Hellenic Statecraft and the Geopolitics of Difference

This book explores competing definitions of Hellenism in the making of the Greek state by drawing on critical historical and geopolitical perspectives and their intersection with difference and exclusion.

It examines Greece's central role in shaping the state system, regional security, and nationalisms of the Balkans, the Black Sea, and the Eastern Mediterranean regions. Understanding the Greek state's social constitution reveals its past and present intentions and strategies as well as local, national, and European notions of security and identity. The book looks at the relation of subaltern communities to state power and the state's ability and willingness to negotiate difference. It also explores how the State's identity politics shaped regional geopolitics in the past two centuries. Chapters present case studies that shed light on the Hellenization of Jewish Thessaloniki, the Treaty of Lausanne's making of Western Thrace's Muslim minority, the role and modes of settlement, urbanization, and 'bordering-as-statecraft' in Eastern Macedonia and Western Thrace, and the politics of erecting the Athens Mosque, the first officially-licensed mosque outside Western Thrace since Greek Independence.

With examples from fieldwork in Greek cities and borderlands, this book offers a wealth of primary research from geographers and historians on the modern history of Greek statecraft and nationcraft. It will be of key interest to scholars and students of political geography, international relations, and European history.

Alex G. Papadopoulos is Professor of urban and political geography at DePaul University. He studies the contestation of urban space in Europe and the United States. His urban work includes studies on Brussels, Saint Petersburg, Istanbul, and Chicago, and his political geographic research entangles 'the urban' in works on SE European geopolitics.

Triantafyllos G. Petridis, Director at the 3rd Secondary School in Athens, Greece, is an educator and independent researcher with degrees in history, archaeology, and political science. He has worked extensively on minority education in Greece, the critical teaching of history, and inter-communal reconciliation based on new pedagogies and curricula.

Routledge Geopolitics Series

Geopolitics is a thriving area of intellectual enquiry. The *Routledge Geopolitics Series* invites scholars to publish their original and innovative research in geopolitics and related fields. We invite proposals that are theoretically informed and empirically rich without prescribing research designs, methods and/or theories. Geopolitics is a diverse field making its presence felt throughout the arts and humanities, social sciences and physical and environmental sciences. Formal, practical, and popular geopolitical studies are welcome as are research in areas informed by borders and bordering, elemental geo-politics, feminism, identity, law, race, resources, territory and terrain, materiality, and objects. The series is also global in geographical scope and interested in proposals that focus on past, present and future geopolitical imaginations, practices and representations.

As the series is aimed at upper-level undergraduates, graduate students, and faculty, we welcome edited book proposals as well as monographs and textbooks which speak to geopolitics and its relationship to wider human geography, politics and international relations, anthropology, sociology, and the interdisciplinary fields of social sciences, arts, and humanities.

SERIES EDITORS

Klaus Dodds, Professor of Geopolitics at the Department of Geography, Royal Holloway University of London, Egham Surrey UK. k.dodds@rhul.ac.uk

Reece Jones Professor of Geography at the Department of Geography, University of Hawai'i at Manoa, Hawai'i, USA. reecej@hawaii.edu

Published

The Geopolitics of Region Building in the Black Sea
A Critical Examination
Yannis Tsantoulis

Hellenic Statecraft and the Geopolitics of Difference
Alex G. Papadopoulos and Triantafyllos G. Petridis

For more information about this series, please visit: https://www.routledge.com/Routledge-Geopolitics-Series/book-series/RFGS

Hellenic Statecraft and the Geopolitics of Difference

Alex G. Papadopoulos
Triantafyllos G. Petridis

Routledge
Taylor & Francis Group
LONDON AND NEW YORK

First published 2021
by Routledge
2 Park Square, Milton Park, Abingdon, Oxon OX14 4RN

and by Routledge
605 Third Avenue, New York, NY 10158

Routledge is an imprint of the Taylor & Francis Group, an informa business

British Library Cataloguing-in-Publication Data
A catalogue record for this book is available from the British Library

Library of Congress Cataloging-in-Publication Data
A catalog record has been requested for this book

ISBN: 978-1-138-49746-7 (hbk)
ISBN: 978-1-351-01870-8 (ebk)
ISBN: 978-1-032-00674-1 (pbk)

Typeset in Times NR MT Pro
by KnowledgeWorks Global Ltd.

Contents

Figures

Tables

Acknowledgements

The idea for this book originated in 2017 while drafting a collaborative article on Greek medievalism in the nineteenth century. The work led us to questions about Hellenic territoriality, identity formation, and geopolitics worthy of greater exploration. Over the next three years, the Hellenic statecraft project evolved a new focus on the geopolitics of difference. Teaming as geographer-and-historian, we brought to bear complementary perspectives on Hellenic historiography and the politics of place and their intersection with difference and exclusion. We had much help. The book has been made possible by the kind assistance of a great number of people and institutions.

Alex G. Papadopoulos

I owe my co-author a great debt for guiding me through some of the most complex and fraught questions of Greek history. His many contributions have made this book possible.

I am grateful for the assistance I received over the course of the project from many scholars, researchers, archivists, and library and museum staff: Dr. Christos Chalkias and Dr. Thomas Maloutas of Harokopio University's Geography Department, Athens, made me welcome and offered me resources as I was setting out for fieldwork in Western Thrace. Dr. Konstantinos Hatzopoulos, Professor of History and Ethnology at Dimokritos University of Thrace extended to me the warmest hospitality, access to the university library, and provided an eye-opening field visit to the historic core of Komotini. Stateside, Dr. Paris Papamichos Chronakis, historian of Modern Greece at Royal Holloway University of London, provided relentless support and encouragement. His scholarship on Thessaloniki's Jewish community was critical to our own work.

I would like to extend thanks to the directors and staff of academic and cultural institutions that made available to me a broad range of bibliographic and archival resources: Aggeliki Giannakidou, director of the Ethnological Museum of Thrace, helped me build a strong foundation on Thrace's history and geography through the museum's rich photographic and material culture collections. Dr. Yorgos Koumaridis, archivist-historian at the Hellenic

Literary and Historical Archive in Thessaloniki, provided critical archival resources on Macedonia's and Thrace urbanization and shared important insights on Greece's politics of toponymy. Mathilde Pyrli, photography archivist at the Hellenic Literary and Historical Archive in Athens, provided guidance on and licenses for illustrations, as did Nopi Ploutoglou, archivist, at the National Map Library in Thessaloniki. Vasileios Nikoltsios, former director of the Museum of Macedonian Struggle in Thessaloniki, made available museum exhibits on Thessaloniki's Great Fire of 1917 and the reconstruction that followed. Athanasios Gouridis, archaeologist and head engineer of Technical Services of the Municipality of Soufli, provided detailed information and expert insights on Soufli's, Didymoteicho's, and Nea Oresteias' urban morphology and the settlement and urbanization of the Greek-Turkish borderlands. I gained important insights on Pomak life by discussing developments in the Village of Echinos with its Community President, Ahmet Kopel. I thank him for his time and generosity. I received much help at the National Historical Museum, the Historical and Folklife Museum of Nea Oresteias, and the Alexandroupolis Public Library. The Historical Archive of the General Staff of the Army was a revelation. I am grateful for the archivists' help identifying reports and correspondence that shed light on the State's blending of borderlands security, refugee resettlement, and regional development during the 1930s Metaxas dictatorship. Konstantinos Daskalopoulos of the Greek Ministry of Foreign Affairs reviewed our narrative on Greek geopolitics for errors of fact. The interpretation of that information is all ours. Dr. Vassilis (Bill) Kappis, Deputy Director of the Centre for Security and Intelligence Studies in the University of Buckingham, kindly reviewed sections on strategy and security and provided important insight on Eastern Mediterranean geopolitics.

Warm thanks goes to Vagelis Papadias, geographer and GIS analyst at Harokopio University, who created several spatial data visualizations that have enhanced the analysis and made our arguments legible. I am grateful to photographer Tasos Vrettos, who granted me an interview to discuss his work on Athens' informal spaces of worship and, especially, for his kind permission to publish three of his photographs. I thank Christos Paridis, reporter and art critic for the weekly LIFO, who kindly provided contemporary photographs of some of Athens' informal mosques and insights on the controversies surrounding the building of Athens' new mosque.

I want to thank Dr. Deborah Brown Kazazis for providing English translations, and Linda Levendusky, Projects Coordinator at DePaul University's Faculty Scholarship Support Center, and Christopher Thomas, Professor of English at Harold Washington College, Chicago, for copyediting the manuscript and helping us streamline the narrative. I thank Sophia Koutsopapa for providing research support in the early stages of the project when I needed it most.

Finally, I would like to thank my spouse, Jeff Lynn, for his loving and unwavering support. We missed much time together to see this project through.

Triantafyllos Petridis

I am grateful to many people who have contributed to this study: Anna Frangoudaki and Efi Avdela for the trust and support they provided all these years and their willingness to address my questions and help meet the challenges of this project; Nelli Askouni for invaluable insight on the best directions for our research and her great advice regarding the project's structure; Alexandra Androusou for making herself available any day, any time to address my questions concerning all topics "Thrace" and more; Thalia Dragona for her encouragement and her assistance in identifying critical bibliographical resources, and Maria Zografaki for providing support and advice on a wide range of topics. I want to extend my thanks to Stefanos Pezmatzoglou, Lampros Baltsiotis, and Leonidas Empeirikos, who shared invaluable information and insights concerning Greece's minority groups and their histories.

I would also like to extend a special thanks to Dr. Moustafa Moustafa, former Member of the Hellenic Parliament (Alliance of the Radical Left: SYRIZA) for introductions to members of Western Thrace's Muslim minority willing to engage with the project and share their experiences and insights. I am grateful to A.K., E.A, E.X., I.M., M.A. and N.E. for agreeing to speak with me extensively and share their memories, concerns, and challenges. Their contributions have been invaluable.

Warm thanks goes to Costas Lykourinos and the rest of the staff of the General Archives of the State in Kavala for their help identifying critical archival material; to Christos Iliadis for his assistance in accessing the archive of The Coordinating Office of Minority Schools when access to it was nearly impossible; and to Olga Katsiani for all her support during my stay in Komotini. Finally, I especially want to thank Spyros Koskinas for his patience and significant contributions to the book project.

Introduction

Hellenic statecraft and the geopolitics of difference

Greece, a Small Power in the European Union's periphery and NATO's southern flank, has played a central role in the geopolitics that have shaped and are shaping the state system, regional security, and ethnic and sectarian communities of the Balkan, the Black Sea, and the Eastern Mediterranean regions. A better understanding of the Greek State's social constitution sheds light on its past and present intentions and strategies and its local, national, and European perspectives. Including geopolitics-from-below as explanation, we gain insight on the relation of subaltern communities to state power, the ability (and willingness) of the state to negotiate difference, and the different roles the State's nationalist- and civil society "turns" have played in regional geopolitics in the last 200 years.

Select case studies ground the constructivist proposition: nationalist "historiography-as-geopolitics," the Hellenization of Jewish Thessaloniki, the making of Western Thrace's Muslim minority by the Treaty of Lausanne, the evolution of the borderland Pomak minority from "suspect other" to Hellenism's "sibling," the role and modes of settlement, urbanization, and bordering as statecraft in Eastern Macedonia and Western Thrace, and, finally, the contestatory politics of erecting the Athens Mosque in 2006–2019, the first officially-licensed mosque outside Western Thrace since Greek Independence.[1]

We explore the constitution of the State and Greek society by ideas and social facts that produced competing ontologies of Hellenism and Hellenicity: *ideas* about national origins, manifest destiny, cultural hierarchy, exceptionalism, privilege, and difference; *social facts* (Durkheim, 1895/1982; 1994; Schmaus, 1994; Shaffer, 2006) expressed as formal or informal norms, processes, and practices, affirm or disrupt the exercise of power and social and territorial control. Difference becomes the crucible in which institutional actors and subjects sort their relationship to state, nation, and liberty. Accepting that difference exists in the state, enables subaltern communities to claim a place in politics and engage in contestatory politics.

In the Greek State's 200-year-long history, the pervasiveness of different ideas and social practices produced, at times, a totalizing effect upon the polity, religion, and perceptions of security – in Marcel Mauss's sense of un *fait social total* (Mauss, 1925/1966). Such instances, very often concerning antagonisms related to difference, reveal the devices and contradictions of Hellenic national statecraft and how the State, during its development, expressed consequential geopolitics of difference.

Our study of Hellenic statecraft is predicated on three main ideas drawn on constructivism (Agnew, 2003, 2017; Greenhill, 2008; Katzenstein, 1996; Ruggie, 1993; Weber, 1994/2009; Wendt, 1992, 1996, 1999). First, we argue that since the creation of the Hellenic State in 1830–1832, Hellenic cultural history and geopolitics become periodically co-constituted to produce different ideas about, and strategies for statecraft and the territorial grounding of a national ideology – statism and nationalism often expressed on different geographical footprints.

In the first decades of its existence, Greece as state-archetype borrowed heavily from the politics and economic-engineering establishments of its "Vienna" Great Power patrons. The State encountered developmental challenges connected to its Ottoman legacy: a society-constitutive role for the Christian Orthodox Church, an Ottoman land-tenure system that required the break-up of large estates and land property-transfers to rural populations, and relatedly, the urgent need for the social uplift of a substantially agrarian national population. It worked to re-establish disrupted commercial networks, build infrastructure to animate a re-territorialized national economy, and, as we argue in the book, build, rebuild, expand, and transform cities, towns, and villages in telling exercises of statist governmentality. Moreover, military preparedness aimed to shield the state from the instability of the Ottoman retreat and Europe's imperial surge, but also served expansionist campaigns.

The success of national statecraft projects in the Balkans was contingent upon the Empire's relations with Europe's Great Powers. The creation of the Greek state in 1830–1832 set in motion a set of expansionist strategies and actions that were, at times, perilous but successful, and at other times strategically misguided and disastrous, and invariably conditioned by Great Power intentions and actions. Nonetheless, considered as a whole, Greece's grand strategy was successful in rolling back the Ottoman Empire and in catalyzing territorial and ethnonational contests in and among ambitious Ottoman successor states.

This materialist statecraft was invested with the energy of a range of nationalist ideations; some generated in the Greek diaspora, others usefully connected to European Enlightenment, neoclassicism, and Romanticism, others still emanating from Ottoman and Mediterranean cosmopolitanism. As members of the European Enlightenment, this intellectual class systematically reworked and reinterpreted the Hellenic past to rally a Greek audience (also to *create* a Greek audience) and to build geopolitical capital with a Western one. It also called back to memory Greece's historical geographic grandeur.

A new narrative about Hellenism would emerge that would animate a sort of Manifest Destiny: the Great Idea. Territorial and cultural primacy were both objectives of foreign and security policies and historical arguments for achieving them: in the case of Greece, the territorial grandeur and civilizational superpower of Hellenism in its Classical, Hellenistic, Medieval Hellenic/Byzantine, and Post-Byzantine ("Latter Hellas") incarnations, become bases for defining the territorial target-list of the new state and justifications for Greece's right to these territories. Admittedly, a nineteenth century Balkan geopolitical playbook, rather than one strictly used by Greece, the re-writing of the region's histories as triumphal ethnic narratives with very expansive and aspirational geographic objectives became the thin edge of the wedge for territorial expansionism across the region.

Initially crystallized as a national historiographic paradigm in the second half of the nineteenth century (a period of pronounced geopolitical anxiety but also nationalist élan and a forward-thrust foreign policy), the triangulation of cultural history, territory, and geopolitics would be significantly redefined four times in the twentieth century: following Greece's defeat in the Greek-Turkish War (1920–1922), as a consequence of the Greek Civil War (1946–1949), after the fall of the Colonels' military dictatorship (1967–1974), and the establishment of a liberal democracy, and, again, after the European Union Treaty of Lisbon entered into force in 2009. This relentless, accelerating reinvention of the state that produced different configurations of social facts, suggests a course toward democratization and Europeanization, albeit both perilous and non-linear.

Second, in our studies of Hellenic statecraft, we were always drawn to stories of settlement, urbanization, and, more specifically, the political contestation of village, town, and city spaces. Phenomena germaine to state-making, state maintenance, and political change, such as territoriality, conditionalities of sovereignty and governmentality, and external and internal borders and bordering, gave new insights on Greek geopolitics and statecraft when intersecting "the urban."

From the very beginnings of the Greek State, urbanism was the forge of both state and nation. The rearticulation of Hellenism as the cultural phenotype of the new state required the rearticulation of "the urban" at all scales – from the built fabric of cities expressed in a national variant of architectural neoclassicism, to village, town, and city plans, and, scaling-up, to the planning of a new, national urban system. Urban morphology as the drawing and transformation of town plans and their fringe belts, the making and changing of built fabric (including ordinary buildings), and the palindrome of building- and land-use grounded and shaped the politics of statecraft and nationcraft, sometimes legibly at the lot-scale, and also shaped it. Athens, Thessaloniki, Nafplion, Komotini, Alexandroupolis, Didymoteicho, Nea Orestias, and expanded towns and new villages built for war refugees on the Greek-Bulgarian and Greek-Turkish borderlands provided empirical grounding to our studies.

Urban regulation was, and in some instances continues to be, the fulcrum for national paradigm shifts: The de-Ottomanization of urban landscapes, the embracing of a European planning canon in urban renewal and the new town planning, the Hellenization of Thessaloniki (in 1913, a cosmopolitan, plurality Jewish city), the explosion of urbanization following the exchange of populations under the auspices of "Lausanne" (a significant feat of Hellenic governmentality under dire conditions), the literal martialing of settlement on the Greek-Bulgarian borderlands in the inter-war period to meet "the peril from the North," and the political contestation of the building of a new mosque in Athens demonstrate how statecraft works in the Greek case when encountering otherness.

The two ideas together – both about the political making of Hellenic space-time and political praxis with territorial meaning – gave rise to a third: the emergent geopolitics of difference and the role of the State in opposing it as contrary to a crystallized definition of Hellenism, or curating it as an expression of a turn to a civil society politics in the Europeanist mold.

The politics of difference defined Greece's national project early on. The origins and identity of nineteenth century Greeks became the subject of racialized interpretations of Hellenic ethnicity, precisely as the time the new state was recognized (1830–1832). Notably, German historian Jakob Philipp Fallmerayer provocatively claimed that the last Hellene fell in the Battle of Chaeronea (338 BCE), where Macedon consolidated its victory over the South. Calling Hellenic history discontinued in centuries past, he inserted himself and the European academy into Greece's national renouveau. In an Orientalist reading of history, Fallmerayer remarks in the *The Fall of the Peloponnesian Greeks and the Repopulation of the Empty Land by Slavic Tribes,* "The race of the Hellenes has been wiped out of Europe. Physical beauty, intellectual brilliance, innate harmony and simplicity, art, competition, city, village, the splendour of column and temple – indeed, even the name has disappeared from the Greek continent's surface... for not even a drop of noble and undiluted Hellenic blood flows in the veins of the Christian population of present-day Greece" (Fallmerayer, 1830/1836).[2]

The Greek academy engaged Fallmerayer's race thesis frontally. The fullest response came in the form of Konstantinos Paparrigopoulos' "continuity of Hellenism" thesis – an exposition of Hellenism's evolutionary arc from Prehistory to the age of nations. Paparrigopoulos' first response came in 1843 and developed over the next 25 years into the paradigm-setting, ethno-poetic work, *The History of the Hellenic Nation.* By crystallizing the making of Hellenic identity, national historiographers, like Konstantinos Paparrigopoulos and his academic and political followers, closed the door to readings of history and geography that would have made place for difference in the Greek story.

The code for his contemporary Greek society was Helleno-Christianity. Despite the inherent contradiction of fusing the Hellenism of polytheistic, same-sexed Greek antiquity with Christianity, Helleno-Christianity as keystone signifier of the nation-state, proved to be a durable social fact with great consequences (Papadopoulos, 2002). In an important sense, it was an act of self-inflicted epistemicide, as Enlightenment and nationalist intellectuals used the academy to define and, ultimately, delimit who and what a Hellene was. In further acts of epistemicide, it helped invalidate, or at least devalue, the cultural complexity and multiplicity of the Ottoman cosmos setting the stage for domestic contests between the charter group and minorities.

Our focus is on geographies of ethnolinguistic and sectarian difference and its racialization in the conduct of politics of national Hellenic statecraft. Therefore, the study's scope of difference is not comprehensive. We decided that the gendering of Hellenism as imagery and as state conduct needs extensive study elsewhere, as does sexual difference. The queering of Hellenism is explicitly visible in popular media and unintendedly parodic when it concerns the extreme-Right (the toughs of the neo-Nazi "Golden Dawn" party like to bathe en groupe in cold streams like their Spartan idols). Tourist placemaking and its uses of the past and of folk culture (Would you like to learn "syrtaki"?) is a promising strand of research that would be a natural extension of our work here.

Contesting the boundaries of "Hellenicity:" Greece's minorities politics

Establishing Helleno-Christianity as the unofficial and, at illiberal times, official passkey into Hellenism made making space for linguistic and sectarian difference politically fraught, and in the contemporary era of liberal European politics, also highly contestable. This stance of state and society on difference was tempered under Prime Minister Eleftherios Venizelos' Liberal Party (1910–1920, 1928–1933). Venizelist liberal policies for Thessaloniki's Jewish community and Western Thrace's Muslim communities were, in part, guided by realist readings of how Great Powers, the ultimate arbiters of Balkan and Mediterranean geopolitics at that time, would have reacted to their exclusion. However, the second ingredient was a deep conviction to the soundness of liberalism as a political philosophy that could neutralize the most pernicious qualities of ethno-nationalism.

A decade after the Balkan Wars, statization and social-territorial Hellenization in the New Lands of Macedonia regressed liberalization. Thessaloniki's Jewish community, which was granted citizenship in the 1920s, saw its rights gradually erode, as the city – once renowned for its cosmopolitanism – became demographically, administratively, and materially Hellenized (Chapter 4). Its near-annihilation in the Holocaust further crystalized the city's national character.

Exclusionary policies were applied to other recognized linguistic and religious minorities, such as the Muslims of Western Thrace. Following the Greek-Turkish War, the Treaty of Lausanne (1923) defined the relationship between the signatory states and the rights of officially recognized minorities in their respective territories. Between 1936 and 1975 (the year Greece's democratic Constitution was ratified), Greece treated its Muslim and allophone populations of Western Thrace as suspects, vulnerable to alleged subversive activities by Turkey in Western Thrace. Although constitutionally Greek citizens, they were *de facto* disenfranchised members of the Hellenic polity. Turkey's long-standing campaign to hinge Western Thrace's Muslim communities to its national imaginary has capitalized on that rift for decades (Chapter 5). Turkey, for its part, in violation of the intent of "Lausanne," engineered in 1955 the expulsion of nearly all the Treaty-protected Greek Orthodox of Istanbul.

In assessing Greece's treatment of its ethnic (per "Lausanne," religious) minorities in Western Thrace, it is useful to consider the polity's socially illiberal character. In 1975, state and Church were not separate.[3] Greece did not have civil marriage, dowries were legal, abortion was illegal as was divorce, and homosexuality, although not criminalized, it counted as a liability in criminal proceedings. Since 1975, Greece has been proceeding along a path to Constitutional liberalization, in part in fulfillment of its obligations to the European *acquis*. The election of Andreas Papandreou's *Panhellenic Socialist Movement* party in 1981, ushered in fundamental changes to civil law, legalizing divorce and abortion, outlawing dowries, and legislating civil marriage. The road to legalization of same-sex civil unions was much longer, becoming legal in 2016. Changes to the definition of citizenship and attendant rights, the annulment of discriminatory legislation and the lifting of regulations that obstructed minority group mobility, and full participation in the economy, were addressed in law in the 1990s and 2000s, eliminating, finally, many injustices against the Muslim communities of Western Thrace.

There is more ground to cover in building a civil state and society in place of (or at least in compromise with) the long-standing national state. The "Lausanne" legacy minority groups enjoy cultural, economic, and political rights that they were denied for decades in the twentieth century. However, as the geopolitics of Pomak identity reveals in Chapter 6, there is still tentativeness on how to expand the definition of "Hellenicity" to include non-Orthodox, non-Greek speaking citizen communities.

Minority groups that are not officially recognized by the State, as in the case of Slavophone communities in Greece's north, harbor unfulfilled non-territorial aspirations. Decades of garrisoning the northern frontier against "the peril from the North" (initially, Bulgaria, but subsequently, also the countries that supported Communist militias during the 1946–1949 Civil War), fostered an ideology of bordering that shaped both borderland communities but also the state itself (Chapters 7 and 8).

The 2018 Prespa Agreement settled the dispute of the name of what is now North Macedonia, and removed Macedonian irridentism as an irritant between the two states. North Macedonia's accession to NATO and its advancement to Candidate Member State by the European Union, further support a turn in domestic politics to recognizing the pluralistic profile of some of the country's regions.

Lastly, since the 1980s, and especially since 2010, Greece became the stopover on the way to Western Europe and, in some cases, the destination of economic migrants and war refugees. Since the fall of Albania's totalitarian regime in 1991, Greece has received hundreds of thousands of Albanian migrants, many of whom are substantially integrated in Greek society and economy. Greece's urban demography is becoming multicultural. The contestatory politics of building the new Athens Mosque (Chapter 9) illustrate dilemmas of geopolitics of difference. Multicultural futurity and a new urban cosmopolitanism challenges the state to test Greek society's commitment to a civil society turn.

Carving the national state out of the Ottoman cosmos: Hellenic cosmopolitanism

Historic and extensive diasporas contributed to the building of the Greek State. Some of the most significant urban centers of Hellenism were outside the boundaries of the 1830 state. Their relationship to the franchised nation was consequential to the shaping of the nation-state. Adding to the tension between national and diasporic communities concerning control over the definition of Hellenicity, was the new relationship of Hellenism to the Ottoman Cosmos it rebelled against. It would, inevitably, involve the exchange of a regional, Mediterranean worldview with a complicated, national Hellenic cosmopolitanism.

Distance is ontologically formative of cosmopolitanism. Distance contours it differently from place to place by relating – in comparative, contrasting, or generative terms – "here" to "there," and by locating opportunities for subject and community interlockedness. Writing about it in ecumenical terms, Barney Warf defines cosmopolitanism as "an ethical, moral, and political philosophy that seeks to uncouple ethics from distance, arguing that each person is bound up with, and obligated to, humanity as a whole" (Warf, 2012: 272). Cosmopolitanism denotes the willingness and ability of persons, groups, and other collectives to elevate and scale-up familiarity, amity, and community to include social and cultural Others, and the structural conditions that make that transition possible and durable. This dilation of the meaning of 'homeplace' to include the world (or some meaningful part of it) would require more than *verstand* of the Other (in Hannah Arendt's terms, objective "knowing" of another or a situation). It would necessitate *vernunft* (reasoning through and understanding in moral terms the "meaning" of the Other's motivations and circumstances) (Arendt, 1977: I, 13).[4]

Arguably, adjectival place-specific qualifications of cosmopolitanism, as "Mediterranean," "Ottoman," or "Hellenic," would undermine this universalist meaning. Nevertheless, there may be good reason for the aporia: David Harvey cautions, "cosmopolitanism ... gets particularized and pluralized in the belief that detached loyalty to the abstract category of "the human" is incapable in theory, let alone in practice, of providing any kind of political purchase even in the face of the strong currents of globalization that swirl around us (Harvey, 2000a: 530). Place attachment, or "rootedness," as Anthony Appiah posits, grounds the politics of difference (and distance) to "the local" and also re-hinges it to "the national."

For Martha Nussbaum (1996), subject agency is the key to understanding cosmopolitanization as a metaphor of concentric circles of affiliation, the ultimate circle encompassing a like-minded ecumene, which sees value in curating a universal, practiced ethics. Such cosmopolitanism, she posits, would be the antipode and the antidote to virulent, exceptionalist nationalism (1994). The spatial configuration of Nussbaum's nesting of affiliation places the Self at the source – at the geographic center – and locates the family, the community, the nation, and the world along a distance gradient. However, borders and bordering scramble what appear to be a straight-forward geography of attachment-at-a-distance. Diasporas are emblematic of how "the distant" and "the different" do not always neatly map one upon the other.[5] Further, if the indexical spatial unit is the Self, the world is our shared home, and its peoples our kin existing, ideally, in a respectful state of peace and equality, why is that ideal continuously and viciously qualified (if not outright undermined or annihilated) by fractious, particularistic, and exclusionary politics?[6] The answer may lie at the intersection of cosmopolitanism and nationalism, on national cosmopolitanism.

To return to Warf's problematizing of cosmopolitanism, sovereigntism and nationalism have been primary responsible for cosmopolitanism's inability to fully scale-up from sites where it was historically, territorially, and sustainably grounded (places in the Eastern Mediterranean region like Alexandria, Thessaloniki, Smyrne, and Istanbul). National cosmopolitanism, as a compromise, can be seen in two ways: first, as Yao Souchou suggests in her study of Malay openness toward non-Malays, national cosmopolitanism can be considered as "a specific strategy of the local to negotiate and absorb new transnational or other universal experiences. Moreover, in [the Malay] case, the cosmopolitan feeling is 'national' because it brings with it a new structure of feeling that encompasses cross-ethnic concerns in the social and economic arena within the nation-state" (Souchou, 2003: 222). Souchou continues that national cosmopolitanism adjusts nationalism's perspective about the "Malay-non-Malay" binary. As that binary relationship is no longer either a proper or adequate basis for understanding the world, cosmopolitanism provides new world-interpretive tools (Ibid, 221).

Building on Souchou's definition, the character and uses of national cosmopolitanism appear contingent upon its social constitutive relationship to a national community's ideal, imagined Self. Its sustainability would depend on the generation of social facts that root it materially and symbolically, and, down the road, map it upon the unconscious and the habitual.

In realist terms, national cosmopolitanism appears contingent upon the national territorial state's interest. It would reflect the power positionalities and aspirations of nationals, the state, and Others as (most often unequal) stakeholders. As such, national cosmopolitanism can work as a statization instrument or strategy of political territorialization with particularly significant implications for the shaping of "the local." Reflexively, local ties filter and root cosmopolitanism back in "the national" (Appiah, 2006).

Turning to the Greek case, pre-Independence, statelessness was not a barrier to cosmopolitanization as the tissue that connected Hellenism's urban agents to the world (commercial networks and maritime trade and transportation) was not contingent upon statehood. Accordingly, Ottoman cosmopolitanism, evidenced in the supra-familial polities of Eastern Mediterranean port cities, drew its vitality from the Empire's multiculturalism and the expanding Greek commercial networks in the Mediterranean. Therefore, that in 1830, the new Hellenic State did not encompass in its territory any such cosmopolitan centers, was no barrier to cosmopolitalization.

The provincialism of Greece's cities, including Athens, was gradually urban-planned away by the Capodistria regime and, subsequently, King Othon's Bavarocracy. Instead, it is the Ottoman and European "Cosmoses" that condition Hellenic cosmopolitanism: a national ideology that drew liberally from European centers of the Enlightenment, emerging ideas on nationality and citizenship, the commercial liberalism of trans-Mediterranean networks curated by shipping concerns in places such as the islands of Andros and Syros (Delis, 2015: 255, 273–274), and the worldliness of diasporic Greek communities – all Ottoman places with efflorescent Greek societies.

The Hellenic ecumene was, therefore, much greater geographically than the new state. It included diasporic places with substantial, or at least influential, Hellenic communities. Distributed on the Mediterranean and Black Sea littorals, some – Istanbul (Constantinople), Smyrne (Izmir), Thessaloniki (Salonica), and Alexandria – are upheld in the popular and academic imaginations as exemplary of supranational Mediterranean and Ottoman cosmopolitanisms. The diaspora was built on amalgams of Hellenic, Jewish, Armenian, Turkish, and other homosocialities and perspectives on the world, thus enriching the definition of a new State-sanctioned Hellenism but also introducing contradictions that will invert the culture-generative relationship between the supranational diaspora and the national metropolis (Figure 1).

The point of inflection that signaled the shift national Hellenic cosmopolitanism came in the second half of the nineteenth century. Cosmopolitanization became increasingly regimented by a nationalist outlook that claimed Hellas's lost Hellenistic and the Byzantine dominions. Enlightenment-inspired historical scholarship on Hellenism was supplanted by a nationalist school of history, defined by historian Konstantinos Paparrigopoulos. The academy's adoption of his "continuity of Hellenism" doctrine, provided the logic for nationalizing the cosmopolitanism of Greek culture outposts remaining in the Ottoman Empire.

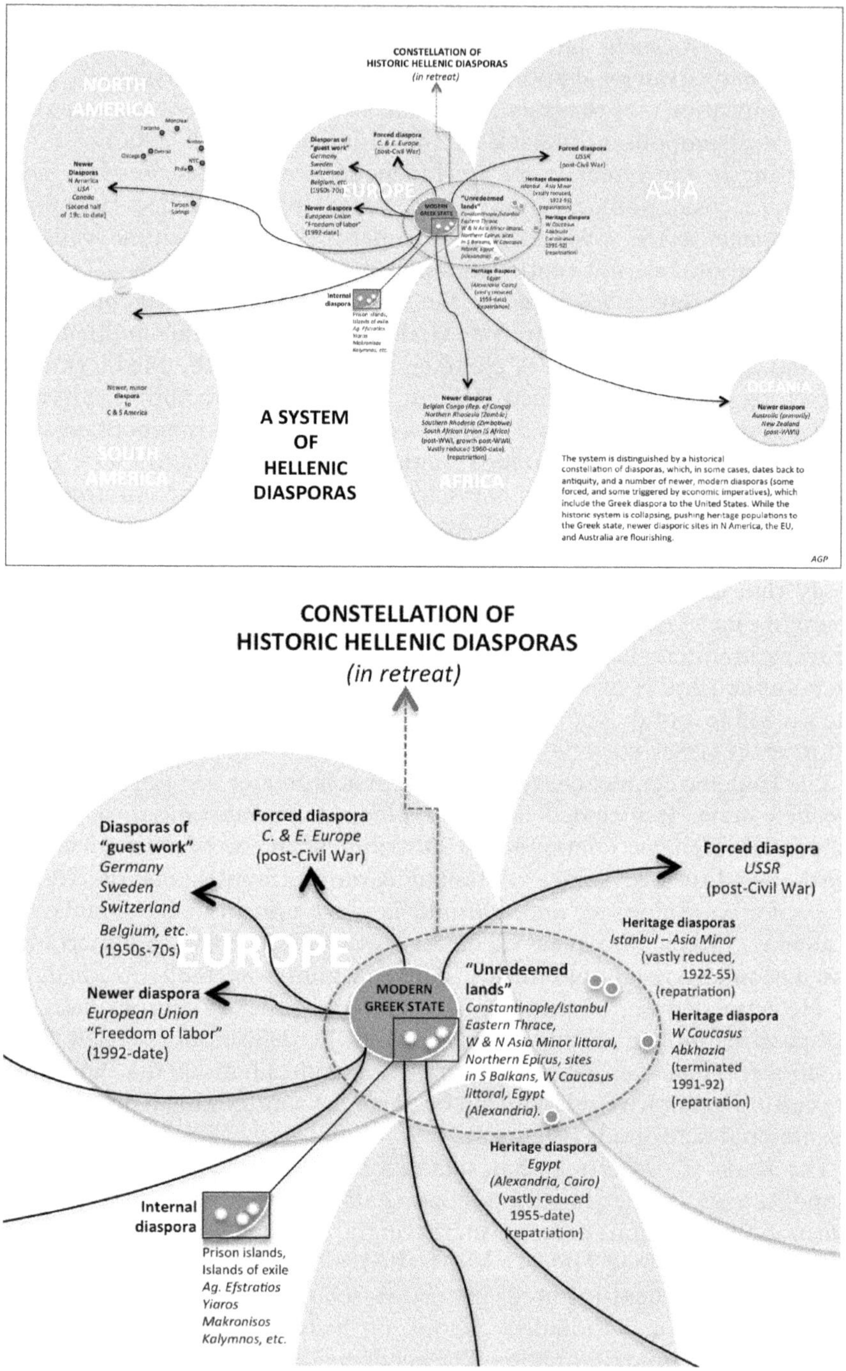

Figure 1 (Above) Global Hellenic diasporas past and present. (Below) (detail): Historical Hellenic diasporas in heritage and "unredeemed" lands (*Source:* authors).

Therefore, Cosmopolitanism was re-forged in the nineteenth century as a particularistic and exclusionary ideology by its adherence to the historical-geographic imaginaries of a "Greater Greece" torn out of the atlases of antiquity and the Byzantine Empire. The thickening of the administrative and security state turned some of that historical imaginary into expansionist strategy.

For Paparrigopoulos and his followers, the world where Hellenes of different eras dwelled was, essentially, Hellenic. Accordingly, Mediterranean cosmopolitanism was re-fashioned as an extension of Hellenism's cultural enterprise and a handmaid of the geopolitics of the expansionist Great Idea. A national Hellenic cosmopolitanism re-territorialized the wellspring of Hellenism from the diaspora to the national homeland. From that fountainhead, a novel, national Helleno-Christian doctrine was transmitted to an emerging set of modern, industrial-era Hellenic diasporas in Western Europe, the Americas, Africa, and Oceania, and propagated by networks of State-supported, Church-managed schools abroad (Zervas and Papadopoulos, 2020).

National Hellenic cosmopolitanism had both organic and politically engineered dimensions. On the one hand, it drew on the supranationalism of Hellenic diasporic culture nested in, and marbled through, the complex multiculturalism of the Ottoman imperial experience. On the other hand, the uplift by Europeans of the Enlightenment of Classical Hellenism as the cultural prototype of Europeanism, re-territorialized the Hellenic experience from Southeastern Europe and the Eastern Mediterranean, to the West.

National Hellenic cosmopolitanism endures to a considerable degree to the present, nourished by a public education establishment and a political culture still in the thrall of Hellenic exceptionalism. Since the era of the European Grand Tour, Hellenism and its study have been instrumentalized in European and American education as core knowledge and cultural touchstone. Classics departments in elite European and American universities long operated as guardians of what critics have deemed a Western and white supremacist episteme.

The effective end of the expansionist geopolitics of the Great Idea in 1922, the emergence of post-Second World War internationalism, the establishment of a robust, ideologically pluralistic democratic system in 1974, and Greece's accession to the European Communities in 1981, have blunted the more extreme expressions of cultural particularism. They have also rendered Hellenic cosmopolitanism less strident and nationalistic. Hellenic cosmopolitanism is expressed in both banal, everyday ways (especially in tourist zones displaying kitsch Greek material culture), and formally, in Greece's international branding strategies in tourism and high-culture sites, such as museums and art spaces.

Less visible is a parallel, subaltern cosmopolitanism of immigrant and refugee communities, primarily in Greece's largest cities and commercial agriculture suburban and rural settlements. *De facto* exclusion from full participation in the economy has compelled Athens's immigrant communities to develop their own co-ethnic and co-religious local networks of support, sometimes backed by civil society organizations, and with links to transnational networks of support. In the case of Athens' Muslim communities, the development of as many as 100 informal mosques, commonly registered as cultural centers, signals the need to mitigate social and political marginality.

Despite a relatively low degree of socio-spatial segregation at the local level, at least in comparison to other European countries, the two cosmopolitanisms meet, most commonly, where the economy mediates (the import/export warehouses in the historic core, the hospitality industry in urban centers and resort towns, and in the agricultural cooperatives that aliment Greece's export-oriented agricultural industry. We will encounter this cosmopolitan dualism later in the book, where it will serve in Chapter 9 as the framework for understanding the contestation of Islamic places of worship in Greece and will provide a critical reading of contestatory politics that built Athens's state-licensed mosque.

"The urban" in Hellenic statecraft: urban morphology and the geopolitics of place

The making of the successor states of the Ottoman Empire in the Balkans required an urban revolution. Urbanization would incubate new, modernizing social relations, new urban forms in which the national state and its steam-industrial economy would propagate, and new hubs for emerging transportation and commercial networks. Ottoman legacy urbanism, its morphology and society, was deemed an obstacle and was targeted by technocrats and ideologues for transformation.

Greece's, Serbia's, and Bulgaria's national projects produced sharp urban-rural cleavages. The city was where the state came together to provide for governmentality. The country may have been the agrarian territorial base of most of the population, but it was politically and functionally dependent on the centers, especially the primate city in the national urban hierarchy. It was in cities where the military-administrative state generated national ideas, where elites formed and interacted, and where people, parcels, and post flowed through. They were the places were diplomatic and commercial intelligence was generated, the polyglot press flourished, financial capital was brokered, commercial and industrial contracts written, and the intellectual enterprises of academics were deliberated.

When dissent would rise in the country, the urban state would intervene to align it with national goals (Roudometof 2000). The contrasting attitudes between urban and rural groups "reveal the articulation of irredentism as a phenomenon associated with the modernization [as in urbanization and 'statization'] of the local national societies … [T]hroughout Europe in the nineteenth century, proponents of nationalism were found everywhere among groups of people whose livelihood depended on the state" (Roudometof, 2001: 165). Accordingly, the main beneficiaries were urbanites. Cities were generators and depositories of state machinery and national memory that promoted the interests of urban elites. Irredentism provided justification for territorial expansionism, which required urbanizing the new lands in terms of the conquering state. So, it went the institutive, generative, cyclical, co-constructive process of tethering "the urban" to statecraft.

In the telling of the stories of national Hellenic statecraft, urban form and process emerged as keystone concepts. Urban form reflected the habitual sphere's material and social circumstances – where and how people live day to day. Urban form also described the places and kinds of spaces that were either centers of administrative, coercive, or symbolic power, or places targeted by the state for control. Urban form also described places and kinds of spaces subaltern communities create, inhabit, socially reproduce in, and resist from.

In her description of colonial Algiers' urban morphology, Zeynep Çelik notes that the colonizer and the colonized lived their lives side by side in public and private spaces where "the metaphorical and the practical, the confrontational and the compromising [clashed]." She continues, "because the particular evolution of Algiers under French rule was so dramatic, the physical city kept its residents constantly aware of political conditions and power relations" (1997: 2).

The evolution of Greek cities was as dramatic as in colonial Algiers, although the urban forms and the processes generated and animated did not fully map on Athens, Thessaloniki, and smaller cities and towns. Urban forms – the town plan that is built lot by lot, street block by street block into block groups and plans, the varied voids and volumes that make up the built fabric, the building – and land-uses in constant flux – materially territorialize and reflexively produce "Greek" regulatory and social change. When and in what time frame did each event of morphological change occur, and at what scale? Who authored it or was instrumental in its making? Who owned it or benefited from it as a piece of exchange, functional, or symbolic-use value? And what did it replace (or, better, displace)? What urban form was created to replace it?

Athens' morphological transformation in the early decades of the Hellenic State's existence prototyped the politics of place that underwrote the geopolitics of difference in later periods and the case studies we cover in the book – Thessaloniki, Xanthi, Komotini, Alexandroupolis, Soufli, Didymoteicho, Nea Oresteias, the Pomak villages of Xanthi and Rhodope prefectures, and the villages on the Greek-Bulgarian and Greek-Turkish borderlands. In the last chapter, we return full-circle to Athens in the twenty-first century.

The making of the Hellenic "urban"

Between the outbreak of the Greek War of Independence and the establishment of the state, several temporary revolutionary-era centers of power emerged, though not quite official capitals. The selection of Hydra and Ermoupolis, in the Saronic and Cycladic islands, respectively, signaled the critical role of maritime merchant elites in the war effort. Between 1827 and 1833, war-related emergencies necessitated the brief removal of the revolutionary government to the islands of Aegina and Poros.

In 1827, the installation of Ioannis Capodistrias as Governor of Greece by the Great Powers set in motion the creation of a formal Hellenic administrative state. A critical act of this government was to select, in 1829, the City of Nafplion as temporary capital of the new state. The consensus was that the war would need to reach its conclusion before a permanent national capital was chosen. Nafplion was an excellent choice: it was well-fortified and served as an important commercial center in the Peloponnese's Aegean Sea littoral. It was also the first testing ground for urban regulations intended to Hellenize (or re-Hellenize) the morphology of Greek cities.

Capodistrias saw the modernization of existing cities and the foundation of new ones as instrumental to the new state's successful organization. Upon arrival in Greece, he assigned the redesigning of five cities, one of which was Nafplion, to French-educated military engineer Stamatios Voulgaris. During the Governor's short tenure (political rivals assassinated him in 1831), new plans were drawn for 19 cities, although they were not fully implemented (Dorovinis, 1985: 288).

Greece's urban modernization continued in earnest under King Othon's successor administration, substantially under the auspices of Western and Central European architects. During Othon's reign (1832–62), 23 cities were redesigned, some to a significant degree, and ten new cities were founded (Kokkou, 1985: 361–63).

The urbanization drive, although modest compared to rates of urbanization in industrializing Europe, was going to transform Greek politics. The renewed "Hellenic city" became the locus of a Western-minded administration and economy, which nurtured broad geopolitical and explicitly territorial ambitions. Rural areas languished. While land reforms, starting with the 1835 "*Law for the Dotation of Greek Families,*" created a class of independent smallholder farmers, the rural economy was not able to produce critical amounts of surplus for regional development. It remained mostly dependent upon the good offices of Athens and the central government. While property relations in post-Ottoman Greece did not resemble those of the Ottoman era, living conditions were not substantially improved. The rural-urban divide made class-based, cultural, and ideological divisions more profound. While the town was becoming "European" and, thus, "Hellenic," the village's morphology remained substantially Ottoman for much of the nineteenth century.

What mattered most was the alignment of the new Hellenic urban imaginary and planning practices with the principles and priorities of neoclassical urbanism as it was expressed in cities like Paris, Vienna, Munich, and Berlin. Regularity, symmetry, monumental perspectives, built fabric with historical narrative qualities (in this case, Hellenic), a spatial-organizational hierarchy of town plan that reflected political-economic relations (in Athens, an urban core that aggregated the most important functions, avenues dotted with the elite's hôtels particulier, a demarcated early industrial zone), and an acknowledgment of the transformative power of embourgeoisement. Admittedly, the application of these principles was uneven across the new urban state. Parts of Athens retained a village atmosphere decades after its elevation to national capital.

Ottoman urbanism as planning and architectural practice, was vilified as "barbarous" (Kalafati, 1995: 279). Aleka Karadimou-Gerolympou notes that in the nineteenth century, the dominant perception was that "Hellenicity" was related to formal, orthogonal town planning and "Turkishness" with the spontaneous urbanism of the pre-capitalist city (1985: 386). Systematic efforts were made to build cities "in a proper Hellenic manner" (Ελληνοπρεπεις πολεις), which would be distinguished by the alignment and broadening of streets, the creation of geometrized public spaces, and the promotion of public health through architecture (Fakoura 2006: 88–89). In contrast, "poorly constructed" Ottoman-era urbanism, was distinguished by an organic town plan of sinuous streets of irregular width and close-knit built fabric of modest dwellings built around courtyards. Many years later Nikolaos Dragoumis described Nafplion as it appeared before its reconstruction, as "entirely Turkish … with streets covered in malodorous mud, the wooden houses … in great disrepair, lacking any architectural style"[7] (1874: 95).

There is a double conceit at work here: first, that Ottoman urbanism and architecture were inferior in aesthetic and functional terms to Western prototypes, and, second, that there was, in fact, a historically-reasoned nineteenth century "Hellenic manner" of designing and building cities, supported by architectural and planning traditions and practices that persisted through the centuries. In the first instance, Orientalist critics of Ottoman urbanism devalued the multicultural, cosmopolitan, ontologies of the Ottoman built environment and material culture, and claimed them as phenomena of an alien, integral Other. According to this view, the Hellenic element nurtured its own built environment traditions and practices aside from the Ottomans. Therefore, the Ottoman built environment, whatever its amenities or diseconomies, was attributable to the occupier alone. Following Independence, their constructions were unwanted landscape intrusions.

In fact, Pallis reports that the Athens Kaza and the town and villages in it enjoyed considerable self-rule. Councils of elders (δημογέροντες), representatives of villages in the surrounding region, the Orthodox Church, through its Athens Metropolitan and its ecclesial presence in the villages, and great monasteries that controlled significant lands and a significant portion of the local economy, had substantial say in town and Kaza affairs (Pallis, 2007: 45-56). Therefore, Athens' urban morphology reflected complex agency that entangled the local Ottoman administrative state, the leaders of *Millets* (in this case, the Orthodox Christian *Millets*), and the broader Turkish and

Millet communities of the Kaza that contributed at a granular scale to the organic buildup of the town. Ergo, no pure urban types present, nascent or established.

The claim to a continuous Hellenic urban morphological and architectural tradition opened the door to the following options for urban preservation or renewal: first, preserve select elements of Ottoman-era urban morphology and vernacular architecture that were recognized as Hellenic (the practice that gained traction among folkloric, cultural preservation societies in the late nineteenth century). Therefore, surviving built fabric was graded as «national» («εθνικο») or not. Second, regard the continuous use of pre-Ottoman-built environment as evidence of an enduring Hellenic tradition in architecture and town building. Lastly, develop a new architectural and urban planning style and practice with clear referents to the historic built fabric of the ancient and medieval Hellas. The state embraced this third option in public projects.

City morphology became the forge of a new national identity. The urban renaissance was not predicated entirely upon economic-engineering and public health concerns or aesthetic preferences. The new regulatory regime prioritized the removal of Ottoman urbanism in politically important cities like the temporary capital City of Nafplion. The «national» standard became the litmus test for renewal through demolition (Fakoura, 2006: 90) – an early version of Haussmannization with expropriation at its core but sans its financial léger de main. New urban fringe belts followed regular or orthogonal town plans in the European mold.

If the temporary Capital City of Nafplion redesign suggested a new, national way of urbanizing the new state, no city's plan and architecture was more scrutinized and made a testing ground of a national material culture than Athens'. The selection of Athens in 1833 as the Capital of the new Hellenic State should not have been a surprise, given the importance of ancient Hellenism for Greek and philhellene intellectuals of the Enlightenment. However, that year, the newly installed Regency Council expressed the desire to, ultimately, place Constantinople in that role. Any other city would only serve as a temporary seat of government (Dimakopoulou, 1998: 114).

By the geopolitics of the time, Constantinople's re-conquest and elevation to Capital of a great Hellenic State were geopolitical fantasies. So, on 6 February 1833, Otto of Wittelsbach, the second son of King Ludwig I of Bavaria, already elevated to the Kingship of Greece, arrived in his new capital city escorted by Bavarian expert advisors. Among the latter was Adolf Weissenburg, who in May 1833 was designated Conservator of Antiquities "responsible for the preservation, as well as the exploration and collection of the Kingdom's archaeological treasures" (Petrakos, 1982: 182). In the following 12 months, significant progress was made in the regulation of cultural heritage. The Royal Edict of 29 April/11 May 1834, which created the Office of Public Economy, mandated registering and recording "remains and monuments" across the State (ΦΕΚ 18, 22 May 1834). The state concurrently passed the first law that explicitly addressed the regulation of archaeological activity (Law 10/22 May 1834).

Works of architecture and sculpture, and material culture and epigraphical remains, were designated "antiquities" and thereby targeted for registration and preservation. The law was explicit in its inclusion of objects from "the most ancient era of Christianity or the so-called medieval era" (Article 111). Eleni-Anna Hlepa notes that Paleo-Christian art and sites were at the forefront of scientific interest in the first decades of the nineteenth century, which provided arguments for their preservation.

Scientific norms for archaeological preservation originated, in the Greek case, either in Bavaria, given the composition of the first Hellenic Government, or more likely Prussia, where architect Karl Friedrich Schinkel was instrumental in prototyping the related authorities in 1815 (Hlepa, 2011: 31). She further remarks that the term "so-called Medieval era" surely included Byzantine, Venetian, and Frankish (Latin Empire of Byzantium) antiquities, but also, possibly, Ottoman ones (Ibid). However, despite the general interest in "the Medieval" and its explicit inclusion in the first archaeological preservation law, the Archaeological Committee (Αρχαιολογικη Επιτροπη), created by the Royal Edict of 6/18 November 1837, could not avert the demolition of a considerable range of monuments of the medieval period writ large. That was especially the case in Athens, where the characterization of the city as national capital supercharged the political signification of archeological remains. It would take another Royal Edict to command "the preservation of the remains of the Medieval Era in Athens in carrying out the new plan of the Capital City..." It makes the critical call for the preservation of Byzantine, Venetian, and Turkish buildings, even if they are interdigitated with either Hellenic or Roman antiquities (which was very often the case). Further, it calls for them not to be demolished without "our explicit permission ... because they augment the picturesque [qualities] of the city (RE 1/19 December 1837) (Figure 2).

As in Nafplion's case, the call to ethnicize the material remains of the past and the works of the present was irresistible. The Ottoman cosmopolis was to be mostly excised from the capital's new Hellenic streetscapes, although, in fairness, the new administrative elites were equally brutal on Byzantium. The Athens Archeological Society was dominated by Classicists who had little interest in Byzantine heritage, and even less in intellectual or political arguments for preserving the remains of the Ottoman era.

The first plans of Athens, by Gustav Eduard Schaubert and Stamatios Kleanthis (1833), Leo von Klenze (1834), and Friedrich von Gaertner (1843) became the scaffold for the development of the city as national capital (Biris, 1933. Initially commissioned by Ioannis Kapodistrias, then continued by King Othon, the first plan, drawn by Gustav Eduard Schaubert and Stamatios Kleanthis, was based on a preliminary archaeological and topographical survey of Athens in 1831, in which they identified much of what the politicization of the archaeological preservation enterprise would later

Figure 2 Edward Dodwell. "Bazaar at Athens." *Views in Greece*, London, 1821, p. 51.

either save and showcase or discard. A Palace, a Royal Garden, a Cathedral Church, a University, a Library, an Academy, all of national scope, coded the city's morphology as a national state project (Figure 3).

Ioannis Travlos remarks that in Athens, out of 129 churches that existed at the beginning of the War of Independence, only 24 survived into the twentieth century. 105 churches were either demolished, remained in ruin, or were substantially transformed into other types of structures (Travlos 1960/2005: 244). Concerning the Ottoman-era built environment, Konstantinos Mpiris notes that the few Turkish structures that were preserved were given new uses:

> The Medrese became a prison; the Fetiye Mosque and the Konak, or the Voyevod's House, were turned into barracks; and the nearby Teke was remissioned as a "Western" [meaning a non-Orthodox Christian] church for the use of foreigners (1933/1960: 4–5).

Successive rearticulations of urban planning codes will take place later, notably, following the Great Fire of Thessaloniki (1917–1920), and the massive influx of exchangees of the Treaty of Lausanne (1923). New town and village plans will be prototyped in the New Lands of Macedonia and Western Thrace in response to the migration crisis. We address such plans and their connection to Hellenic geopolitics of difference in Chapters 4 to 9.

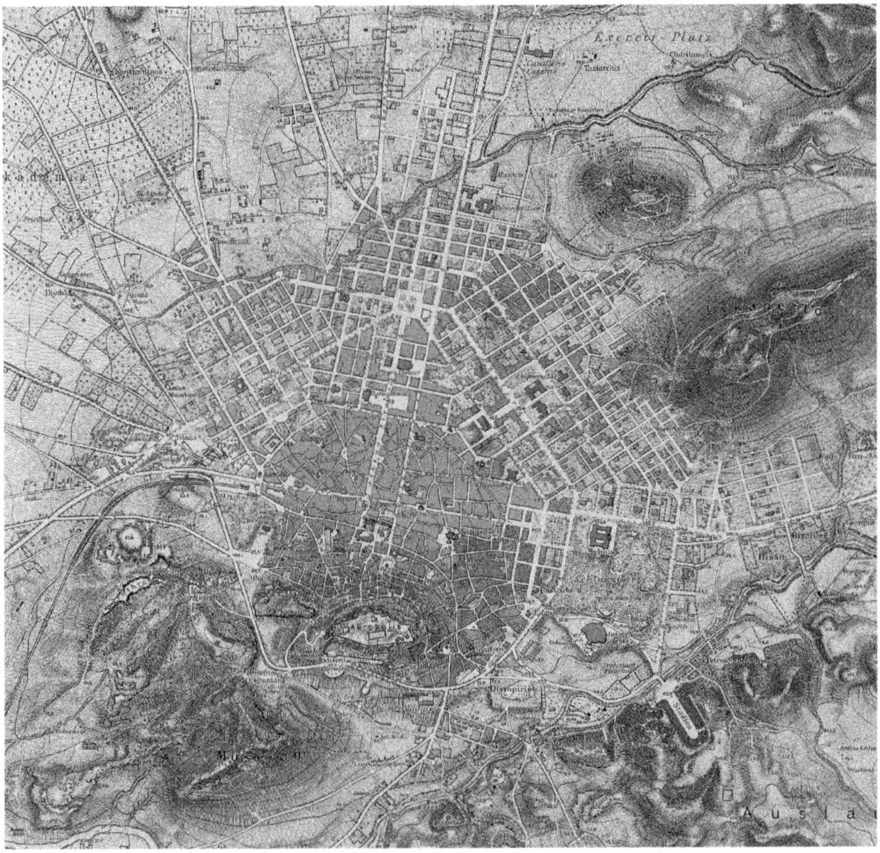

Figure 3 The 1875–1877 plan of Athens by Johann A. Kaupert (1895 edition) visualizes how early plans were translated into built forms. Despite interventions, the historic core's organic plan is still substantially intact.

Source: Heidelberg University Library Digital Collections.

The discourse of power and the geopolitics of difference

Power in various guises conditions the State's perception and management of difference in the process of statecraft: As state power entangled in the domestic political sphere at the national, sectional, prefectural, and local scales. As a Small Power engaging contiguous Small Powers in regional competition. As a Small Power instrumentalized by Great Powers and international organizations (League of Nations, NATO, the EU), achieving geopolitical ends of mutual interest, or more grimly, solely serving Great Power aims. As cultural power, leveraging (or wielding) the signifiers of Hellenism and Orthodox Christianity for domestic and international audiences in national state-making and identity politics. Also, as punitive state power directed at subaltern communities during Greece's time as national security state. And more recently, as state power conditioned by Europeanization and engaging difference in liberal, contestatory politics. We start in Chapter 1 with an exploration of the relationship of power geopolitics to the geopolitics of place and difference. That discussion of power should inform the granular studies of Hellenic statecraft in the chapters that follow.

Notes

1. Our empirical coverage of geopolitics of difference in Greece is not comprehensive. The massive migration of Albanians in the 1990s warrants separate treatment, as do the Greek-Albanian borderlands. Difference works through politics in distinct ways in globalizing places like Mykonos, Santorini, Corfu, and Rhodes. More spatially granular studies, like our work on the City of Didymoteicho, would be necessary to test our model of how contestatory politics work at the local scale, while entangling and constructing meaning at other scales of political engagement.
2. The first volume of the *History of the Morea in the Middle Ages (Geschichte der Halbinsel Morea während des Mittelalters),* subtitled *The Fall of the Peloponnesian Greeks and the Repopulation of the Empty Land by Slavic Tribes (Untergang des peloponnesischen Hellenen und Wiederbevölkerung des leeren Bodens durch slavische Volksstämme)* took direct aim on European Enlightenment and Romanticist interpretations of Hellenism and its continuity.
3. The current Constitution continues to recognize the Christian Orthodox Church's historic and continuing significance in Greek society, short of calling it the official State religion. The Constitution guarantees complete freedom of religion for "known" religions, a qualification that undermines the state's commitment to social and political liberalism and complicates relations with communities the religions of which are not Constitutionally accounted for.
4. Hannah Arendt's work on the moral sources and consequences of political responsibility spans *Eichmann in Jerusalem* (1964) and *The Life of the Mind* (1977). The binary of *verstand* and *vernunft* takes new significance in the information age.
5. To paraphrase the Stoics and Warf, individuals construct and are constructed by progressively distant, nested circles of compassion within networks of mutual obligation and assistance. These circles link body-space and individual agency to "the local," to region, to "the national," and ultimately to "the international," "the interstate," and "the global" (Warf, 2012a: 276).
6. Skeptics and critics of cosmopolitanism understand the concept and the practice as the product of liberal Western moral philosophy, evolved into a pragmatic and visible internationalism of collective action intergovernmental organizations. Cosmopolitanism is described as a project of whiteness, capitalism, Western-type social democracy, and muscular statism. David Harvey's characterization of cosmopolitanism as potentially another geographizing of banal evil, draws on Hannah Arendt's work on Adolf Eichmann's trial and political responsibility to posit "what if Arendt's characterization of evil has some subterranean connection to the banalities of "dwelling," of "place" and of "heimat" as social constructs essential to the human condition?" (Harvey, 2000a: 545). Our attachment and rootedness to place, scaled up to spaces of state- and nationcraft (even in the pre-modern sense of the Greek city-state),

can be found to be "the fount of all prejudice, aggression, and evil" (Ibid). The implication here is that cosmopolitan geographies often embed the forces and practices that bring forth abstract cosmopolitanism's demise.

7. «[Ο]λως τουρκική, τας μεν οδούς είχε στενάς, ανωμάλους και βορβορώδεις, τας δε οικίας ξυλοκτίστους, πολυ-θύρους, σεσαθρωμένας και πάντη αρρύθμους.»

Sources

Agnew, J., Checkel, J. T., Deudney, D., Mitzen, J., and Guzzini, S. Symposium on Stefano Guzzini's (ed.) "The Return of Geopolitics in Europe? Social Mechanisms and Foreign Policy Identity Crises." *Cooperation and Conflict* 52, no. 3 (2017): 399–422.

Agnew, John. *Geopolitics: Re-visioning World Politics.* London: Routledge, 2003.

Appiah, Kwame Anthony. *Cosmopolitanism.* New York: Norton, 2006.

Arendt, Hannah. *Eichmann in Jerusalem: A Report on the Banality of Evil.* London: Penguin Publishing, 1964.

Arendt, Hannah. *The Life of the Mind.* New York: Harcourt Brace Jovanovich, 1977.

Çelik, Zeynep. *Urban Forms and Colonial Confrontations: Algiers Under French Rule.* Berkeley: University of California Press, 1997.

Delis, Apostolos. "A Mediterranean Insular Port-City in Transition: Economic Transformations, Spatial Antagonism and the Metamorphosis of Landscape in Nineteenth-Century Hermoupolis on the Island of Syros." *Urban History* 42, no. 2 (May 2015): 225–245.

Durkheim, Émile. *The Rules of the Sociological Method and Selected Texts on Sociology and its Method.* Edited with an introduction by Steven Lukes; translated by W. D. Halls. New York: The Free Press, (1895)/1982.

Durkheim, Émile. "Social facts". In *Readings in the Philosophy of Social Science.* Edited by Martin, Michael; McIntyre, Lee C. pp. 433–440. Boston, MA: MIT Press, 1994.

Fallmerayer, Jakob Philipp. *Geschichte der Halbinsel Morea während des Mittelalters [History of the Morea in the Middle Ages*], Volume I *Untergang des peloponnesischen Hellenen und Wiederbevölkerung des leeren Bodens durch slavische Volksstämme* (The Fall of the Peloponnesian Greeks and the Repopulation of the Empty Land by Slavic Tribes). Stuttgart, 1830/1836.

Greenhill, Brian. "Recognition and Collective Identity Formation in International Politics." *European Journal of International Relations* 14, no. 2 (2008): 343–368.

Harvey, David. "Cosmopolitanism and the Banality of Geographical Evils." *Public Culture* 12, no. 2 (Spring 2000a): 529–564.

Harvey, David. *Spaces of Hope.* Edinburgh: Edinburgh University Press, 2000b.

Karadimou-Gerolympou, Alexandra. *Επανασχεδιασμός και ανοικοδόμηση της Θεσσαλονίκης μετά την πυρκαγιά του 1917* (Re-planning and reconstruction of Thessaloniki after the Great Fire of 1917). Doctoral dissertation, Polytechnic School, Department of Architecture and Engineering, Aristotle University of Thessaloniki, 1985.

Katzenstein, Peter J. ed. *The Culture of National Security: Norms and Identity in World Politics.* New York: Columbia University Press, 1996.

Mauss, Marcel. *The Gift: Forms and Functions of Exchange in Archaic Societies.* London: Cohen & West, 1925/1966.

Nussbaum, Martha. *For love of Country: Debating the Limits of Patriotism*. Boston, Mass.: Beacon Press, 1996.

Papadopoulos, Alex G. “Mapping ‘Romeic’ and ‘Hellenic’ Same-Sex Desire: Articulating Heteropatriarchy and Male Homosexuality in Contemporary Greece.” *Antipode* 34, no. 5 (2002): 910–934.

Roudometof, Victor. “The Social Origins of Balkan Politics: Nationalism, Underdevelopment, and the Nation-State in Greece, Serbia, and Bulgaria, 1880-1920.” *Mediterranean Quarterly* 11, no. 3 (2000): 144–163.

Roudometof, Victor. *Nationalism, Globalization, and Orthodoxy*. New York: Praeger, 2001.

Ruggie, John Gerard. “Territoriality and Beyond: Problematizing Modernity in International Relations.” *International Organization* 47, no. 1 (Winter 1993): 139–74.

Schmaus, Warren. *Durkheim’s Philosophy of Science and the Sociology of Knowledge: Creating an Intellectual Niche*. Chicago: Univ. of Chicago Press, 1994.

Shaffer, L. S. “Durkheim’s Aphorism, the Justification Hypothesis, and the Nature of Social Facts.” *Sociological Viewpoints* (fall 2006): 57–70.

Souchou, Yao. “After the Malay Dilemma: The Modern Malay Subject and Cultural Logics of ‘National Cosmopolitanism’ in Malaysia.” *Journal of Social Issues in Southeast Asia* 18, no. 2 (October 2003): 201–229.

Warf, Barney. “Nationalism, Cosmopolitanism, and Geographical Imaginations.” *The Geographical Review* 102, no. 3 (July 2012): 271–292.

Weber, Cynthia. *Simulating Sovereignty: Intervention, the State and Symbolic Exchange*.Cambridge Studies in International Relations. Cambridge: Cambridge University Press, 1994.

Wendt, Alexander. “Norms, Identity and Culture in National Security” (with Ronald Jepperson and Peter Katzenstein). In *The Culture of National Security*, edited by Peter J. Katzenstein, pp. 33–75. New York: Columbia University Press, 1996.

Wendt, Alexander. “Anarchy is What States Make of it: the Social Construction of Power Politics.” *International Organization* 46, no. 2 (1992): 391–425.

Wendt, Alexander. *Social Theory of International Politics*. Cambridge, UK: Cambridge University Press, 1999.

Zervas, Theodore G.; Papadopoulos, Alex G. “Creating Greeks and Greek-Americans: Geographic and Educational Identity Constructions at the Socrates and Koraes Greek-American Schools.” *European Education* 52, no. 1 (2020): 16–32.

Δημακοπούλου (Dimakopoulou), Χαρίκλεια. «Το Ναύπλιον Εδρα της Διοικήσεως (Κυβερνήσεως) (1823-1834)» (Nafplio, seat of administration (government)), *Ναυπλιακά Ανάλεκτα* 3 (1998): 112–125.

Δραγούμης (Dragoumis), Νικόλαος. *Ιστορικαί Αναμνήσεις* (Historical reminiscences) τ. 1–2. Αθήνα: Ερμής 1874/1973.

Δωροβίνης (Dorovinis), Βασίλειος. «Ο Σχεδιασμός του Ναυπλίου κατά την Καποδιστριακή Περίοδο (1828-1833). Η Ειδική Περίπτωση και τα Γενικότερα Προβλήματα» (The urban plan of Nafplion during the Capodistria era (1828-1833)). In Οντέτ Βαρόν-Βασάρ (ed.), *Νεοελληνική πόλη. Οθωμανικές κληρονομιές και ελληνικό κράτος* (Πρακτικά Διεθνούς Συμποσίου Ιστορίας, Αθήνα 26-28 Σεπτεμβρίου 1984, Ερμούπολη 29-30 Σεπτεμβρίου 1984), τ. 2. Αθήνα: Εταιρεία Μελέτης Νέου Ελληνισμού – Μνήμων, 1985: 287–296.

Καλαφάτη (Kalafati), Ελένη. «Η Πολεοδομία της Επανάστασης: Ναύπλιο, 1822-1830» (The Revolution's urban planning practices: Nafplion, 1822-1830), *Τα Ιστορικά* 2 (1984): 265–282.

Καλαφάτη (Kalafati), Ελένη. «Κράτος, Δημοτική Διοίκηση και Οργάνωση του το 19ο αιώνα» (State, municipal governance and its organization). In Οντέτ Βαρόν-Βασάρ (ed.), *Νεοελληνική πόλη. Οθωμανικές κληρονομιές και ελληνικό κράτος* (Πρακτικά Διεθνούς Συμποσίου Ιστορίας, Αθήνα 26-28 Σεπτεμβρίου 1984, Ερμούπολη 29-30 Σεπτεμβρίου 1984), τ. 2. Αθήνα: Εταιρεία Μελέτης Νέου Ελληνισμού – Μνήμων, 1985: 367–373.

Καραδήμου-Γερολύμπου (Karadimou-Gerolympou), Αλέκα. «Σχεδιασμός και Ανάκτηση του Χώρου της Πόλης. Χαρακτήρας της Πολεοδομικής Παρέμβασης του Κράτους κατά τη Μετάβαση από την Οθωμανική στη Νεοελληνική Πόλη» (Planning and recovery of urban space. State intervention in the transformation of the Ottoman to a neo-Hellenic city). In Οντέτ Βαρόν-Βασάρ (ed.), *Νεοελληνική πόλη. Οθωμανικές κληρονομιές και ελληνικό κράτος* (Πρακτικά Διεθνούς Συμποσίου Ιστορίας, Αθήνα 26-28 Σεπτεμβρίου 1984, Ερμούπολη 29-30 Σεπτεμβρίου 1984), τ. 2. Αθήνα: Εταιρεία Μελέτης Νέου Ελληνισμού – Μνήμων, 1985: 381–395.

Καρύδης (Karidis), Δημήτρης Ν. «Αθήνα – Αττική στον Πρώτο Αιώνα Οθωμανικής Κατοχής: Η Σχέση Πόλης – Υπαίθρου» (Athens in the First Century of Ottoman occupation: the rural-urban relationship). In Οντέτ Βαρόν-Βασάρ (ed.), *Νεοελληνική πόλη. Οθωμανικές κληρονομιές και ελληνικό κράτος* (Πρακτικά Διεθνούς Συμποσίου Ιστορίας, Αθήνα 26-28 Σεπτεμβρίου 1984, Ερμούπολη 29-30 Σεπτεμβρίου 1984), τ. 2. Αθήνα: Εταιρεία Μελέτης Νέου Ελληνισμού – Μνήμων, 1985: 50–53.

Κόκκου (Kokkou), Αγγελική. «Η πολεοδομική ανασυγκρότηση στην περίοδο 1828-1843. Κρατική πολιτική και πραγματικότητα» (Urban reconstruction, 1828-1843. Realities of state policy). In Οντέτ Βαρόν-Βασάρ (ed.), *Νεοελληνική πόλη. Οθωμανικές κληρονομιές και ελληνικό κράτος* (Πρακτικά Διεθνούς Συμποσίου Ιστορίας, Αθήνα 26-28 Σεπτεμβρίου 1984, Ερμούπολη 29-30 Σεπτεμβρίου 1984), τ. 2. Αθήνα: Εταιρεία Μελέτης Νέου Ελληνισμού – Μνήμων, 1985: 361, 363.

Μπίρης (Biris), Κωνσταντίνος. *Τα Πρώτα Σχέδια των Αθηνών. Ιστορία και Ανάλυσίς των* (The first plans of Athens. Their history and analysis). Αθήνα: Τύποις Κ. Σ. Παπαδογιάννη, 1933.

Πάλλης (Pallis), Γεώργιος. *Τοπογραφία του Αθηναϊκού Πεδίου κατά τη Μεταβυζαντινή Περίοδο. Οικισμοί, Οδικό Δίκτυο και Μνημεία* (Topography of the Athenian plain in the post-Byzantine era. Settlements, road network and monuments). Θεσσαλονίκη: Εκδοτικός Οργανισμός Π. Κυριακίδη, Αριστοτέλειο Πανεπιστήμιο Θεσσαλονίκης-Κέντρο Βυζαντινών Ερευνών, 2009.

Παπαρρηγόπουλος (Paparrigopoulos), Κωνσταντίνος. *Ιστορία του Ελληνικού Εθνους, από Αρχαιοτάτων Χρόνων μέχρι των Νεωτέρων* (History of the Hellenic nation, from the most ancient to recent times), τ. 1–5. Αθήνα: εκ του τυπογραφείου Νικήτα Πάσσαρη, 1865-1874.

Πετράκος (Petrakos), Βασίλειος. *Δοκίμιο για την Αρχαιολογική Νομοθεσία* (Treatise on legislation concerning archaeology). Αθήνα: Ταμείο Αρχαιολογικών Πόρων και Απαλλοτριώσεων (Δημοσιεύματα του *Αρχαιολογικού Δελτίου*, αρ. 29), 1982.

Τραυλός (Travlos), Ιωάννης. *Πολεοδομική Εξέλιξις των Αθηνών* (Athens' urban planning evolution). Αθήνα: Καπόν, 2005.

Φακούρα (Fakoura), Αλίκη. *Από την Οθωμανική στην Ελληνική Πόλη. Οι Λειτουργίες της Πόλης του Ναυπλίου (1715-1833): Κοινωνία, Οικονομία και Διοίκηση* (From Ottoman to Hellenic city. Nafplion's city services (1715-1833)), Μεταπτυχιακή διπλωματική εργασία. Αθήνα: Πάντειο Πανεπιστήμιο Κοινωνικών και Πολιτικών Επιστημών, Τμήμα Πολιτικής Επιστήμης και Ιστορίας, 2006 (file:///C:/Users/User/AppData/Local/Temp/Fakoura%20A1.-1.pdf Accessed on 7 August 2020).

Χλέπα (Hlepa), Ελένη-Άννα. *Τα βυζαντινά Μνημεία στη Νεότερη Ελλάδα. Ιδεολογία και Πρακτική των Αποκαταστάσεων, 1833-1939* (Byzantine monuments in Modern Greece. Ideology and conservation practices, 1833-1939). Αθήνα: Καπόν, 2011.

1 Hellenic statecraft in power politics

The limitations of hard power and the promise of collective action

The power geopolitics of Greece's founding and territorial expansion provides critical insight into the making of its geopolitics of difference. Greece's commitment to Hellenization has read difference as a real or perceived challenge to its security and territorial integrity. The blurring of boundaries between the ontologies of ideational Hellenism and material state security has traditionally constructed linguistic and sectarian difference as a threat, or at least, as evidence of geopolitical disorder.

As the state grew territorially, racialized, linguistic, and sectarian differences, once across the border, were enfolded inside it. Thus, newly acquired lands reflected complex demographic, sectarian, and linguistic complexities of the Ottoman Empire – *un fait social total* that compelled difference's reading by the state in security terms and animated its Hellenization machinery. Internal bordering would become a standard domestic security containment tactic. The border would be as much a construct of the state – a reflection of its geopolitical anxieties and a locus of military and police investment – as a cause for reorganizing the state. Greece's security leitmotif, at least until the fall of the Colonels' Junta in 1974, is distilled into two geopolitical constructs: "The peril from the North" (Ο εκ Βορρά κίνδυνος) – Bulgaria, or a more amorphous Slavic adversary, and "the peril from the East" (Ο εξ ανατολών κίνδυνος) – Turkey, which remains to this day Greece's sole territorial threat.[1]

Greece's security imaginaries were built on perceptions of its capabilities, assessments of its importance to Great Powers, appraisals of the relative strength and aims of potential regional allies and adversaries, and domestically, on the degree of support for a forward-thrust strategy of territorial expansion, and, importantly, on how ethnolinguistic and sectarian minorities figured at the time in the national security equation.

Newly incorporated territories acquired through war or treaty – the Ionian Islands in 1863, Thessaly in 1881, parts of Macedonia and Epirus in 1913, Western Thrace in 1919, and the Dodecanese Islands in 1947 – necessitated a recalibration of these security imaginaries. They also required the rebalancing of Greece's grand strategy and the reordering – too often ineffectually – of its finances. In other words, territorial expansion came at the cost of considerable structural adjustment.

Did Greece's power measure up to its strategic ambition? It depends. The Greek State – a Small Power – was made possible in 1821–1832 by Ottoman decline and Europe's imperial surge, its *de facto* sovereignty conditioned by the degree dictates of other, more powerful states, determined who governed it and how (Murphy, 1996; Agnew, 2005). That condition of absolute dependency in the first decades of its existence, gave way to outsize foreign and security policies and a matching ambition.

However, Greece was not the Empire's sole rebellious adversary. Territorially ambitious successor states – while enjoying Great Power patronage themselves – created new regional balance of power configurations. The result was two nested and interacting systems of states: one regional, including the rump Ottoman Empire and its Small Power successor states, and a more geographically comprehensive system that includes Great European Powers. For the latter, the regional system was reducible to a "theater of conflict," a "mosaic of potential junior allies or adversaries," "tradeable territories-as-diplomatic-coin," and a context for "rulemaking and regulation." The question is how effectively Greece, at the time a bellicose, expansionist, anti-status quo Small Power, could instrumentalize this playbook that limited it as much as its immediate regional adversaries?

The measure and classification of state power has been much debated (Fox, 1959; Keohane, 1969; Katzenstein, 1985; Handel, 1981, 1990; Mearsheimer, 2001, 2018; Hoyt, 2017; Long, 2017; Maass, 2009). The magnitude of the aggregate of military, economic, and diplomatic power is, in many cases, the full measure of the state. Pairings of "Small" or "Weak," and "Power" or "State" denote lesser or diminished, although not inconsequential, ability to influence the international system. To that last point, the proliferation of statehood after the Second World War, the emergence of multilateralism, and regional integration projects (notably European unification) has empowered Small States, especially in economic and environmental matters. As Joseph Stalin allegedly claimed, "There is quality in quantity." The great number of Small States has become consequential in international relations.

While the conventional view holds that a Small Power (or a Small State), like Greece, cannot pursue a substantially independent foreign and security policy in opposition to the wishes of Great Powers, a more nuanced approach finds otherwise. First, it is not only how much power a Small State like Greece has but also how it exercises it (Steinmetz and Wivel, 2010: 7). Second, *relative* power compared to other Small Power adversaries is a more meaningful measure of capability, as is the manner of its application in the region. Hence, while any power asymmetry is a strategic consideration, Michael Handel notes that power "is best measured not against all other countries, but in relation to [a Small Power's] neighbors, and by the degree to which the strength at its disposal matches its national goals and ambitions" (Handel, 1981: 52).

Stephen Walt posits that Small Powers are animated by their position in a "balance of threat," the threat measured by the aggregate of the adversary's geographic extent and proximity, demographic profile, economic and military strength, its offensive capabilities, and its offensive intentions. He continues that it is the *immediate* (meaning *proximate*) threats that drive the [Small Powers'] security policy. The context of that threat map is how they gauge their involvement in larger-scale conflicts with neutrality, alliance, and bandwagoning some of their options (Walt, 1987). Accordingly, Greece's historically outsized foreign and security policies should be gauged against the regional balance of threat or opportunity, since for Bulgaria and a diminished Ottoman Empire, Greece was, indeed, the threat source in 1912–1913, 1917–1919, and 1920–1922.

The effectiveness of Greece's pre-1922 forward-thrust foreign and security policies drew on three conditions. First, the alignment of Greece's strategy with Great Power interest and intent. In that case, Greece would gain "derivative power." Handel defines it as the output of "[t]he diplomatic art of the weak states...to obtain, commit, and manipulate, as far as possible, the power of other, more powerful states in their own interests" (Ibid). Great Powers would assess Greece's role as logistically important (e.g., Thessaloniki as Entente entrepot during the Gallipoli campaign) and gauged its military or naval contributions as consequential (for their assistance in enforcing the Entente's occupation of Thrace, and since the Cold War, the defense of NATO's southern flank). Therefore, any action Greece took would complement their own. Second, Greece was able to engage in exercises of self-help and balancing through collective security agreements, even when not explicitly directed by Great Powers. Allied with other Small Powers, Greece achieved considerable territorial and political gains in Epirus and Macedonia in the case of the Balkan Wars. The third condition was economic. Greece's ambitious foreign and security policy agenda required an equally robust economic base. Specifically, sustained state capability was contingent upon the balancing of the extravagant costs of a forward-thrust strategy and the need for economic modernization and social development.

Slow to industrialize, Greece's national economy was based on primary sector activities (some of them export-oriented), the merchant marine, and commerce. A point of inflection for the national economy was the premiership of the reformist Harilaos Trikoupis (1882–1885). He prioritized large-scale infrastructure projects that propelled economic modernization: the expansion of railroads, the Corinth Canal to facilitate commercial shipping, and the launch of the Lake Copaïs land reclamation project to increase arable land, carry out effective land reform, and address rural poverty. The Greek economy remained under-resourced and bankruptcy-prone, which again made the state dependent upon international economic patrons – some Great Power or Western financial consortium.

State-driven economic development and structural adjustment – the geography of infrastructure investment, access to credit, the regulation of property markets, business licensure – created a scaffold for geopolitics of difference. Illiberal Greek governments practiced exclusion of borderlands minorities from the goods of statism and modernization into the late twentieth century. In our studies of Western Thrace's Muslim Minority (its Turkish, Pomak, and Roma communities) and the Greek-Turkish borderlands in the Evros Prefecture, we describe how geopolitical anxiety and a closed-system Hellenic nationalism and exceptionalism produced a security ecosystem resistant to change, even after democracy was reestablished in 1974. Since the 1990s, measurable change has arrived as liberalization in a Europeanist mold that has gradually taken root and has started mitigating the effect of structural exclusion. Our study of Pomak villages in the Xanthi and Rhodope Prefectures observed both general and granular impacts of economic development and structural reforms promulgated and financed by the state and, in part, by the European Union (EU).

Geopolitical anxieties: state and post-state perspectives

Europe's security map is more than the aggregate of threats – real or perceived – to its constituent states. The EU's threat map, which now includes climate change and pandemics, is sensitive to contingencies of distance. In security terms, the EU triangulates to the world across all scales: from the European subject and the Member States, to transboundary, "neighborhood," regional, continental, and global threats that may be state- or non-state-based, may be conventionally territorial, or spatially complex, such as dynamic networks of illicit trafficking.

National and collective organizations cooperate to deliver European security. Member States of NATO and the EU (acting through its intergovernmental Common Foreign and Security Policy [CFSP] and its integrated Common Security and Defense Policy, the European Defence Agency, FRONTEX – the EU's Border and Coast Guard Agency – and EUROPOL) serve regional security and policing demands by different means and at different scales. Their scope extends beyond the geographical boundaries of their membership and includes coordinated interdiction of the trafficking of terrorism, humans, narcotics, and arms. Importantly, the intergovernmental legal basis of the EU's security arrangements allows Greece to act in the defense of its territory as necessary.

In Europeanization, small Member States appear to have additional options available beyond traditional balancing and "bandwagoning." In the best case, their diplomacy might bind more powerful Member States to policies that contribute to their security (Inbar and Sheffer, 1997; Mosser, 2001; Steinmetz and Wivel, 2010). More grimly, they can rely on self-help.

The resilience of the EU's security commitment to its Member States and the operability of its instruments are currently being tested on the Union's borders by increased migration. As a frontline border state, Greece has called on the EU to help manage the migration from the Middle East,

North Africa, West Asia, and Sub-Saharan Africa at its borders. It has an active immigration policy that is the subject of some controversy, generating vigorous legislative debate and review and continuous comment by the press. State-EU cooperation extends to the coordinated interdiction of terrorism, human, narcotics, and arms trafficking.

The record and perceptions of threats to the state and nation coexist with (and within) the institutional, substantially intergovernmental European security ecosystem. Historical experiences with adversaries, war chronicles and eyewitness accounts, and the memorialization of past conflicts in the habitual sphere (as in the war memorials in Evros' borderlands villages) and in school texts, constitute a world of information and ideas that channel geopolitical anxiety. Such ideational constructs may or may not map upon rationalist assessments of threat and peril. Or perhaps they once did but no longer do. Two such constructs had a durable impact on Greece's geopolitical discourse: "The peril from the North" and "The peril from the East."

"The peril from the North"

Greece no longer faces state-based threats on its northern frontier. Its neighboring states have been joining NATO and the EU: Bulgaria acceded to NATO in 2004 and the EU in 2007, Albania and North Macedonia acceded to NATO in 2009 and have official EU candidate country status, and Greece and North Macedonia settled the issue over the name of the former Yugoslav Republic of Macedonia in the 2018 Prespa Agreement. Outside bilateral relations in the immediate neighborhood, Greece manages non-state challenges, such as immigration, on its behalf, and on behalf of the EU.

While the end of the Cold War and Europeanization have made "the peril from the North" strategically irrelevant, it remains a formative "code" of Greek geopolitics (Gaddis, 1982; Huliaras and Tsardanidis, 2006: 465). When the code is also a norm, it is constitutive of society and actors. The earliest formulation dates to the national awakening of Slavs in the Ottoman Empire in the nineteenth century. As geographically ambiguous warning, it has influenced security policy and domestic political and social cohesion. It was an intellectual tool for statecraft, inflected by popular imaginaries of the state's relative position in the world (Dijkink, 1998: 293). The specter of danger supplied an emotive charge.

Until 1913, the only power sharing a land border with Greece was the Ottoman Empire. Following the Congress of Berlin in 1878, however, it was clear to the Great and Small Powers that future territorial retrenchments were likely to occur in the Balkans at the expense of the Ottomans. Alongside guerrilla warfare against the Empire, a covert war would break out in the 1890s in Ottoman Macedonia among Bulgaria, Greece, and Serbia, paving the way for military action. On 8 October 1912, the Balkan League, comprising of Bulgaria, Greece, Montenegro, and Serbia, invaded the Ottoman Empire. The aims of this First Balkan War were to roll back the Ottoman's European dominion and partition Epirus and Macedonia among them. Unsatisfied with its territorial gains, Bulgaria launched the Second Balkan War against Greece for control of Macedonia on 29 June 1913, losing to the combined forces of Greece, Romania, Serbia, and Turkey.

Following the conclusion of the Balkan Wars, the Treaty of Bucharest in August 1913 awarded Greece considerable lands in Epirus and Macedonia. Bulgarian resentment over its territorial losses found an outlet in the First World War. Greece and Bulgaria again warred, resulting in Bulgaria's loss of Western Thrace to Greece, including its access to the Aegean and Mediterranean Seas. The deep enmity of these contests shaped "the peril from the North" security imaginary with resonance in both international and domestic security concerns. Here too is located Greece's historical geopolitical anxieties over the challenge to the State and Hellenicity posed by Slavophone Greeks in Macedonia and Western Thrace.

At the onset of the First World War, Bulgaria's strategic location between Austria-Hungary and the Ottoman Empire offered the Entente a range of military opportunities. The British Admiralty was divided in its assessment of the relative geopolitical value of Greece and Bulgaria. Great Britain was particularly intent on Bulgarian neutrality and, even more so, on recruiting Bulgaria to join the war effort, even if it meant pressuring Greece to make territorial concessions. For example, the Allies offered Bulgaria Greek territory, including the City of Kavala and its Eastern Macedonian hinterland, as an incentive and reward for joining the Entente Alliance. Serbian territory was similarly objectified. Both the Greek and Serbian governments rejected such dealings outright (Silberstein, 1970: 139–140).

Winston Churchill, as First Lord of the Admiralty, was particularly keen on the Bulgarian scenario. After British forces failed to capture the Ottoman forts protecting the entrance to the Dardanelles in February 1915, Churchill blamed the reversal on Britain's failure to recruit Bulgaria into the alliance (Robbins, 1971: 583–534). He wrote, "... other circumstances will intervene in the Balkans unless we can gain Bulgaria to our cause or attack Constantinople before the end of September [1915] without her, and these other circumstances may be fatal to the issue of the war and disastrous in a peculiar degree to Great Britain" (Gilbert, 1966: 1093). Countering that view, Greece's steadfast treaty commitment to Serbia, based on their common understanding of Bulgaria as a regional threat, shielded it from Britain's most predatory imaginings of a strategic understanding with Bulgaria (Bataković, 2004: 64, 66–67).

It was, ultimately, Bulgaria that rejected the Entente's offer and sided on 14 October 1915 with the Triple Alliance. It aspired to secure considerable territorial concessions at the expense of Greece and Serbia to realize some version of a Greater Bulgaria (Hall, 2011: 301; Christensen, 1997: 65–67; Christensen and Snyder, 1990: 138–139; Kenneth Waltz, 1979: 67, 165–169). "Chain ganged" to Germany, Austria-Hungary, and the Ottoman Empire, as the junior partner of an alliance not of its making, Bulgaria operated fitfully on the Balkan front without significant military cooperation with its allies.

Bulgaria's decision to join the war on 14 October 1915 on the side of the Triple Alliance was geopolitically countercultural. On the one hand, Bulgaria was intent on defeating, if not destroying, Serbia to settle their long-standing competition for Macedonia. But on the other hand, Bulgaria was joining its historical adversary, the Ottoman Empire, in a war against an alliance that looked to Bulgaria's great patron, the Russian Empire, as a critical participant. The poor state of Russo-Bulgarian diplomatic relations at the time partly explains this seeming paradox. Bulgaria believed that Russia had ignored its claims to Macedonia in the 1913's Bucharest treaty conference, effectively ceding Eastern Macedonia to Greece (Hall, 2011: 301–303).[2] In that diplomatic context, the Triple Alliance's territorial incentives for joining were both attractive and extensive. However, frankly, so were the enticements offered by the Entente.

Greece had been garrisoning British and French expeditionary forces in Thessaloniki since 5 October 1915. It officially joined the war on the side of the Entente on 17 June 1917, 5 days after the abdication of Germanophile King Konstantinos I. Once again Eleftherios Venizelos assumed Greece's premiership. Capitalizing on his reputation as a diplomat and statesman, he aligned Greek foreign and security policies to those of the Allies, thus insolating Greece from a direct attack by Bulgaria.

At the end of the First World war, Bulgaria irretrievably lost Western Thrace. The chance of reacquiring the Macedonian lands it had conceded at the Treaty of Bucharest in 1913 also evaporated. A strategy of aggressive Bulgarization of Western Thrace started in 1913 and went underground during the joint French-Greek occupation. By terrorizing and displacing Turkish and Greek rural communities, Bulgarian irregulars aimed to maintain a Bulgarian presence there in the hope of winning some accommodation of Bulgarian interests, such as the declaration of an Autonomous State of Western Thrace. Chapter 5 will show how adroit policies to protect the Muslim Minority (the Pomak community being also Slavophone or Pomakophone) on the part of the Venizelos government and demonstrations of good administration on the part of the Greek Occupation administration, persuaded France, the League of Nations' Mandate Power, to support Greek annexation of Western Thrace in 1919 (recognized by Bulgaria in 1923).

"The peril from the North" security discourse would see several permutations in the decades that followed. In the inter-war period, Bulgaria will remain a regional antagonist of Greece, Yugoslavia, and Turkey, who all signed the Balkan Pact in 1934 to contain it (later amended in 1938, and abolished by Nazi Germany in 1941). The dictatorial Ioannis Metaxas regime (1936-1941), built massive fortifications on the Greek-Bulgarian borderlands in anticipation of an invasion "from the North." Nazi Germany, allied with Bulgaria, breached that hard line of defense and occupied Greece during the Second World War. In the Greek Civil War that followed in 1945–1949, "the peril from the North" will be cast as Greece's revolutionary Left, allied with Albania, Bulgaria, and Yugoslavia, and, ultimately, the Soviet Union (Figure 1.1).

During the Cold War, the defense of Greece and Turkey's borders became a collective security responsibility of NATOs Southern flank. Greece reasonably calculated at the time that with NATO's brawn on its side, it is Turkey – "the peril from the East" – that represented a real threat to its

Figure 1.1 Poster of the Rightist Popular Party used in the elections of 1946 at the onset of the Greek Civil War. "Hellenes, beware! Those who truly love Greece vote Popular Party." The Communist Party (KKE) is laying down the carpet for Slavic-Communist invaders. The tower is Thessaloniki's signature monument. There is no ambiguity to the geopolitical signing: Greece is under threat, and the peril comes from the North.

Source: Wikimedia – In the public domain.

territorial integrity and warranted its full security attention. Geopolitical concerns increasingly plagued Greek-Turkish relations throughout the post-Cold War era.

"The peril from the East" and "the Muslim arc"

First as the Ottoman Empire, then as the Republic, Turkey has historically been considered the legacy "peril from the East," notwithstanding interludes of amicable relations and cooperation. Since the creation of the Greek State in 1830–1832, Greece and Turkey have fought four wars: in 1897, 1912–1913, 1917–1918 (the First World War), and 1920–1922. All sprung from Greece's aspirations for a Greater Greece encompassing most of Eastern Thrace and legacy parts of Asia Minor. Its defeat in the Greek-Turkish War of 1920–1922 was in part the result of strategic and tactical miscalculations and in part because Greece went to war against the wishes of the Great Powers. As in war-making, so in peace-making, the Great Powers this time working bilaterally and through the League of Nations influenced the contours of the resulting Treaty of Lausanne – an enduring, critical international agreement shaping Greek-Turkish relations and a touchstone of Greek geopolitics of difference.

Following the Second World War, membership in intergovernmental, multilateral organizations – including NATO – created new outlets for hard and soft power politics. The United States replaced Great Britain as the Cold War hegemonic power in the region. Early NATO members – Italy, Greece, and Turkey – constituted the alliance's southern flank.

Greece and Turkey's joint membership, however, did not denote consistent alignment with the aims of the alliance. Both concerned themselves with domestic security and regional priorities, including the disposition of Cyprus' decolonization from Great Britain. Starting in 1954, Greece and Turkey turned antagonistic over competing security priorities. Their actions concerning Cyprus threatened the operational integrity of the alliance in the Eastern Mediterranean. In 1974, they found themselves at the brink of war over Cyprus.

Dissonance between national and NATO actions persisted to the close of the Cold War. A 1985 US Senate hearing entitled "Europe Without America" laid out the strategic dilemma of the US in stark terms:

> Senator PRESSLER [SD-R]: [D]o we truly have a "crisis of will?" Is there something structurally obsolete in NATO; does it need to be changed? I also have a second question on a subject that we have not touched on the soft underbelly of NATO, the Greek – Turkish conflict in the Aegean that has been called NATO's weakest link. Do you believe that NATO's southern flank is adequately protected?
>
> ...
>
> Senator MATHIAS [MD-D]: I would prefer the chairman's description of it as the southern flank. NATO does not have a soft underbelly.

Mr. RAVENAL [Georgetown University, School of Foreign Service]: You are referring to the defection, practically speaking, of Greece and the weakening of the southern flank because of the continuing altercation and the diversion of defense resources by Greece, particularly against Turkey, and to some extent by Turkey, I suppose, as well?

Senator PRESSLER [SD-R]: Yes.

(United States Senate Hearing, 19 September 1986: 66–67).

The two states came close to clashing in 1974 over Cyprus and in 1996 concerning the Island of Imia. In both situations, the United States played a critical mediating role. As new crises over territorial and maritime disputes unfold, US leadership will once again play a critical role. Its absence will test the will of the EU's Common Foreign and Security Policy (CFSP) and Common Security and Defence Policy (CSDP).

The Muslim arc

Amidst the tumultuous events of 1991, a year that saw the dismantling of Soviet Union and beginning of collapse of Yugoslavia, a new geopolitical code appeared in the Greek press to eclipse the traditional idea of "the peril from the East." A new interpretation of the old "peril from the North" was introduced: "the Muslim arc," with Turkey as its anchor in the east. The mass influx of Albanians to Greece (and Italy) following the fall of Albania's totalitarian regime, also in 1991, ignited fears of an invasion. Muslim populations in the successor states of Yugoslavia, notably the former Yugoslav Republic of Macedonia (now North Macedonia), Bosnia, and, later, Kosovo, were seen as potential targets of Turkish diplomacy. To the east, Bulgaria's Muslim minority represented approximately 8 percent of the national population, two-thirds of whom identified as ethnically Turkish. It was also considered a potential catalyst that would align Bulgaria with Turkey over domestic politics. From the conservative Greek security perspective, the concern was over the emergence of a "Turkish" as opposed to an "Muslim arc" (Papathemelis, 1994). Huliaras and Tsardanidis posit that Greeks feared Turkey's manipulation of the arc to create conditions suitable for Muslim secessionist movements. In Greece's case, Western Thrace – a Greek territory with about 100,000 Muslim inhabitants – would be the likely target of Turkish diplomacy (2006: 465–472).

It was a case of adapting the traditional geopolitical code of an eastern peril to fit the realities of a post-Cold War world. The long record of competition, if not enmity, between Greece and Turkey made the fear a potent imaginary. Although, as popular geopolitical discourse has faded since the late 1990s, Turkish rhetoric about championing Muslim minorities in the Balkans can only rekindle the "arc" discourse.

The Muslim arc code may be considered an expression of geopolitical paranoia. In its most extreme articulations, it prognosticates the end of time.

However, the Turkish counterpart of that code – a forward-thrust, status quo-disrupting Neo-Ottoman geopolitics – is evident. Addressing an audience of Justice and Development Party (AKP) members in Sakarya, Turkey, on 13 March 2018, President Erdoğan translated the code to imagine a geography of unredeemed lands currently inhabited by ethnic Turks in neighboring Balkan and Middle Eastern countries:

> From this magnificent place I personally want to congratulate my brothers in Bosnia and Herzegovina, Kosovo, Albania, Macedonia, Serbia, Western Thrace, Crimea, Bulgaria and Romania … "We send our greetings to all the victims and oppressed brothers of ours in Sarajevo, Skopje, Xanthi, Komotini, Kardzhali, and Mostar. We share our cordiality with these brothers whose souls and eyes are turned to Turkey, for those who pray for the success of Turkey … Every time I say it – these cities are physically located in the borders of other countries, but they are part of our spiritual boundaries. The meaning of Turkey does not fit within 780,000 kilometers. The half of our hearts are in Istanbul, Diyarbakir, Trabzon, Antalya, Izmir, and the other half is Aleppo, Kirkuk, Jerusalem, Sandzak, and Bukhara.

The subject fits Erdoğan's domestic political discourse and, on first reading, can be interpreted as political rhetorical myth-weaving. It suggests that all Muslim communities in the region are ethnically Turkish, part of a historic diaspora yearning for stronger ties with Turkey. However, it also posits that Muslim heritage writ-large in the region, regardless of a community's ethnolinguistic, or otherwise defined identity, is durably connected to the Ottoman legacy and the Empire's Caliphal authority established in 1517. According to this speech, the Turkish national project under AKP rule is mapped upon the Ottoman Caliphate.

The Bulgarian Foreign Ministry responded to the speech by pointing to the perilous subjectivity of cultural bordering and affirmed the status quo:

> Regarding the comment made by the President of the Republic of Turkey, Recep Tayyip Erdoğan, referring to Kardzhali, the Ministry of Foreign Affairs of the Republic of Bulgaria would like to point out that:
>
> - Any country with a rich and long history could claim that its spiritual and cultural borders go beyond the physical ones.
> - If the political rhetoric in a country for some reason opts to hark back to the past, it can, in nearly every case, outline boundaries that depart from the present ones.
> - In reality, today's state borders have been affirmed by international treaties, recognized by all states and are not subject to questioning or revising.

> Bulgaria and the Republic of Turkey, enjoying excellent bilateral relations, are no exception to these agreements.
>
> (Communiqué of the Ministry of Foreign Affairs, Republic of Bulgaria, 13 March 2018).

Any of the countries referenced – Greece's Prefectures of Xanthi and Komotini have Muslim minorities – could have responded similarly. The immediate effect of statements of this sort act as accelerant within Turkey's nationalist circles and roil up neighborhood politics. The chief question becomes to what degree and by what means does such rhetoric generate actionable policy and strategy. Concerning the spin that the Neo-Ottoman discourse gives to "the Muslim arc" code, Europe-minded governments have good cause to resist Turkey's diplomatic pull in the region.

A structural realist perspective on the Greece-Turkey balance of power

Greece and Turkey are of significant geopolitical value to the United States and NATO. Middle Power Turkey is actively restructuring itself into Great Power status to pursue regional dominance, seemingly unrestrained by NATO's norms and obligations. Since 1974, Small Power Greece has been reorienting its grand strategy toward Europe with an understanding that it would sustain concessions to sovereignty. At the same time, it has strived to build as credible a military deterrent as its resources and democratic political culture would allow.

By the metrics of the US structural realism, Turkey appears geopolitically stronger than Greece in terms of hard power. That has been a constant even when Turkey ignores US direction. Henry Kissinger's statement on the relative importance of the two countries was representative of the common US stance toward Turkey during the Cold War:

A discussion on the direction of US policy concerning the evolving 1974 Greek-Turkish crisis over Cyprus (Washington, 26 September 1974):

> …
>
> Senator Fulbright: Why not take it to the UN. Then we wouldn't have all the responsibility.
>
> President Ford: The Greeks and the Turks both trust us.
>
> Secretary Kissinger: Giving it to the UN is a pro-Turk move because the UN cannot do anything [about Turkey's second offensive in Cyprus], and the situation would freeze as it is. If we move away from the Turks, the Soviet Union will probably move toward them. Turkey is more important to the Soviet Union than Greece …
>
> (Memorandum of conversation, Kissinger papers, Box CL 124)

… and by the function of a bipolar balance of power, Turkey was more critical to the United States than Greece.

If the United States could not be reliably counted as a counterweight to Turkish expansionism, could the EU? Despite the *acqui*'s stated commitment to the collective security of both countries, a similar calculus appears to in fact be at work in the EU-Turkish relations. Although the EU already extends itself diplomatically to shield Greece and its own interests from external challenges, under current law, EU military assistance to Greece will not be forthcoming in the event of a conflict between Greece and Turkey.

However, if the EU, also, is an unreliable source of security for Greece, individual EU Member States, operating beyond the limits of the CFSP, participate in the regional security system in pursuit of their own interests and factor into Greece's security calculations. In other words, conventional deterrence, bilateralism, and its political armature remain at the center of Greek strategy.

The strategic asymmetry has required Greece to recalibrate its international relations. Also, since its accession to the European Communities in 1981, treaty commitments to the *acqui* contributed to progressive change in its politics of difference.

With strong support from France, a fellow NATO member, Greece applied for membership to the European Communities (EC) in 1976. That same year, the EC proposed the adoption of a "Common Declaration by the European Parliament, the Council, and the Commission concerning the respect of Human Rights" (COM (76) 374 Vol. 1976/0109) requiring all members to respect and protect human rights in addition to their commitment to pluralistic government. The Declaration was integrated into the *acqui* in 1977. The Commission's opinion on Greece's membership application pointed to deficits in these areas (Dagtoglou, 1984: 149). Alignment with Europe's standard of democratic government and civil society was a prerequisite to accession to the EC and the only way Greece could secure European support for its post-Junta political rehabilitation and economic development.

Greece's membership in the EU serves as a soft power multiplier. Arguably, membership also increases Greece's hard power profile. The EU's intergovernmental CFSP and its integrated CSDP are aimed at "strengthen[ing] the EU's external ability to act through the development of civilian and military capabilities in Conflict Prevention and Crisis Management" through "external action" (Articles 42–46, Treaty of European Union). By building out its strategic and operational capacities to address security and humanitarian crises, the EU builds up its overall defensive capacity and advances a sense of One Europe with One set of Borders. By expanding its autonomous defense capabilities, Europe etches out a hard-power profile.[3]

Security burden-sharing between Member States and the Union scrambles the logic of state size, strength, and geographic position as hard corollaries of state power. It would require significant expansion and development of the *acqui*. There are, however, handholds in the *acqui* for powering-up Europe's Common defense. For example, FRONTEX, the EU's Border and Coast Guard Agency, is no equivalent to a European Army guarding EU's borders. Built on the concept of Integrated Border Management (IBM), it was designed to enhance capacity and leaves primary responsibility for border security to the Member States. Increasingly central to the EU's global strategy, it is presently capable of granular interventions in border hotspots. Such policy articulations, such as the IBM may allow crossovers and spillovers that would further institutionalize defense at the European level.

CFSP and CSDP allow Member States "which see their common defence realised in NATO" to defer to NATO. Were the United States to retreat from its primate position in NATO, CFSP and CSDP would likely need to be redefined to fill the power vacuum. That would propel the institutional EU and the Member States into greater prominence in logistical and military planning and preparedness, articulated as a new call for strategic autonomy. This is not an unprecedented call. The Saint-Malo Summit (1998) that negotiated the shape of the CSDP germinated ideas about reduced dependency on the United States and NATO, In 2003, in general opposition to European involvement in the Iraq War, a somber discussion started about what that concept may mean in policy terms. By 2016, three different conceptual visions of strategic autonomy had emerged: autonomy as responsibility, autonomy as hedging and autonomy as emancipation (Fiott, 2018: 1-8). The damage done to the US-EU relationship by the Trump administration may not be long-standing, however, the EU was shown in 2017-2020 tangible arguments in support of some version of strategic autonomy, short of emancipation, For its part, the Biden administration will go to great lengths to rebuild confidence among its European allies that the United States is fully committed to NATO and European security. American pushback on calls for European strategic autonomy is a certainty, should they advance to action.

The delicate balance sought between EU and NATO security commitments is disrupted when states belonging to these organizations are also competitors. Security relations between Greece (a member of both NATO and the EU-CFSP) and Turkey (a NATO member) are an example. The probability of military conflict between them either by strategic design or accident has increased under President Recep Tayyip Erdoğan. The reasons for their geopolitical enmity most critical to current and future relations include the unresolved Cyprus issue (started in 1954, it remains the most intractable issue), the Aegean littoral dispute, minority-related issues (referring here to Western Thrace's Muslim community), foreign policy and security dogmas and doctrines, enemy stereotypes, virulent nationalism (at least on the part of extremist parties), and the curating of security-state personas (Heraclides, 2019: 4–5). The challenge is keeping the competitive relationship from turning adversarial, and when it turns adversarial to keeping the peace.

From a structural realist perspective, Turkey's strategic dogmas of the last 30 years and its record of substantial investment in its armed forces, offer evidence that it aims to become a regionally dominant, if not a global power. Described in the 1990s as a "coercive regional power" in hardline realist terms, Turkey appeared to de-escalate the tone and substance of its aims in the 2000s, when membership in the EU was central to its foreign policy. The emerging Euro-Asianist turn of its foreign policy was cast as "soft" (Öniş and Yilmaz, 2009: 7–8) and consistent with Europeanization (Tüysüzoğlu, 2014: 89). Euro-Asianism was articulated as a forward-thrust strategic vision: the Doctrine of Strategic Depth announced in 2001 by Ahmet Davutoğlu, Erdoğan's Prime Minister and Foreign Minister.

Described as Neo-Ottoman in scope and texture, this doctrine urged Turkey to resuscitate its historical ties to the Balkans, the Middle East, North Africa, and the Eurasian interior in order to best represent its interests as a Eurasian state actor. Initially portrayed as "zero problems with neighbors," the foreign and security policies underlying the doctrine are anti-status quo, hence disruptive of established international law, intrusive, and expansionist, and an irritant to neighborhood relations. Moreover, since 2016, Turkey has downgraded, if not compromised, its strategy of Europeanization as relations with the EU have deteriorated to the point that "de-Europeanization" of Turkish foreign policy may be more apropos (Heraclides, 2019).

Departing from their roots as a secular republic, Turkish governments under the Premiership and then the Presidency of Erdoğan have instrumentalized Islamic, Pan-Turkish, and Pan-Turanist ideologies to form the bases for a hegemonic claim over the Middle East and the Central Asian Turkic belt (Taşpinar, 2011: 1–2; Murinson, 2006: 945). That strategic posture was sharpened further through the "Blue Homeland Doctrine" introduced by Turkish Admiral Cem Gurdeniz in 2006 on behalf of the Government. The primary objective was regional naval dominance, nominally as a NATO partner, but in realist pursuit of national security interests.

In 2009, Turkey reaffirmed energy security as central to its developing economy. The well-documented coordination between the Ministry of Foreign Affairs and the Ministry of Energy and Natural Resources (MENR) describes Turkey's efforts to bolster its energy security by diversifying supplies and suppliers. MENR's vision for 2023 (also the hundredth anniversary of the Turkish Republic) rests on five economic and sector-specific strategic pillars: (1) energy supply security, (2) Turkey's regional and global effectiveness in the field of energy, (3) the environment, (4) natural resources, and (5) corporate structure (Bilgin, 2015: 73). The links between energy, foreign policy, and hard-power geopolitics may be apparent but not explicitly spelled out in strategic dogma or MENR planning documents and legislation. How Turkey plans to fulfill the first two pillars of Vision 2023 raises questions and potential[?] alarm about regional stability.

Establishing naval superiority and eventual dominance over the Black, Aegean, and Eastern Mediterranean Seas, would allow Turkey to control continental shelf fossil fuel exploitation, gas pipelines, and the liquified gas tanker traffic in service of its energy security and still broader geopolitical objectives. These actions directly implicate and threaten the interests of Greece, Cyprus, Israel, and Egypt within their legally-delimited Exclusive

Economic Zones, and increase security costs for all. The second targeted "envelope" includes the Caspian, Red, and Arabian Seas, the Persian Gulf, Saudi Arabia, the Arab Gulf States, Iran, and, importantly, the other Euro-Asianist power, Russia (Ibid: 78–79). Such foreign policy élan predates the United States' gradual retreat from the region since 2017, and it may reasonably be assumed that the power vacuum caused by US inaction has further emboldened Turkish and Russian strategy aims.

Since the 2001 Erdoğan-Davutoğlu pivot, Turkey has been code-switching from "US partner" to "regional competitor." Under such circumstances, old standards, such as assessments of the relative geopolitical value of Greece and Turkey as assets to US strategy, do not entirely make sense anymore. Turkey is building diplomatic, security, military procurement, and financial network links that have, partially, unhinged it from US oversight. A possible pivot in the United States' position on Greek-Turkish relations may occur when the United States recognizes that its primacy in the Eastern Mediterranean region is challenged not only by Iran and Russia but also by Turkey. By operating as an anti-status quo power in the broader region (in Iraq, Israel, Iraq, Libya, Syria, and the Eastern Mediterranean), Turkey directly impacts US interests, taxes US military assets, and undermines US naval primacy. By saber-rattling with EU Member States, Greece, and Cyprus, Turkey introduces stressors in the EU security ecosystem that have resonances on the EFSP, NATO, and by implication, on the EU-US security relationship. This is not necessarily good news for Greek security. Any reduction of US influence on Turkey introduces uncertainty and instability in the region. The EU has yet to exhibit any interest in balancing Turkey by any means other than financial.

The new security demands of Turkey's forward-thrust geopolitics situate Turkey and Greece at odds over economic and territorial issues. Turkish doctrine, actions, and démarches are antagonistic to the economic and territorial security interests of Greece – and by extension the EU. Under President Erdoğan, the Turkish government has made revisionist border, territorial, and continental-shelf claims in the Aegean and the Eastern Mediterranean Seas (the latter implicating Cyprus, Egypt, Israel, Libya, and, ultimately, the United States' naval primacy). It questions the terms of the Treaty of Lausanne, opening the door to potentially calamitous political and territorial claims in the region, specifically in the Balkans, Syria, and Iraq. It continues to instrumentalize Western Thrace's Muslim Minority rights as diplomatic leverage and a means of unsettling Greece's domestic politics and social cohesion. Nearly daily violations of Greek and Eurocontrol airspace by Turkish fighter planes – now visible to the EU's FRONTEX – threaten military escalation. Its unwillingness to negotiate the reunification of Cyprus in good faith perpetuates the territorial division of an EU Member State that primarily hurts the population in the Turkish-occupied sector. Lastly, it uses Greek-heritage religious monuments and institutions in Turkey, such as UNESCO-protected Hagia Sophia, and the

Halki Patriarchal School, as pressure points and bargaining pawns. As the scope of the European Defence Agency (EDA) and the CFSP expands, these transgressions are no longer, singly, Greece's concern.

The Europeanization of Greece's domestic politics and its regional geopolitics should be understood not as a replacement for but as an addition, support, or extension to its self-help posture. If neither the United States nor the EU is willing to contain Turkish expansionism, Greece will need to actively seek other security partners and develop its own regional balance of power geopolitics, while maintaining maximum military capabilities.

In a Colonel Blotto world

> A dominating strategy:
>
> There are ten locations with respective value $a_1 < \ldots < a_{10}$. Player i ($i = 1, 2$) is endowed with n_i soldiers ($n_i < 10$) and must allocate them among the locations. To each particular location, he can allocate no more than one soldier. The payoff at location p is a_p to the player whose soldier is unchallenged, and $-a_p$ to his opponent, unless both have a soldier at p or no one has, in which case the payoff is zero to both. The total payoff is obtained by summing up local payoffs … (Moulin, 1986: 30–31)

The classic Colonel Blotto game lays bare the hard contours of the application of hard power at "the local" and scales upward. Of course, strategy matters and produces different outcomes beyond this demonstration of a dominating strategy equilibrium. The broader claim here is that statecraft and the domestic and transnational geopolitics of difference are partly conditioned by, and grounded in regional hard-power geopolitics. If the constructs of "the peril from the East" and "the Muslim arc" are sufficiently imaginative and emotionally charged to move public perception and influence government policy, so are the concrete facts of the military balance of power.

Turkey enjoys significant military superiority over Greece. That limits Greece's deterrence strategy based strictly on hard power. Yet, the physical determinants of probable zones of contention between Greece and Turkey – in Thrace, the Aegean Sea, and the Eastern Mediterranean – make outcomes uncertain, regardless of Turkish arms superiority. Further, Turkey's wide-ranging Euro-Asianist commitments may require Turkey to disburse its hard power across several conflict zones in Syria, Northern Iraq, and Libya, in addition to its domestic campaign against the Kurdish militia on its own soil. However, the Turkish-Greek balance of power increasingly favors Turkey's offensive capabilities.

Greece's defense expenditure is outsized for its economy and surpasses the NATO guideline of 2.0 percent of gross domestic product (Tables 1.1 and 1.2). Maintaining that level of spending was especially burdensome during the leanest years of the Eurozone crisis, but domestic support for continuing to fund defense programs mostly unchanged was strong across political parties.

Table 1.1 Defense expenditure of NATO countries – billions USD (2012–2019).

Budget year	*2012*	*2013*	*2014*	*2015*	*2016*	*2017*	*2018*	*2019*
Greece	5.63	5.31	5.23	4.52	4.63	4.75	4.85	4.84
GR real GDP	*202*	*196*	*197*	*196*	*196*	*199*	*202*	*207*
Turkey	13.89	14.43	13.59	11.96	12.65	12.97	14.15	13.92
TR real GDP	*710*	*771*	*810*	*860*	*887*	*953*	*978*	*952*
NATO Europe	264.52	269.44	270.43	235.34	237.27	250.62	275.97	284.04
NATO E r. GDP	*15,526*	*15,640*	*15,945*	*16,282*	*16,603*	*17,058*	*17,372*	*17,568*

Source: Press release. NATO Public Diplomacy Division (25 June 2019).

Cooperation and collective security are fundamental to the enhancement of armed forces. Greece is committed to making qualitative improvements in armaments and operational capacity in alignment with the guidelines of NATO and the EU CFSP. Greece National Defense Procurement Policy in encoded in Law 3978/2011, which transposes European Directive 2004/18/EC and Defence and Procurement Directive 2009/81/EC. The Law fully aligns national policies and regulations to the European Defence Agency's common standards: quality and environmental management, security clearance, security of supply, supply chain management, payment to contractors, e-procurement, and the resolution of disputes (EDA, 2020: 40–44). Military services integration based on an EU rubric ties Greece's hands and represents loss of sovereignty but also confers tangible material and political advantages.

Bilateralism is at least as critical to Greece's strategic dogma as European multilateralism. Cooperation at the Community level overlays bilateral diplomatic relations, some of which are long-standing and vital to Greece's military preparedness. Greece purchases arms and military technology from many of the world's top suppliers, including France, Germany, Italy, the United Kingdom, Belgium, and Israel. The EU is also an emergent arms supplier (Eurofighter, Eurocopter). Greece's commitment to purchasing or leasing arms of significant value consolidates these relationships, and the diversification of its portfolio shields it, to a degree, from political pressure.

Greece's top strategic source of arms, however, is the United States for reasons that include a desire to match the quality of US arms in Turkey's arsenal. Turkey's procurement of the F-35 fighter jet has been derailed in

Table 1.2 Defense expenditure as a share of GDP and annual real change – Percent in 2015 USD prices (2012–2019).

Budget year	*2012*	*2013*	*2014*	*2015*	*2016*	*2017*	*2018*	*2019*
Greece	2.29	2.21	2.21	2.30	2.38	2.34	2.23	2.24
Turkey	1.59	1.52	1.45	1.39	1.46	1.52	1.85	1.89
NATO Europe	1.55	1.52	1.47	1.45	1.46	1.48	1.53	1.58

Source: Press release. NATO Public Diplomacy Division (25 June 2019).

the US Congress because of its purchase of the S-400 Russian anti-aircraft system. These procurements, should they be completed, would significantly increase Turkey's regional air superiority (*Financial Times*, 2019: online).

Suppliers compete for lucrative long-term contracts that include in-country co-production and subcontracting. Since Greece returned to a democracy in 1974, France and Greece have invested considerable political capital in their alliance. Greece's tenders for two intermediate-sized frigates have attracted bids from France and the United States. That includes one from the naval shipbuilding company *Naval Group*, which produces France's FDI/BELH@RRA frigate in collaboration with the French Defense Procurement Agency (DGA) (*Naval Technology*, 2020: online). The United States has counter-offered four *Lockheed Martin* Multi-Mission Surface Combatant (MMSC), including all weapon systems and missiles, for approximately the same cost of EUR 2 billion. The company commits to building the vessels in Greek shipyards, under its supervision (*Naval News*, 2019: online). Greece accepted the French offer, adding the two BELH@RRAs to its fleet.[4]

Infrastructural improvements benefit NATO's allies and extend NATO's own operational reach. Moreover, the United States has long-maintained basing access in military bases in Greece established by bilateral agreement. The base in Souda, Crete, is an important anchorage and supply base for the USN's Sixth Fleet. The United States has expressed interest in the privatization and expansion of the Port of Alexandroupolis. Its ambassador to Greece, Jeffrey R. Pyatt, called it "a crucial link to European energy security, regional stability, and economic growth … [I]t makes sense that the United States and Greece have chosen here to work together to advance our shared security and economic interests (Press release, 13 September 2019).

Defense of the homeland and diplomacy form the basis of current Greek strategic dogma, both headlined by commitments to the respect of international law and human rights. The defense pillar is organized around military and logistical preparedness for homeland defense and for participation in NATO and CFSP actions, as defined by their respective collective governance bodies. The diplomacy pillar commits the state's political and diplomatic resources and good offices to improving relations and fostering economic development in the neighborhood and the broader region. Greece supports the status quo, the inviolability of borders, and the resolution of disputes by peaceful mediation through international courts. Its defense dogma is mapped on to that of the *acqui*, in alignment with EDA, but also permits an independent self-help strategy. Compared to Turkey, Greece is less sovereign (like all Member States); it is committed to institutionalist, collective action schemas that prescribe means and methods of military management, but is otherwise unrestrained from catering to its territorial integrity. Should deterrence fail, a long, multiple-front defensive war promises grave consequences. However, in the Cold War spirit of France's "force de frappe," Greece appears to be prepared to exact a high price for them.

Beyond military defense, Greece's grand strategy significantly matches the EU's *Six Priorities for 2019–2024*: (1) a European Green Deal (making it the first climate-neutral continent) with significant linkages to Common transportation and energy policies; (2) a digital age Europe (investing in technological innovation for economic competitiveness and popular empowerment); (3) a people-centered economy (still neo-liberal but committed to social fairness), (4) a stronger Europe in the world (calling for "strengthening Europe's unique brand of responsible global leadership"); (5) promotion of a European way of life (calling directly for "a Union of equality"); and (6) a new push for European democracy (nurturing and strengthening institutions and civil society, while building firewalls to protect European democracy from extremism) (European Commission, 2019).

The EU's goals for a Green Deal and global economic competitiveness are tied directly to Greece's plans for investment in its transportation and energy infrastructures. Greece's goal is to make Greece, and the broader partnership it has forged with Cyprus, Israel, and, lately, Egypt, a major energy hub and clearinghouse for Europe and the Middle East. Actions taken in pursuit of this goal include The Energy Triangle (2010), a natural gas extraction plan created by Cyprus, Greece, and Israel; the Eastern Mediterranean Gas Forum (2019) with Cyprus, Greece, Israel, Jordan, Italy, and the Palestinian Authority; and, most[?] importantly, the EastMed gas pipeline agreement (2020) between Greece, Cyprus, and Israel, among others (Tziarras, 2019: 5–10). This extensive energy diplomacy is founded on bilateral agreements establishing maritime boundaries between the partners and demarcating their respective exclusive economic zones (EEZ).

Turkey disputes the substance of the agreements, pitting the partnership's plans in direct competition with the Turkish energy security strategy and, by extension, The Blue Homeland doctrine. Added to disputes over the continuing division of Cyprus, the undermining of "Lausanne" and the politicization of Greece's minorities, the delimitation of national waters in the Aegean Sea, air space violations, the weaponization of war refugees at Greece's land border with Turkey and Greece's littoral islands, and incendiary rhetoric, Greek-Turkish relations are at the lowest point since 1954–1955 or 1974. How Turkey's strategy impacts the interests of Greece's partners and what their responses might be would demonstrate the resilience of Greece's energy and security strategies.

The Europeanist 'turn'

Conventionally, distance and time favor the defense, as the offensive power needs first to reposition its forces. However, Greece's prospects in a Colonel Blotto world are unfavorable. The doctrinal distance between Greece's and Turkey's dogmas and the means for applying them is considerable. For a Small Power, violation of international law can be catastrophic, as the Cyprus crisis of 1974 (and the Arab-Israeli War of 1948 before it) have made clear.

An abject lesson about the limitations of hard power in regional geopolitics came in 1974. On July 15th of that year, a coup d'état by Greece's military Junta toppled the democratically elected government of Cyprus, led by Archbishop Makarios III. It installed a short-lived puppet regime intent on declaring the unification of the island republic to Greece – calling for "ENOSiS" (Union) – a rogue application of an older nationalist imaginary of a Greater Greece that extremist nationalists had revived in 1954.

This unlawful action contravened the terms of the 1960 Treaty of Guarantee, the international agreement that underwrote Cyprus' decolonization and independence from Great Britain. The treaty made Great Britain, Greece, and Turkey the guarantors of the new state's sovereignty. The Junta violated Article II, Paragraph 2 of the Treaty whereby "Greece, Turkey, and the United Kingdom ... undertake to prohibit, so far as concerns them, any activity aimed at promoting, directly or indirectly, either union of Cyprus with any other state or partition of the Island." Article IV, Paragraph 2 allowed unilateral action: "In so far as common or concerted action may not prove possible, each of the three guaranteeing Powers reserves the right to take action with the sole aim of re-establishing the state of affairs created by the present Treaty."

Greece's violation of Cyprus' sovereignty and, by extension, the Treaty, provoked Turkish intervention for the declared purpose of establishing the status quo *ante*. The Treaty is vague on acceptable means for achieving that goal and does not preclude military intervention. Turkey invaded and occupied 3 percent of the north part of the island, precipitating a crisis that removed the coup leaders and brought down Greece's Junta. The return of Cyprus' internationally-recognized government nominally established the status quo *ante*, satisfying the rationale for Turkey's intervention as guarantor. One month later, Turkey engaged in a second phase of military action that increased its occupational sector to 37 percent of the island. Neither Great Britain nor the United States sought correction action against Turkey for its violation of Article II, Paragraph 2 of the Treaty. Since then, the United Nations has maintained a peacekeeping mission on the island where occupation continues 46 years later, despite repeated international efforts for reunification. Although Turkish Cypriots' historical grievances have been recognized and incorporated into a range of proposed agreements for reunifying the island republic, Turkish occupation of the island's north has precluded a solution.

Greece's primary strategy is the principled rules-based engagement with the world through bilateral and multilateral institutions and agreements. It remains strongly committed to maintaining as great a military deterrent as possible in the spirit of self-help, in tandem with an investment in deeper European military and diplomatic activism through institution-building. Critically, however, the EU is neither legally able within the *acqui* nor willing to apply hard power to defend its southern flank Member States, Greece and Cyprus. Continuing deference to NATO provides Brussels and

Strasbourg with some diplomatic cover. That bleak finding demands that Greece continue providing maximally for its own territorial security.

In this grand-scale geopolitics, issues of difference appear diminished. However, they are to a substantial degree constitutive of Greece's geopolitics and political culture. An example is the consequences following the imbrication of minority policies to security and economic development in Macedonia's and Western Thrace's borderlands. The policies and their adverse outcomes for minority populations have thrown in sharp relief the stark contours of political, social, and economic exclusion, posing the dilemma Greece's political parties and society must face: what kind of a republic does Greece want to be.

The *acqui* provides a legal road map for how to achieve liberal goals but offers no solutions or strategies for addressing the potential effects they might have on a country's international relations beyond the EU's fold. These decisions still rest in the hands of national executives. Since Greece's accession to the EC in 1981, and especially after the creation of the EU in 1993, Greece's obligation to conform to the *acqui* has led to the progressive liberalization of nationalist perspective on difference – a critical contribution the country's civil society turn. However, democratic reversals in Hungary and Poland and surges of nationalist and extreme rightist parties in the EU warn that civil society gains are reversible. Greece's challenge became how to abandon the garrison state mentality and practices that deemed civil rights incompatible with national security and embrace a liberal government paradigm without compromising external security.

Notes

1. Bulgaria is now Greece's EU partner, but Greece's extreme Right still uses the construct to demonize the European Union as a scourge (from the North) more dangerous than neo-Ottoman Turkey.
2. As was standard practice in diplomatic negotiation since the Congress of Berlin in 1878, Bulgaria provided ethnographic maps of the region, which indicated that the majority of the population was Bulgarian (however, that was deduced from Ottoman sources) as well as Turkish. In contrast, the Greek element was both minor and spatially limited to clusters along the Aegean Sea littoral zone. That claim is illustrated in the 1912 so-called Professors' map of Bulgaria-in-the-Balkans' ethnography. It is attributed to Anastas Ishirkov, Lyubomir Miletich, Benyo Tsonev, Yordan Ivanov, and Stoyan Romanski and was published by Pettermanns in 1915. Notably, Ishirkov is considered the founder of geographic science in Bulgaria.
3. What are the options for small states like Greece? Per liberal intergovernmentalism, Europeanization is shaped in a two-step process: first, Member State governments aggregate domestic elite and constituency interests. In the second step, they are introduced to the intergovernmental branch – the Council and non-EU-based circuits of bargaining and negotiation – which make decisions that reflect the relative power and influence of Members. In that view, the EU's supranational organs are secondary to the process (Moravcsik, 1998). With options limited to balancing and "bandwagoning," small Member States

would participate in the building of the Union in ways that may, at best, bind large Member States to policies that are beneficial to them (Inbar and Sheffer, 1997; Mosser, 2001; Steimetz and Wivel, 2010). Since that influential articulation, some attempts have been made to find common ground among rationalist schools (neorealism, neoliberalism, and rational-choice institutionalism). Mark Pollack (2000) suggests that the theories converge where they identify fixed preferences and rational behavior among all actors in the EU as important. They also focus on how Member governments engage European institutions that constrain and channel their behavior. Institutionalists interpret the policy-making as substantially visioned and applied by the supranational Commission, the Parliament, and the European judiciary. Constructivists claim that ideations-turned-norms and social facts imbricated into constituency beliefs and interests are more important than rationalist frameworks.

4. Now all procurement contracts have to comply with the *acqui*, which attempts to balance neoliberalist commitments to market efficiency, public finances, and Member State/Union strategic priorities. Thus far, concerning non-EU suppliers, such as the United States, NATO's primacy over the EDA serves as a political equalizer.

Sources

Agnew, John. *Hegemony: The New Shape of Global Power.* Philadelphia, PA: Temple University Press, 2005.

Agnew, John. "No Borders, No Nations: Making Greece in Macedonia." *Annals of the Association of American Geographers* Vol. 97, No. 2 (2007): 398–422.

Bataković, Dušan. "Serbia and Greece in the First World War: An Overview." *Balkan Studies* Vol. 45, No. 1 (2004): 59–80.

Bilgin, Mert. "Turkey's Energy Strategy: Synchronizing Geopolitics and Foreign Policy with Energy Security." *Insight Turkey* Vol. 17, No. 2 (2015): 67–82.

Christensen Thomas; Snyder Jack L. "Chain Gangs and Passed Bucks." *International Organization* Vol. 44 (1990): 137–68.

Christensen, Thomas J.; Snyder, Jack L. "Multipolarity, Perceptions, and the Tragedy of 1914." *International Studies Quarterly* Vol. 55, No. 2 (2011): 305–308.

Christensen, Thomas J. "Perceptions and Alliances in Europe, 1865-1940." *International Organization* Vol. 51 (1997): 65–97.

Dagtoglou, P. "The Southern Enlargement of the European Communities." *Common Market Law Review* Vol. 1, No. 21 (1984): 149–162.

Dijkink, Gertjan. "Geopolitical Codes and Popular Representations." *GeoJournal Vol.* 46 (1998): 293–299.

European Commission (EU). *6 Commission Priorities for 2019-2014* (2019). https://ec.europa.eu/info/strategy/priorities-2019-2024_en (accessed on 26 June 2020).

Fiott, Daniel. "Strategic Autonomy: Towards 'European Sovereignty' in defence?" *European Union Institute for Security Studies (EUISS) Brief 12* (2018): 1–8.

Fox, A.B. *The Power of Small States: Diplomacy in World War II.* Chicago, IL: University of Chicago Press, 1959.

"FTI Medium-Size Frigates." *Naval Technology* (January 2020) (accessed on 30 June 2020) https://www.naval-technology.com/projects/fti-medium-size-frigates/#:~:text=The%20French%20Navy's%20medium%2Dsize,displacement%20of%20approximately%204%2C500t.).

Gaddis, John L. *Strategies of Containment.* Oxford: Oxford University Press, 1982.

Gilbert, M., ed. *A Century of Conflict, 1850-1950: Essays to A. J. P. Taylor.* London: Hamish Hamilton, 1966.
Hall, Richard, C. "Bulgaria in the First World War." *The Historian* Vol. 73, No. 2 (2011): 300–315.
Handel, Michael I. *Weak States in the International System.* New York: Frank Cass, 1981.
Heraclides, Alexis. "Greek-Turkish Relations and Conflict." Edited by Heraclides, Alexis; Alioğlu Çakmak, Gizem. *Greece and Turkey in Conflict and Cooperation: From Europeanization to De-Europeanization.* London: Routledge, 2019: 3–12.
Heraclides, Alexis; Alioğlu Çakmak, Gizem. *Greece and Turkey in Conflict and Cooperation: From Europeanization to De-Europeanization.* London: Routledge, 2019.
Hoyt, Timothy. *Military Industry and Regional Defense Policy: India, Iraq and Israel.* London: Routledge, 2017.
Huliaras, Asteris; Tsardanidis, Charalambos. "(Mis)understanding the Balkans: Greek Geopolitical Codes of the Post-communist Era." *Geopolitics* Vol. 11 (2006): 465–483.
Inbar, Efraim; Sheffer, Gabriel, eds. *The National Security of Small States in a Changing World.* London: Frank Cass, 1997.
Katzenstein, Peter J. *Small States in World Markets.* Ithaca, NY: Cornell University Press, 1985.
Keohane, R.O. "Lilliputians' Dilemmas: Small States in International Politics." *International Organization* Vol. 23, No. 2 (1969): 291–310.
Long, Tom. "Small States, Great Power? Gaining Influence Through Intrinsic, Derivative, and Collective Power." *International Studies Review* Vol. 19, No. 2 (June 2017): 185–205.
Maass, Matthias. "The Elusive Definition of the Small State." *International Politics Vol.* 46, No. 1 (2009): 65–83.
Mearsheimer, John J. *The Tragedy of Great Power Politics.* New York: W.W. Norton & Company, 2001.
Mearsheimer, John J. *Great Delusion: Liberal Dreams and International Realities.* New Haven: Yale University Press, 2018.
Moravcsik, Andrew. *The Choice for Europe: Social Purpose and State Power from Messina to Maastricht.* Ithaca: Cornell University Press, 1998.
Mosser Michael W. "Engineering Influence: The Subtile Power of Small States in the CSCE/OSCE." In *Small States and Alliances.* In E. Reiter and H. Gärtner (eds.), Heidelberg: Physica, 2001: 63–84.
Moulin, Hervé. *Game Theory for the Social Sciences.* New York: New York University Press, 1986.
Murinson, Alexander. "The Strategic Depth Doctrine of Turkish Foreign Policy." *Middle Eastern Studies* Vol. 42, No. 6 (November 2006): 945–964.
Murphy, Alexander B. "The Sovereign State System as Political-Territorial Ideal: Historical and Contemporary Considerations." In T. J. Biersteker and C. Weber (eds.), *State Sovereignty as Social Construct.* Cambridge, U.K.: Cambridge University Press, 1996: 81–120.
(NATO) "Defense expenditure of NATO countries – billions USD (2012–2019)." Press release. NATO Public Diplomacy Division (25 June 2019).

(NATO) "Defense expenditure as a share of GDP and annual real change – Percent in 2015 USD prices (2012–2019)." Press release. NATO Public Diplomacy Division (25 June 2019).

Öniş, Ziya; Yilmaz, Şuhnaz. "Between Europeanization and Euro-Asianism: Foreign Policy Activism in Turkey during the AKP Era." *Turkish Studies* Vol. 10, No. 1 (2009): 7–24.

Papathemelis, Stylianos. "The Aims of Greek Foreign Policy in the Balkans." In P. Tsakonas (ed.), *Greek Foreign Policy. Domestic and International Considerations*. Athens: Odysseas, 1994.

Pollack, Mark A. *International Relations Theory and European Integration*. Brussels: EUI Working Paper RSC No. 55, 2000.

Robbins, Keith. "British Diplomacy and Bulgaria 1914-1915." *The Slavonic and East European Review Vol.* 49, No. 117 (Oct., 1971): 560–585.

Silberstein, Gerard E. *The Troubled Alliance: German-Austrian Relations, 1914–1917*. Lexington: University Press of Kentucky, 1970.

Steinmetz, Robert; Wivel, Anders, eds. *Small States in Europe. Challenges and Opportunities*. London: Routledge, 2010.

Taşpinar, Omer. "The Three Strategic Visions of Turkey." *Center on the United States and Europe at Brookings. US-Europe Analysis Series*, No. 50 (8 March 2011): 1–5.

Tüysüzoğlu, Göktürk. "Strategic Depth: A Neo-Ottomanist Interpretation of Turkish Eurasianism." *Mediterranean Quarterly* Vol. 25, No. 2 (2014): 85–104.

Tziarras, Zenonas, ed. *The New Geopolitics of the Eastern Mediterranean. Trilateral Partnerships and Regional Security*. Nicosia, Cyprus: Friedrich-Ebert-Stiftung and Peace Research Institute Oslo, 2019.

"US Trying To Counter France's Frigate Offer To Greece With 4 MMSC." *Naval News* (31 October 2019) (accessed on 30 June 2020 https://www.navalnews.com/naval-news/2019/10/u-s-trying-to-counter-frances-frigate-offer-to-greece-with-4-mmsc/).

Walt, Stephen M. *The Origins of Alliances*. Ithaca, NY: Cornell University Press, 1987.

Waltz, Kenneth. *Theory of International Politics*. New York: McGraw-Hill, 1979.

"Why Turkey's S-400 Missile Purchase Angers the US." Financial Times Online, 12 July 2019 (accessed on 30 June 2020 https://www.ft.com/content/8d03f422-a270-11e9-a282-2df48f366f7d).

Wivel, A. "From Small State to Smart State: Devising a strategy for influence in the European Union." In R. Steinmetz and A. Wivel (eds.), *Small States in Europe: Challenges and Opportunities*. Farnham, England: Ashgate, 2010: 15–30.

United States Senate. "Crisis of Will in the NATO Alliance," Hearing Before the Subcommittee on European Affairs of the Committee on Foreign Relation, Ninety-ninth Congress, First Session, on US Policy East Europe, West Europe, and the Soviet Union, September 19, 1985.

2 Historiography as geopolitics of statecraft

The Byzantine turn and the crafting of a modern Hellenic identity

The state made possible

The Greek War of Independence (1821–32) resulted in a plausible post-Ottoman state when the retreat and ultimate collapse of the Ottoman Empire were far from apparent.[1] The new state's territorial, ideological, and governance characteristics revealed the increasing importance of European imperialism in the region and the transformative power of European nationalism married to military, economic, and political modernization. In the course of the Long Nineteenth Century, and after intellectual and political contestations, Greece's elites configured modern Hellenic identity and gave shape to the ideological and territorial contours of its geopolitics. One such contestation, about the historical-geographic character and durability of Hellenism, was catalytic for identity politics, and its outcomes are still discernable in Greek public discourses. The resurgence of nationalism today calls for a fresh look at arguments for, and the structure and implications of national exceptionalism and its relationship to geopolitics. The rise of a wrathful new nationalist politics in Greece, animated by both the Left and the Right, has implications for the stability of the Balkans, the European Union, and, importantly, Greece's fragile relationship with the Turkish Republic.[2]

The post-Vienna Ottoman retreat

The possibility of creating an independent Hellenic State in 1830–32 was made manifest, in part, because of power disparities between the Ottoman Empire and the Great Powers – especially the Russian Empire, Great Britain, and France – and in part as a result of narrow but effective nationalist agitation within the Ottoman Empire by its subject people (the latter encouraged and animated by European elites). Concerning the geopolitics of the post-Napoleonic era, the Ottoman Empire's most significant threat came from the Russian Empire. Russia continued to pursue its long-standing geostrategic goal of access to open seas, and, more narrowly and importantly, of controlling the Bosphorus, which links the Black

Sea to the Mediterranean. It continued to assault Ottoman supremacy in the Balkans, the Black Sea, and the Caucasus throughout the nineteenth century. On the landward side, Russian imperial ambitions in the Balkans were checked by Austrian aspirations. The Hapsburg Empire wished to extend its frontier southward at the expense of a relatively weaker Ottoman Empire, which was preoccupied with the containment of Balkan nationalist revolutions. The thalassocracies of Great Britain and France found benefit in the creation of a small and dependent Hellenic State in the Eastern Mediterranean littoral but not in the destruction of the Ottomans, who were seen as an essential bulwark against Russian expansionism. A fine geopolitical line had to be walked. On 20 October 1827, in a decisive naval action in the Bay of Navarino against the massed Ottoman and Egyptian fleets, Britain, France, and Russia threw their weight behind the Greek War of Independence and determined its outcome.

This major reconfiguration of the map was inconsistent with the geopolitics of the 1815 Congress of Vienna, which disapproved of security actions that destabilized the post-Napoleonic balance of power. Two factors qualified the Vienna rules: Russia was relentless in its territorial expansion, and the Ottoman world was increasingly becoming a relative power vacuum and an opportunity target. Then, Vienna rules did not apply to vulnerable Ottoman borderlands and parts of the Ottoman Empire's Mediterranean littoral that was especially vulnerable to naval power projection. Therefore, in conventional realist and balance of power sense, international system conditions were consistent with an attempt to wrest lands off the Ottoman Empire to craft one or more successor states that would be structurally dependent and politically and diplomatically beholden to one or more Great Power patrons. Great Power strategies intended to reduce Ottoman power, first, did not include the complete dismantling Ottoman lands in Europe, North Africa, and the Middle East, and second, visioned different roles in the international security system for a diminished Turkey and the constellation of successor states that would emerge from its lands. The landward powers – Austria and Russia – were directly interested in the territorial expansion of their empires in Europe and increased their diplomatic and economic influence upon the successor states.

In contrast, British and French interests on the future of Ottoman lands were shaped, first, by global colonial strategies that would benefit from control of key territories in the Eastern Mediterranean (most notably, Egypt and Cyprus), and second, by their desire to deny Russia structural and territorial benefits in the region. Ultimately, a new regional balance of power was in the making by 1821, the year the Greek War of Independence broke out. It will take its final form in 1922 with the Republic of Turkey's creation, and the crystallization of borders in the region until the end of the Cold War would revise the Balkan map again following the dismantlement of Yugoslavia.

While the historical geopolitics of the Ottoman imperial state's dismantlement are well known, less is known about the geopolitics of difference and the geographies of statecraft and nationhood of the fledgling Hellenic State in the nineteenth century. Its small territory and near-complete dependence on Great Power patronage describe the plausible, barely defensible, Small Power. What distinguishes Hellenic statecraft from other such exercises in the Balkans in the age of nationalism is the potency of spatializing (and weaponizing) history to support and catalyze the State's territorial ambitions. Greek politicians, academic historians, and intellectuals of statecraft amalgamated and elevated cultural traditions (that nominally belonged to an expansive, broadly defined historical Hellenic universe) into a cogent imaginary of a modern Hellenic State. They employed a grand idea of a modern-Hellas-as-inheritor-of-ancient-and-medieval-Hellas as a basis for thickening the state, shaping its governmentality, and engaging in territorial expansionism.

Such cultural strategies with political ends were (and continue to be) part of the well-known nation-state-making playbook in the Balkans. The complex architecture of the historical argument of cultural "continuity," how it is politicized, and how it is grounded and territorialized distinguishes the Greek case. Moreover, for political and cultural geographies, in particular, the Greek case is further distinguished by its internal contradictions and its liminal qualities: neither East nor West, but both; the new state as a geopolitical instrument (and weapon) of the West in the East; both ancient and modern; burdened to live up to idealized expectations of outsiders (as was the case with philhellenism and its eventual decline); both pagan and Christian; both an arc of Western civilization and a European periphery; demographically modest locally and nationally, at times spatially interdigitated with others but also expressive of a vast diaspora.

Exceptionalism as nationalism

What it is to be a "Hellene" in the twenty-first century is still tethered to intellectual constructs and statecraft strategies of the nineteenth century. We explore here the role that Greece's "national historiographer," historian Konstantinos Paparrigopoulos' major work, *The History of the Hellenic Nation, from Ancient Times to the Present*, played, first, in shaping the way Greeks, in the newly created Greek State, read the region's past, and second, in articulating ideas that animated the State's foreign and security policy-minded elites. Paparrigopoulos' historical thesis promotes the idea of an unbroken, continuous, durable, and vital Hellenism, which originated in the Classical Era. Concerning "the medieval," Paparrigopoulos, alongside historian Spyridon Zambelios, is substantially responsible for resuscitating Byzantium as a "Hellenic Middle Ages." His argument that the Hellenic Middle Ages were as formative of nineteenth-century Hellenism as Classical Greece was, redefined the teaching of Greek history in the era of state formation and expansion. Accordingly, Byzantium, and its attendant value system and geopolitical culture – and its territorial footprint straddling Eurasia – became a fundamental source of political- and cultural-geographic signification.

Importantly, the reintegration of Byzantine history into nineteenth century political and cultural discourses illuminated the paths between understanding Byzantine spatiality and using it in pursuit of the national interest. The same can be said about the geopolitical usefulness of Orthodox-Christian spatialities to the Hellenic State: the expansive and dynamic geographical footprint of Orthodox Christian kingdoms and empires of Southeastern Europe and Asia Minor, and the significant material culture that branded and blazoned landscapes across the Eastern Mediterranean and the Black Sea regions. Paparrigopoulos' hard-baking of Byzantinism and Christian Orthodoxy into both the identity of the new state and the definition of its historical mission and national interest, would have significant implications on Greece's relationship to Great Powers, especially concerning Russian imperial ambitions in the region. Between 1860 and 1922, Paparrigopoulos and his milieu, which included well-connected political elites, mobilized the Hellenic part of this medieval space-time in the service of forward-thrusting, expansionist foreign and security policies. Accordingly, historiography became geopolitics and strategic dogma. Historiography became the fulcrum of statecraft by constructing a national Hellenic logic contrary to the realities of the Balkans' racialized and cultural difference.

In that same period, in Europe and the United States, Hellenism was co-opted, integrated, and digested into an historiographic argument for Western mastery and "continuity." It allowed John Stuart Mill to epigrammatically claim that "the battle of Marathon, even as an event in English history, is more important than the battle of Hastings." By claiming Hellenism, and also Romanitas, for their own, the Great Powers weaponized them in pursuit of empire. The Helleno-Roman legacy became the touchstone for an ethic and practice of Western supremacy and, by extension, white supremacy.

Hellenism's geographic expansiveness in Paparrigopoulos' thesis provided an intellectual basis and political justification for the pursuit of the Great Idea starting in mid-nineteenth century: a grand state that incorporated extensive Balkan and Asia Minor territories deemed patrimonial. Often, these lands were inhabited by either majorities or pluralities of Greeks, or by Orthodox Christians doctrinally subsumed by Hellenism. As of 1860, the Great Idea as a communication, if not strategic doctrine, justified military campaigns. When the wars were successful, they extended the geographic footprint of the State to Thessaly, Crete, Epirus, Macedonia, and Thrace, and after the conclusion of the First World War, briefly, to the coast of Asia Minor. the Great Idea was, ultimately, extinguished by the defeat of the Hellenic State in the Greek-Turkish war of 1919–1922. The Treaty of Lausanne of 1923 established the border and regulated ethnoreligious community relations between the two adversaries, putting an end to Greece's territorial ambitions and transforming it into a small, *status quo*-oriented power.

Challenges of the Treaty of Lausanne by Turkey intensified in 2017–2020. The "Neo-Ottoman" security doctrine that has been honed by the administration raises the specter of geopolitical crises in Thrace and the Aegean Sea littoral, and the Turkey-Syria-Iraq triangle. Calling former Ottoman imperial territories, including Greek Thrace and the Aegean Islands, "the lands of our heart," Erdoğan abandons the Davutoglu Dogma of "no trouble with neighbors" and resuscitates qualitatively similar arguments to the ones that once justified the Hellenic State's anti-*status quo* Great Idea, thus increasing the likelihood of conflict in the region and beyond (Yanik, 2011: 80–81).

The contest between enlightenment and romanticism: resuscitating "the medieval" and Byzantium

Konstantinos Paparrigopoulos worked counter to the prevailing scholarly tradition of Greek Enlightenment, which looked at Classical and Hellenistic Greece as sources of signifiers of modern Hellenic identity. Without intending to diminish the signifying role of Hellenic and Greco-Roman antiquity, or undermine the prestige it lent to the new state, Paparrigopoulos, an agent of Greek romanticism, rehabilitated the Byzantine *qua* medieval past as "the Hellenic Middle Ages" and wove it into a grand narrative of the greatness of the Hellenic nation. It is a topic of continuous debate whether that historiographic turn – which lionized the Orthodox religion and placed it at an equal footing as Classicism – derailed Greece's intellectual commitment to European Enlightenment values. To wit, conventional modernization-as-Europeanization did not occur, and Greece's idiosyncratic modernization with European and post-Ottoman Balkan features took a circuitous route: commitments to Westernization and all its attendant characteristics (secularism, a civil constitution of society, liberalism, and industrial capitalism). Instead, Paparrigopoulos attempted to sequence the ancient and medieval claims on Hellenism – no matter how heterodox – to construct (or contrive) a claim to an ethnic, now national, Helleno-Christian, or more precisely, Helleno-Orthodox heritage. To rationalize a concatenation of a Hellenic antiquity that led to a Hellenic medieval period, he needed to render Orthodox Christianity into a normative pillar of nationalism and statecraft for Greece's mid-nineteenth century Europeanist political and intellectual milieu. He accomplished this delicate task by constructing a two-spirited Hellenic Middle Ages, articulating the life force and temperament of both Byzantine and Classical and Hellenistic traditions. This historiographic foundation crystallized the nation's historical-geographic imaginaries and animated the state's drive for territorial expansionism.

Paparrigopoulos started to publish his extensive *History of the Hellenic Nation* in 1861. It would prove to be his most significant work and establish him as Greece's "national historiographer." The work's three, then five volumes, would be published over several years and completed in 1874. His *History* had its starting point in prehistory and the legendary beginnings of Hellenism in antiquity, and concluded with the creation of the Hellenic State in 1830. His decision to divide the text of the first edition into "books" (15 in all) following ancient practice, communicates his desire to produce more than a piece of scholarship but something akin to a *Gesamtkunstwerk* (a total

Table 2.1 Bibliometric outline of Konstantine Paparrigopoulos' *History*

Tome	*Book*	*%*	*Hellenism*	*Period*	*Number of chapters*	*Number of pages*	*Total pages*	*%*
1α	1	33,3	(First)	Prehistoric era	2	1-115	718	22,1
	2			First historic era		117-315		
1β	3		Ancient	Mede or Persian Wars	4	1-191		
	4			The Peloponnesian War	3	194-300		
	5			Decline of First Hellenism	3	301-408		
2α	6	13,3	Macedonian	Philippos - Alexandros	2	1-135	314	9,7
	7			[Hellenistic era]	3	137-316		
2β	8	6,7	Christian	Roman dominion	4	5-265	260	8
3α	9	26,7	Medieval	Gradual Hellenization of the eastern state	6	1-281	1133	34,8
3β	10			Reformation [Iconoclasm]	7	5-311		
4α	11			The Macedonian Dynasty	6	5-308		
4β	12			The Komnenoi – The Crusades	7	5-249		
5α	13	20	New	Dawn of Hellenism	9	3-424	829	25,5
5β	14			Great trials of New Hellenism. Internal structure & organization.	8	6-229		
6α	15			The great revolution and the first political establishment of the new Hellenism.	6	11-196		

piece of work) about Hellenism (Table 2.1). He extensively edited the work in 1885–1887, reordering some of the content, writing new sections, and, importantly, "improving" his language by adopting a stringently archaizing style. It would be essential to note here that to construct a new Hellenic state and craft a durable, and historically meaningful identity, intellectuals of statecraft created a modern linguistic idiom that was significantly based on Attic Greek of the fourth century BCE. Therefore, not the content alone, but the idiom in which it was delivered had great political significance. The audiences for this work included Greek elites, Greek and foreign scholars, and through the popularization of his work over decades, all Hellenes in metropolitan Greece and beyond. With the expansion of public education, Paparrigopoulos' "continuity" theory became pervasive. Pavlos Karolidis compiled a posthumous edition of the *History* in 1932, which extended the chronological coverage of the work to 1930, but left intact the thesis as it had evolved by 1874. The Paparrigopoulos-Karolidis' *History* still resonates across Greece's K-12 public education establishment; parts of it (or at least its ideas) are included in history curricula today. Thus, it is still playing a role as a source of Greek historical-geographic imagination 150 years after its inception.

In his main thesis, Paparrigopoulos adopted and promoted the "continuity" of the Hellenic nation from antiquity to the Byzantine period, its endurance during the centuries of Ottoman dominance, and its ultimate efflorescence on the first half of the nineteenth century. "Continuity" as the fundamental phenomenon is organized and catalyzed by one protagonist: "Hellenism." Its essential qualities are transmitted from one period to another "through the sacrament of the transfiguration of Hellenism from its ancient form to its new" (P 1932, 2b, 95).

He counted five Hellenisms, one succeeding the other: the first was the ancient Greek (or Hellenic) civilization, which faded gradually following the defeat of Athens by Sparta in the Peloponnesian War (a very problematic historical rupture that was greatly exploited by his polemicists – especially Jakob Philipp Fallmerayer). A "Macedonian Hellenism," a "Christian Hellenism," a "Medieval Hellenism," and finally, a "'Νεος'/Late or Modern Hellenism" followed it (Table 2.2). Intentionally or not, Paparrigopoulos placed his argument in the perennial nationalist camp by defining a patrilineal genealogical framework that attempted to explain the morphogenetic association of these manifestations of Hellenism. Except for the first Hellenism, which had the privileged position of progenitor-father, the rest

Table 2.2 Paparrigopoulos' five Hellenisms. The *History* of forefathers is written from the perspective of the "son"

Five Hellenisms	*Generation*	*Optic from origin*	*Optic from mid-19c*
First Hellenism Prehistory-Battle of Chaeronea in 338 BCE	**Ancient Hellenism**	Father	Great-great-grandfather
Second Hellenism Ascendancy of Macedon to the Battle of Actium in 31 BCE	**Macedonian Hellenism**	Son	Great grandfather
Third Hellenism Rise of Roman Empire to fall of Rome in 476 CE	**Christian Hellenism**	Grandson	Grandfather
Fourth Hellenism Fall of Rome to the sack of Constantinople by the Fourth Crusade in 1204	**Medieval Hellenism**	Great-grandson	Father
Fifth Hellenism Palaiologan Byzantium, to the fall of Constantinople, ending with the formation of the new state in 1830)	**«Νεος» Late/ Modern Hellenism**	Great-great-grandson	*Son*

were arrayed along the thread of history as civilizational offspring. The psychoanalytic character of the framework renders the *History*, not only as content but also as act of writing it, into an intimate affair, a family affair, which further promoted the fundamental idea of the project, which was continuity.

Further, neither Paparrigopoulos nor, later, Karolidis, claimed that the fifth Hellenism had reached its full expression or foretell what its closure would signify. Nor do they impose any restriction (territorial or other) upon a new "son" who is to usher a Sixth Hellenism down the line, one which would both capture the spirit of its age and "sacramentally" embody the durable forms and practices of Hellenism's past.

Importantly, Paparrigopoulos claimed that Greece would experience figurative (mystical transcendence), rather than literal, deaths, as his model is genealogical. Ontologically durable as "ethnos," Hellas does not transit from one Hellenic civilizational type to the next by starting in a larval stage and slowly growing and developing across centuries. Paparrigopoulos' conceit is bolder still as each descendant civilizational offspring emerges in full panoply, as did Athena out of Zeus' head. There are two points in that temporal trajectory where political rupture with the past forms would present critical thresholds for Hellenism: the Roman conquest of 146 BCE and the dissolution of the rump Byzantine Empire – the Fall of the imperial capital of Constantinople in 1453. In the case of the former, Paparrigopoulos starts the narrative on *Hellenic Christianity* by claiming that the Hellenic nation (ethnos) was placed for the first time under the dominion of foreigners following the Fall of Corinth. He then proceeds to establish the exceptional character of the ethnos:

> Commonly, nations, whose political autonomy has been interrupted, are considered as not having a history. This is not so because political life is the only object of history, but because, for the most part, absent of political life, both social and intellectual life wither away. Such a thing has never occurred in the case of the Hellenic nation, which is especially so during the dominion of Rome. The Romans were neither able to remake us into their image, nor destroy our political institutions… [T]hus the Roman dominion was not able to intermix with the riverine sterling vortices [that describe] the Hellenic nation […δὲν ἴσχυσε ν' ἀναμιχθῇ μετὰ τοῦ αργυροδὶνου ἑλληνικοῦ ἔθνους].

Paparrigopoulos' metaphor here is extraordinary not only for the skillful application of a literary loan from Homer, which he quotes in full, but also for proposing a general maxim about civilizational intercourse. Concerning the loan, Paparrigopoulos, throughout his monumental work, makes magisterial use of ancient and medieval primary sources, both in his practice of professional historiography, but, we would also claim, as a counterargument to critics, like Fallmerayer, who had earlier claimed that "…[p]hysical beauty, intellectual brilliance, innate harmony and simplicity, art, competition, city, village, the splendor of column and temple – indeed, even the name has disappeared from the Greek continent's surface." Weaving Homer *logos* into his historical discourse about nationhood, he draws (mends?) the line between antiquity and the modern era.

... οἳ περὶ Δωδώνην δυσχείμερον οἰκί' ἔθεντο, οἵ τ' ἀμφ' ἱμερτὸν Τιταρησσὸν ἔργα νέμοντο ὅς ῥ' ἐς Πηνειὸν προΐει καλλίρροον ὕδωρ, οὐδ' ὅ γε Πηνειῷ συμμίσγεται ἀργυροδίνῃ, ἀλλά τέ μιν καθύπερθεν ἐπιρρέει ἠΰτ' ἔλαιον (Homer, Iliad: 750 B)

... those who had built their homes around Dodone of the wretched winters, and those who cultivated the land in the vicinity of the lovely Titarissos, which pours its beautiful waters into the Pinios River, not intermingling them with the Pinios of the silvery vortices, [Tatarissos'] waters flow on top, like olive oil, because [Tatarissos] is a tributary... (authors' translation)

To illustrate the general maxim about the meeting and mixing of civilizations, he chose a physical geographic metaphor: Hellas is constructed as the world-dominant cultural interlocutor as the main stem of a river system. Hellas is imagined as the grand Pinios River of the silver vortices. Rome is imagined as the Titarissos, which may be mightier, militarily, and with cultural goods to contribute, but still only a tributary. As a general geopolitical maxim, contact between dominant and non-dominant cultures may produce a spectrum of mixes, one flowing on top of the other as waters of different composition.

Paparrigopoulos continues, "However, we do not suggest that the nation remained unchanged. We only suggest that it was not Romanized [the reference here is to Rome's failure to impose Latin as the dominant language]. Many other changes did occur. These changes began during the last years of its ancient 'life.'" The operative term here is βιος (life). The life of the second, Macedonian Hellenism is coming to a close in Corinth in 146 BCE, and new life is simultaneously breathed into the third, Christian Hellenism that picks up, also, in Corinth. Paparrigopoulos models the second point of rupture, which occurs between 1204 and 1453, and the third, final one, of the Greek Revolution, in the same manner, by admitting to change but remaining steadfast on the durability and assimilationist capacity of Hellenism, as well as its unquantifiable transcendental spiritedness, which animates them.

The fourth, "Medieval Hellenism" as civilizational fulcrum

Our focus is the fourth, "Medieval Hellenism," and its relationship to the construction of the Modern Hellenic political-territorial and national identity. Paparrigopoulos called it «Η μεσαιωνική ημών ιστορία»: "our medieval history."

According to Paparrigopoulos the terms "Byzantium" and "Byzantine" are inappropriate for three reasons: first, the peoples of the Empire never called themselves by that name; second, the term does not describe in any sense the organizational or ideological make-up of the state, regime, or society; and third, the term and its derivations are an invention of Western

Europeans, and it is through them that the terminology became adopted by the Greeks. In Paparrigopoulos' view, given the syllogistic framework of durable yet evolving Hellenisms, the proper term would be "Medieval Hellenism." The vexing fact that Byzantines called themselves Romans and their Empire Roman as late as the eleventh century does not trouble Paparrigopoulos, who points out significant cultural practices – language among others – that preserved Hellenicity. While until the tenth century, the label "Hellene" connoted idolatry, its persistence, for Paparrigopoulos, pointed to the unbroken string of history, while also pointing back to the first and second Hellenisms. Paradoxically, perhaps, the very association of Hellenism with suspected idolatry in the tenth century becomes, for Paparrigopoulos, trace evidence of continuity, even when it complicates the definition of truly "Hellenic" in Medieval Hellenism. After the eleventh century, the air clears when the label "Hellene" starts to lose its pagan religious associations. In fact, by the end of the Palaiologan period, prominent scholars, such as Pletho Gemistos, proudly called themselves Hellenes.

For Paparrigopoulos, "Byzantine history" is the history of "Medieval Hellenism" (which is not the same as a conventional "medieval history of Hellenism"). He variously (and at times evasively) called it «*η ιστορία ημών*» (our history), «*η πάτριος ιστορία*» (the history of forefathers), and «*η μεσαιωνική ημών ιστορία*» (our medieval history). He was equally evasive when he spoke about the state and its institutions, which he called by names other than Byzantine: «*Η Ανατολική Αυτοκρατορία*» (the Eastern Empire), η «*αυτοκρατορία της Ανατολής*» (the Empire of the East), το «*μεσαιωνικόν του ελληνισμού κράτος*» (the medieval state of Hellenism), το «*χριστιανικόν της Ανατολής βασίλειον*» (the Christian Kingdom of the East), το «*Ανατολικόν κράτος*» (the Eastern Kingdom, an appellation also favored by pre-Revolutionary intellectual Rigas of Velestino), το «*κράτος της Ανατολής*» (the State of the East), το «*μεσαιωνικόν ημών κράτος*» (our medieval state), η «*ημετέρα μεσαιωνική βασιλεία*» (our medieval Kingship), η «*μεσαιωνική ημών μοναρχία*» (our medieval monarchy), το «*εν Κωνταντινουπόλει κράτος*» (the state founded/located in Constantinople). Importantly, he does not shrink from using the term «*ελληνικό κράτος*» (Hellenic state), or «*Ελλάς*» (Hellas) as a synonym, especially when he discusses events in the eleventh century and later. Also, importantly, it is impossible to discount the significance of "East," as a geographical signifier, especially in a period when the ideological contours of "East" and "West" were becoming very much apparent as a result of both Russian expansionism and the emergence of Orientalism.

Paparrigopoulos refers to Byzantine emperors and their rule as «*εν Κωνσταντινουπόλει μονάρχαι*» (the monarchs in Constantinople), «*οι βασιλείς της Κωνσταντινουπόλεως*» (the Kings of Constantinople), η «*εν Κωνσταντινουπόλει βασιλεία*» (the Kingship [based] in Constantinople), «*η εν Κωνσταντινουπόλει μοναρχία*» (the monarchy [based] in Constantinople], «*η εν Κωνσταντινουπόλει κυβέρνησις*» (the Constantinople government), «*ο αυτοκράτωρ Κωνσταντινουπόλεως*» (the Emperor of Constantinople), and, again, for the period that starts in the Eleventh Century, we encounter characterizations such as «η *ελληνική αρχή*» (the Hellenic authority), «ο *αυτοκράτωρ των Ελλήνων*» (the Emperor of the Hellenes), «ο *βασιλεύς των Ελλήνων*» (the King of the Hellenes), and, very pointedly, «οι *μονάρχαι του ελληνικού έθνους*» (the monarchs of the Hellenic nation). Pararrigopoulos used the term «*βασιλείς του Βυζαντίου*» (Kings of Byzantium) only once, but from the context, we can deduce that he used it as a geographical designation, as Constantinople was built in part on the trace of a pre-Classical Greek colonial city bearing the placename "Byzantion." Lastly, he often referred to the subjects of the emperor as «*Έλληνες*» (Hellenes), his army as «*ελληνικός στρατός*» (Hellenic army), and the territories of the Byzantine Empire as «*ελληνικαί χώραι*» (Hellenic lands).

Regarding its chronology, "Medieval Hellenism" started with the fall of Rome in 476 CE, at a time the West was culturally and socially fragmented by the arrival and settlement of Germanic peoples. In contrast, he claimed that the East exhibited cultural cohesion founded on the linguistic dominance of Greek and Christianity along the Eastern Mediterranean littoral.

The tension between a racially continuous heartland (mainly in the lands of the newly independent state) and a vast desired hinterland of lands and people who claimed to be historically Hellenized had to be resolved. Paparrigopoulos' *History* emerged as a lengthy discourse on the reclaiming, in word and historical argument, of vast territories north of the lands of Classical city-states and the Hellenic "racial" continuity of their inhabitants. In his posthumously published work, *The Most Teachable Findings of the History of the Hellenic Nation* (Τα διδακτικώτερα πορίσματα της ιστορίας του ελληνικού έθνους), in 1899, his "continuity" thesis and the construct of Helleno-Christianity had crystallized in their most explicit and succinct form. Following the defeat of Hellas' city-states, Macedonia – now Hellenized, per Paparrigopoulos – became the beacon of Hellenic language and culture across the Eastern Mediterranean and as far away as the interior of West Asia. The death of Alexander the Great and the partition of his Empire by the "diadohoi" (the successors) into a constellation of Hellenic kingdoms do not arrest the geographical diffusion of Hellenism, nor does the subsequent Roman conquest of these territories. The Hellenistic kingdoms that endured until their conquest by Rome only confirms, according to Paparrigopoulos, the commonsensical character of the geographical diffusion of Hellenism – commonsensical because of its locational geography and the exceptional character of a superior, Hellenic culture: "The Hellenization of Asia Minor was readily possible. Its entire coastal region was Hellenic since a very long time, and once Alexander the Great conquered the interior, it was logically expected (φυσικῷ τῷ λόγῳ) that Hellenism would be established throughout that peninsula" (Paparrigopoulos, 1899: 17). Concerning the Roman conquest of Hellenistic Asia Minor, Syria, and Egypt, Paparrigopoulos claims that *Romanitas* – the essential qualities and practices of Roman culture – did not leave a durable imprint upon them. Although he recognizes that the territorial administration was Latinized, he claims that "Romanism [Roman civilization and its ways] was never able to accomplish in the East what Hellenism accomplished: it never succeeded in transmitting to these lands its language, its norms, and its character" (Ibid: 39). In this manner, with the lands and the populations Hellenized, the Eastern Mediterranean littoral was claimed for Greek culture, though not necessarily claimed as a Hellenic population.

The argument, however, was not complete until Hellenism was permanently paired to Christianity. The space-time coincidence of a Hellenized Palestine and the birth of Christianity was "the work of divine providence" (*το έργον της θείας πρόνοιας*) (Ibid: 40). Accounting for the numerous political reversals the Hellenes experienced under Roman rule, he marvels that the "idea of the diffusion of Hellenism" would endure: "How can one not see the finger of Providence in all these reversals, especially when one considers the end result. We can imagine the difficulties the diffusion of Christianity would have encountered had it not been for the timely diffusion of Hellenism" (Ibid: 40–41). Paparrigopoulos claimed that Christianity would be fluidly diffused across the Eastern Mediterranean and be rendered legible to its population only once it was articulated to the Greek language. Engaging in anti-Semitic speech either by ignorance or by willful omission, Paparrigopoulos dismissed Hebrew as a language unworthy of Christianity: it was fragmented into numerous dialects and spoken by a few people in Judea. He claimed that only when scripture was transmitted in the Greek language – an exceptional language capable of the most brilliant and philosophical expression – did Christianity accede to its proper communication plane: "Scripture required more than pronouncement; it required interpretation, [in] a language capable of explaining and elaborating the high truths and meanings of the new religion: a language that is most perfect, most structurally refined, most philologically adept ["*γλώσσης τελειωτάτης, τεχνικωτάτης, φιλολογικωτάτης*"], without which one cannot conceive how [Christianity's] miraculous establishment would take place" (Ibid: 41).

Ergo, a claim was made upon all the Christian lands of the Eastern Mediterranean since they would not be Christian in the first place if they were not Hellenic in language, if not culture, first. Furthermore, the two temporally staggered geographical diffusions of Hellenism and Christianity would change the world, thus cementing Hellenism's (and the Hellenic State's) exceptional position in world history: "It was a great event for all of humanity, great for the fortunes of the Hellenic nation, because from that point on, the entirety of our social, moral, and intellectual being flows out of it" (Ibid: 41). The marriage of Hellenism to Christianity may have occurred soon after the Macedonian, "Asianic" Hellenism ends at the Battle of Actium (31 BCE) and while Rome dominated the region. However, the implications of that union, for Paparrigopoulos and other Byzantium revivalists among the ranks of nationalist historians, would become evident during the fourth, "Medieval Hellenism." Per Paparrigopoulos, it effectively ended in 1204 with the sacking of Constantinople by the Fourth Crusade, and gradually faded away over the next two centuries, giving way to expressions of "Νεος" or "Late Hellenism" ("Modern," *sensu stricto*, is problematic here, so we use the Greek term). Therefore, "Medieval Hellenism" and "Late Hellenism" would grind against one another until the Fall of Constantinople in 1453, one leaving the stage while the other was arriving.

"Medieval Hellenism" was massively important to his overall argument. He wrote 1,133 pages on it, which represented 34.8 percent of his grand *History*. The 728 years it covered captured the contentious and simultaneously creative encounter and amalgamation of pagan/pre-Christian and Christian Hellenisms, all of which occurred in the cauldron of Byzantine imperial statecraft. For Paparrigopoulos and, by extension, for many Greek nationalists starting around the middle of the nineteenth century, the gambit was to nationalize three millennia of history of the Hellenes in a manner in which it could serve the Greek state interest and help crystallize a prestigious and convincing national identity for its citizens. At a time of great geopolitical instability in the Eastern Mediterranean, the Balkans, and the Black Sea regions, in significant part because of the growing structural weakness of the Ottoman Empire, perennialist nationalist arguments like Paparrigopoulos' could turn "history" into "geopolitics," that is, into ideas of territorial expansionism at the expense of the Ottomans and other successor states of the declining Empire.

"Medieval Hellenism" was, thus, defined by structural characteristics of Byzantium that the Modern Greek state would have liked to transubstantiate (and done so "sacramentally," as Paparrigopoulos mysteriously claims).[3] The prestige and geographic expansiveness of "empire" (importantly, a Christian one) were wished into the state's national project and distilled into a statecraft practice. These characteristics were "Hellenism," "Christianity," and a "strong, centralized authority." Paparrigopoulos observed great interdependence among them: by the end of the fifth century, "Christian Hellenism" – the third of five Hellenisms – already enjoyed the support of imperial authorities. At the same time, the Empire itself (in the East) relied substantially on the stabilizing resources of Hellenism. Palace administrations mobilized Christianity, «την ηθικην αυτου δυναμιν» (its moral power or potency), as connective tissue among the peoples of the Empire. Each one of these elements – Imperial governance, Hellenism, Christianity – relied upon the other two for support.

Paparrigopoulos claimed that the "Hellenic" dimension in "Medieval Hellenism" is substantially based on three factors: language, cultural-more-than-racial continuity of specific populations who inhabit historically Greek territories, and cultural practices that can sometimes be traced back to the first Hellenism. Ethnological paradigms based on linguistic, race, and folk custom typologies were in the mainstream at the time and as late as the 1930s when the Karolidis edition was published. Specifically, he argued that "Medieval Hellenism" was defined by the dominance of the Greek language in the Eastern Mediterranean and the Black Sea regions, established during the second, Alexandrian and Hellenistic "Macedonian Hellenism." He noted that his language argument was not undermined by the much later Hellenization (and de-Latinization) of imperial administration, during the reign of Emperor Heraclius – ca. 700 CE. In fact, it is geopolitically instrumental for Greek nationalists, to territorialize Hellenism from Sicily to Alexandria, and from the southern coast of Crete to the Crimea, based on the historical geography of language. Ergo, it does not matter that it would take almost two centuries after the Fall of Rome for Constantinople palace bureaucrats to adopt Greek in the work environment.

The signification of the "East" in Paparrigopoulos' thesis influences Greece's East-thrusting geopolitics. Paparrigopoulos claimed that in the lands conquered by Alexander and his successors (all in the East, spanning Asia Minor, Persia, and Bactria to the Indus River, and from Upper Egypt to Mesopotamia), a special kind of Hellenism emerged – an "Asianic" one – that can be understood as both subsidiary to "Macedonian Hellenism," but also durable beyond the chronological confines of the "Macedonian Hellenism." While some territories, such as Egypt, Syria, and Palestine, became incorporated into the Arab Empire during the seventh century CE, and were gradually Arabized and Islamized in great measure, Hellenic elements persisted (and mere traces persist today, as the remaining 1,500-strong Greek community of Alexandria attests).

Yet, Paparrigopoulos' argument about an "Asianic Hellenism" still flourishing into the "Medieval Hellenic" period is not exemplary of an European mid-nineteenth century romantic flirtation with the East, nor was it Greek-flavored orientalism.[4] The preoccupations, anxieties, and desires of Greek elites were different, and, despite their eagerness to embrace, understand, and capitalize upon the classical and Hellenistic past, their gaze upon the land and people of the new state was exploratory and critical in its discernment of what was "Hellenic." Modern Greece's relationship to the East *qua* Ottoman World was brutally evident in an every-day, everyplace manner in the lands of the new state, and, importantly, in the "unredeemed lands" across the Arta-to-Volos border that separated it from the Empire in mid-nineteenth century. The material culture and built environment of the Ottoman Empire, the rural landscapes upon which Ottoman land-tenure and occupancy patterns persisted to the 1880s (McGrew, 1985), and the spatial segregation of religious groups in cities into quarters, were durable signifiers of the "East" in the new state. Reclaiming (at least some of it) and transforming by re-Hellenizing it and incorporating it to the new state was the aspiration of perennial nationalists like Paparrigopoulos and Zambelios. We can then concede that Greece's Eastern ambitions and Europe's Oriental ones had similar practical, aspired outcomes that involved the continuing dismemberment of the Ottoman Empire. This subsidiary cultural-linguistic claim of "Medieval Hellenism" upon the eastern lands of the defunct Byzantine Empire, which in Paparrigopoulos' time were entirely under Ottoman rule, lays the ground for others, in Europe and in Greece of the late nineteenth and early twentieth centuries, to imagine these lands (re)incorporated either in the colonial empires of Great Powers, or the territorially puny but greatly ambitious Modern Greek State. All of the yearnings and dreams – about Constantinople and lands east of the Straits of the Hellespont – will turn into dust in 1922, when a million and a half Greeks were forced out of Eastern Thrace and Asia Minor by Turkey.

Furthermore, while lands were being lost to Arabs in the East as of the seventh century CE, Paparrigopoulos notes that in Southeastern Europe the Byzantines (or better, "medieval Hellenes") were able, to a degree and for some time, to manage the settlement, pacification, and incorporation into the Empire, of Slavic peoples, mainly through Christianization. The encounter of Hellenes and Slavs in Byzantine lands was significant to Paparrigopoulos and other nationalist scholars, as the extensive Slavization of the Balkan region defined the competing nationalisms and aspirational geographical boundaries of the successor states to the Ottoman Empire. Importantly, it animated Great Power interests and ambitions in the region, as we argue in the next section. Once the Ottoman Empire showed signs of collapse in the beginning of the nineteenth century, the triangulation of "race" (blood), "land," and "statecraft" defined much of the geopolitical contestation among Greeks, Serbs, Bulgarians, Romanians, Montenegrins, and Albanians. So, Paparrigopoulos went to great lengths arguing that substantial populations in the lands occupied by the ancient city-states continued to exhibit *qualified* racial continuity (claiming that mixed populations were becoming Hellenized by the vital and spirited Hellenic culture), while at the same time noting the significant demographic presence of "Hellenized" others in those same lands. An argument for axiomatic racial continuity was unwinnable as racial character was not just impossible to substantiate, but it was quack science from the start. The politics of racial purity was so incredibly fraught that the most infamous of Greece's opponents in Europe, Jakob Philipp Fallmerayer, challenged the Greeks to produce even a single "real" Greek. For Fallmerayer, Greece had essentially died with the defeat of Athens by Sparta at the end of the Peloponnesian War.

So, enduring Greek language and Greek or Hellenized "race" populations served as the two main pillars of "Medieval Hellenism." The third pillar consisted of cultural and social norms that linked past Hellenisms, even the "parent"-First Hellenism, to the Medieval one: the organizational structure and communication culture of the Christian Church was where Paparrigopoulos located that third pillar. He would claim that the institutions and concepts of society and self of antiquity are Christianized, re-signified, and reforged in the smithy of a strong centralized imperial administration. He would claim that the Christian Church, institutionally, and Christian dogma, ethically, sheltered in liturgical practices, concepts and norms traced back to "the ancient Greek spirit": a predisposition for democracy, for social equality, for social justice and the protection of the weak, and the use of "logos" and rhetoric as a medium and means of governing. He further noted that ancient Greek influences, transmitted through philosophy, existed in both Roman and Byzantine law. He alleged that the Athenian Ecclesia – the congress of citizens in Athens' direct democracy – shared essential cultural and administrative characteristics with the Christian ecclesia as a community of equals before God, and Christian Synods where bishops' standing and role evoke the office of archon in Classical Athens. Finally, and admittedly we might consider this a big stretch, Paparrigopoulos drew a parallel between the Parthenon of ancient Athens and the Church of Hagia Sophia in Constantinople, which he found share a common concept of divine wisdom, despite the differences between the religions that built them – an idea that must have played very well in both the national capital and the provinces. We can distill from these contorted imaginaries that Paparrigopoulos instrumentalized 'anachronism' as a historiographic stratagem. Less beholden to scholarly fidelity than service to state and nation, Paparrigopoulos proffered to his national audience a template for resolving historical and geographic incongruities in Modern Greece's master narrative of Hellenism.

Figure 2.1 Awaiting Hellas' resurrection by Chicago-based lithographer, P.D. Vafakos, 1910.

Source: National History Museum.

For Paparrigopoulos, "Medieval Hellenism's" greatest achievement is the Iconomachy (717–848 CE), which he describes as Hellenism's period of Reformation, drawing parallels to the West's sixteenth century experience. Church and Palace factions competed for much more than the status of icons in worship. The outcome may have, ultimately, favored liturgical convention rather than reform, since the icons were restored in churches, but, per Paparrigopoulos, the struggle produced great social, administrative, and cultural changes nonetheless. It reduced Church privileges and limited the influence of monasteries, which were constraining imperial authority. It opened the way to legislation to protect small farmers, especially in so-called stratiotic lands, or borderlands protected by farmer-soldiers. Consequently, with good governance, the Komnenian Emperors maintained a robust and stable empire into the beginning of the eleventh century. Of course, the fatal blow will come from the West in 1204 and will commence the 249-yearlong countdown to the material collapse of the Empire.

Given the three-pillared support of "Medieval Hellenism" in Pararrigopoulos' model, a much-weakened imperial administration pillar served "Medieval Hellenism" poorly in its closing decades. Further, the pillar of the Church was in crisis, and demands mounted for compromises with Catholic Rome over military support from the West. Nevertheless, per Paparrigopoulos, the Hellenic pillar flourished, as imperial subjects started to embrace their Hellenicity when the sway of the institutional church was in some doubt. The European Renaissance's Byzantine links are found in the structural failure of "Medieval Hellenism," which liberates "the Hellenic" from the clerical box. A "New," "Late," and eventually "Modern" Hellenism, the Great-Great Grandson of Hellenisms, will glide out of the dying body of its progenitor full of the memories, norms, practices, and yearnings of Athens, Alexandria, Nicaea, and Constantinople, right back to Athens, though an Athens of the nineteenth century.

In Paparrigopoulos' *History of the Hellenic Nation, from Ancient Times to the Present*, historiography becomes a political instrument in the hands of statist elites that benefit from centralization, militarization, and territorial expansionism – or at least ideations and yearnings for territorial expansion. The realization of the imagined homeland required further centralization and militarism. Paparrigopoulos' historiography provided a blueprint (and a roadmap) to nationhood and, by extension, national statecraft, by providing identity signification anchors. In practical geopolitical terms, it supported statecraft by identifying an aspirational target list for territorial acquisitions. He was a critical contributor to the articulation of the Great Idea – a neo-Byzantinist imagining of a Greater Greece that would have Constantinople as its capital. The dream never came to be, but the traces of these yearnings are still visible in Greek political discourse about Greece's past and its future in Europe's Southeast.

Notes

1. We date the establishment of the Greek State 1830-1832. On 3 February 1830 (London Protocol), the Great Powers, Britain, France, and Russia, agreed to its establishment as a free, sovereign state. Under pressure from the Powers, the Greek State and its boundaries were formally recognized by the Ottoman Empire in the Treaty of Constantinople of 21 July 1832. The new boundary ran from Arta, in the west, to Volos in the east.
2. Even when prescribed and managed within a state's constitutional order, the new wave of nationalisms in the United States, the United Kingdom, The Netherlands, France, and the Turkish Republic, to name a few, are often toxic and virulent, and always outwardly exclusionary in their crystalizing the national group in opposition to others. While nationalist revivals define the new politics of a significant number of states and communities across world regions - especially after the global economic crisis of 2008 - European nationalist revivals, cutting across Europe from Scotland to Turkey, are especially destabilizing of the international system. Greece, as a frontline state of the European Union, has felt the brunt of migrant and refugee waves primarily from North Africa, Syria, Iraq, and Afghanistan. It is experiencing its own nationalist turn and surge of the extreme Right, the latter in the guise of "Golden Dawn", the third largest political party in Greece in a constellation of eight parties currently sitting in the Hellenic Parliament. The global financial crisis of 2008 manifested itself in Greece in 2010 as a massive debt crisis. *Golden Dawn* resurfaces in Greek politics in 2009 after a 13-year hiatus. Within three years it increased its electoral reach by 2,300%, its anti-globalist, anti-neoliberal, and anti-Europeanist positions catapulting it to great political significance: National Parliament electoral results for Golden Dawn published by the Ministry of the Interior (http://ekloges.ypes.gr/): 0.07% (1996); 0.29% (2009); 6.97% (2012/a); 6.92% (2012/b); 6.28% (2015/a); 6.99% (2015/b). European Parliament election: 9.40% (2014). These shifts are, in part, precipitated by widespread opposition to globalization, and in part by the perceived neoliberal character of the European integration project (Bistis, 2013: 35-36).
3. "Sacramentality" as the engine of movement across Hellenisms was readily accepted by political and church elites for its demagogic qualities but emerged as a weak point easily undermined by Paparrigopoulos' scholarly opponents.
4. [4]Konstantinos Paparrigopoulos raised a strong Greek voice in European academic disciplinary circles in the second half of the 19th C, Greek artists took at least as long to discern their own Hellenic thematic and iconographic individuality gradually. To start, unlike their Western European counterparts, Greek artists in the new state were less interested in reimagining idealized peopled landscapes of ancient or Hellenistic Greece (as in the works of Lawrence Alma-Tadema and John William Godward). However, they favored interpreting battlescapes and events of the Greek War of Independence, and sometimes after such events were imagined by Western European painters (most notably Eugène Delacroix). The most important artistic interpreters of the nationalist past were painters Nikolaos Gyzis, Nikiforos Lytras, and Theodoros Vryzakis, all three members of Athens' "Munich School." Their work exemplified the influence of Greece's Bavarian monarchy and the thrall of European romanticism in which Greece's elite artists expressed both historical and everyday subjects in a Western stylistic manner. Emblematic of this artistic intercourse between the West and Greece, is Nikolaos Gyzis' painting "Μετά την καταστροφή των Ψαρών" [After the Destruction of Psara], ca. 1896-98, National Gallery, Athens. Painted more than 70 years after the July 5, 1924 event of the annihilation of the Island of Psara's civilian population by

Ottoman forces, it depicts, in a manner evocative of Théodore Géricault's *Le Radeau de la Méduse* [The Raft of the 'Medusa,' 1818-19, The Louvre], survivors of the massacre on a rowboat battling the waves. The work also readily invokes Eugène Delacroix's *Scène des massacres de Scio* (The Massacre at Chios, 1824, The Louvre), which imagines a scene of the systematic destruction and deportation of the Greek island's population in the spring and summer of 1822. Little known outside of Greece, such *salon* works addressed, on the one hand, the small urban, sophisticated, and substantially westernized society of Athens, and, on the other, lent their visual narrative force to illustrations in Greek school books and scholarly publications. Paparrigopoulos' later Karolidis editions of the *History* and every Greek government-sponsored textbook ever since has made frequent use of such imagery.

Sources

Παπαρρηγόπουλος (Paparrigopoulos), Κωνσταντίνος. *Ιστορία του Ελληνικού Εθνους, από Αρχαιοτάτων Χρόνων μέχρι των Νεωτέρων* (History of the Hellenic nation, from the most ancient to recent times), τ. 1-5. Αθήνα: εκ του τυπογραφείου Νικήτα Πάσσαρη, 1865–1874.

Παπαρρηγόπουλος (Paparrigopoulos), Κωνσταντίνος. *Ιστορία του Ελληνικού Έθνους* (History of the Hellenic nation). *Η πρώτη έκδοση: 1853*. Κ. Θ. Δημαράς (Dimaras) (ed.). Αθήνα: Εστία, 1999.

Παπαρρηγόπουλος (Paparrigopoulos), Κωνσταντίνος; Καρολίδης (Karolidis), Παύλος; Αναστασιάδης (Anastasiadis) Γεώργιος Ο; Μουτσόπουλος (Moutsopoulos), Νικόλαος. *Ιστορία του Ελληνικού Εθνους: Από Αρχαιοτάτων Χρόνων μέχρι Σήμερα* (History of the Hellenic nation, from the most ancient times to the present). Αθήνα: Αλέξανδρος, 1930-1932.

Παπαρρηγόπουλος (Paparrigopoulos), Κωνσταντίνος. *Τα Διδακτικώτερα Πορίσματα της Ιστορίας του Ελληνικού Εθνους* (The most teachworthy findings from the history of the Hellenic nation), (πρώτη έκδοση: Αθήνα, 1899). Αθήνα: Πελεκάνος 2005.

3 Race, identity, territory

Contesting the ethnological profile of the Balkans

The Hellenicity of the newly liberated Greeks became the subject of scholarly inquiry in Europe and the new Hellenic State. The discernment of their genealogical and cultural heritage had implications beyond the academy. Supporters of the view that nineteenth-century Hellenes were, indeed, the descendants of ancient Hellenes were much more inclined to support the defense, if not also the expansion, of a Hellenic State. Those who opposed such a view supported either that the "race" of the ancient Greeks was long gone or, more worryingly for the new State, that the majority of the region's inhabitants were a mixture of Albanian and Slavic peoples deserving of territorial rights. That constituency was also more likely to champion the geopolitical aspirations of other successor states of the Ottoman Empire (especially the aspirations for a Great Bulgaria and a Greater Serbia). All parties – the Great Powers and the Ottoman successor states – worked up historical, geographic, linguistic, and ethnographic analyses as means for arguing for their opposing positions.

The national awakening benefitted greatly from the surge of philhellenism buoyed by Western European and American elites already before the Greek War of Independence (1821–1830). The West saw the liberation of "Hellas" from Ottoman rule as a Western moral victory. Hellas represented a cultural-intellectual tradition that the Europeans had appropriated, interpreted, and codified during the Renaissance and the Enlightenment. The Revolution provided the context to weaponize and wield it against a non-European antagonist. The heroism of the besieged and the death of Lord Byron during the siege of Messolonghi (1824) and their coverage in the British popular press were exemplary of the pro-Hellas fever that had overtaken the West. Alexis Politis remarks that, eventually, Europe's idealized imaginings of the resurrection of ancient Hellas, and its expectations of the Hellenes themselves, did not match reality. Fatigue of revolution set in: "No, these illiterate Orientals may be quaint, but could not, normally, claim the heritage of Classical Greece – the ideal standard of every European bourgeois society" (Politis, 2008: 26; Politis, 1993: 33).

Practicing purism in the new state did not, ultimately, benefit the Hellenes. Dissent among the ranks of Greek intellectuals themselves over who the

Hellenes were in antiquity complicated the effort to make a formidable and compelling argument for greater territorial claims. The traces of the Enlightenment and academic Classicism were manifested among influential intellectuals and civil servants of the new state. In 1803, Adamantios Koraes writes, "those who were the most culpable among the Hellenes allowed themselves to be corrupted by the gold of the Macedonians" (Koraes, 1803: 6). Two years before the start of the War of Independence, the publisher of the Paris-based journal *Melissa* (it is unclear whether it was Konstantinos Nikolopoulos or Spyridon Kontos), writes: "What happened then? [as the result of divisions among the Hellenes] The Macedonians enslaved the fatherland (την πατριδα) of Miltiades. And later? The Romans devoured it. And finally? Alas, my hand trembles, and my quill tumbles [from my hand]" (Melissa, 1819: 100). After the Revolution, Koraes reprises the argument even more forcefully, equating Macedonians, Romans, and Ottoman Turks as masters of Hellas: "The Macedonians ... worked consistently toward the abolition of liberty [of Hellas] and its enslavement to Macedon ... Following the Macedonian despotate, [the Hellenes] were subjugated by the Romans, and finally, by the Turks" (Koraes in Ginis, 1948:54-55).[1] Koumanoudis railed against Byzantium, which, he claimed, diminished Hellenic thought "by not a little," since the Byzantines had distanced themselves from "healthy philosophical pursuits and had surrendered to scholasticism." He writes, "the Hellenic element did not rise during the Byzantine period because neither liberty nor education were awoken."[2] A distaste for Byzantium threatened to crystallize the historical-geographic profile of Hellas and the Hellenes south of Mountain Olympus and Thessaly's Tempi Valley. That inconvenient and ultimately intellectually dubious definition would be systematically and durably dismantled by major historiographers of the mid-century, Zambelios and Paparrigopoulos in the "continuity thesis" and in the articulation of the Helleno-Christianity concept.

For Paparrigopoulos and the elites of the new Hellenic State, the argument over the continuity of Hellenic history and civilization was, in part, predicated upon the claim of the historic continuity of the Hellenes as a discernable ethno-demographic entity. History, geography (including enduring albeit decayed and altered landscapes of the Hellenic past), "race" (το γένος), and later ethnicity (η εθνικότης), religion, and language were rallied to help crystallize the definition and the nineteenth century profile of the Hellenes and their communities. Importantly, the Hellenic profiling of the land and its people provided a potent raison d'état to a state with expansionist aspirations. Most fundamentally, however, Paparrigopoulos' extensive historiographical work had a concrete objective the argument that he and his compatriots were verily Hellenes and their geopolitical claims were worthy of satisfaction by the Great Powers.

The scholarly and policy conundrum was how to curate, reinterpret, and present the Hellenic history of, putatively, Hellenic lands to audiences (domestic and foreign) that were operating in a world structurally framed outside the Ottoman experience: a tradition of statecraft and political territoriality that pointed back to the Peace of Westphalia, a tradition of humanism and rights that emanated out of Western experiences of revolution in England, the United States, and France, and an experience of economic and technological (industrial) change that, again, had entirely Western referents. Scholars of Greek Enlightenment, such as Paris-based Adamantios Koraes and Constantinople (then Athens)-based Neofytos Vamvas, contributed to the liminal character of Greek letters and sciences in the early state, serving as brokers of Western thought in the Ottoman successor state of Hellas. Their Enlightenment credentials biased them to read Hellenic history and geography from a Classicist perspective. Hellas under the Macedonians, the Romans, and the Byzantines were still Hellas, but they diminished and transformed it.[3] Hellas in its indexical state was Hellas of Athens, Sparta, Thebes, and Argos. History was, thus, constructed as a hierarchy, with Ancient Greece at the zenith. Importantly, Greek Enlightenment also defined the geographies of Hellenism along the same lines: real Hellas was the mosaic of territories controlled by autochthonous Hellenes, organized in city-states, to the south of the Kingdom of Macedonia. By the mid-century, the Enlightenment position on the character of Hellenism became both intellectually untenable and (geo)politically inconvenient.

If Paparrigopoulos' and Zambelios' contrivance of a "medieval turn" was consequential for constructing Helleno-Christianity as the keystone of the new identity, it remained to be seen how both the construct and the identity it supported could be operationalized in the foreign and security policies domain. In domestic policy, its application in public education was to have profound and durable implications, becoming the organizational concept for curricula ranging from basic reading to literature, history, geography, and religious instruction (Zervas, 2017).

The challenge for national intellectuals like Paparrigopoulos, who had influence in the government and the national government itself, was how to make the concept of Helleno-Christianity and the continuity thesis both legible and acceptable abroad. They desired to use them as basis and justification for Helleno-centric resolutions to security and foreign policy challenges and opportunities related to the Ottoman retreat and the emergence of a post-Ottoman system of successor states. The approach was not realist, although attempts were made, as we argue below, that support of Greece's moral historical standing on territorial issues in the Balkans amounted to good geopolitics, advantageous to at least two of the Great Power: Britain and France.

In the course of building a cogent ethnonationalist master narrative based on biology, history, and language, the proponents of the theory of continuity were put on the defensive by Western European scholars. The latter questioned the genetic and cultural relationship between ancient and modern Hellenes. What argument could Paparrigopoulos put forth that countered undeniable facts on the ground? Much of the material culture of antiquity and Byzantium (of Medieval Hellenism) lay in ruins;

Hellenic toponyms across large swaths of territory were supplanted by Albanophone and Slavophone ones; outside commercial and ecclesiastical elite milieus, Greek patois was spoken alongside other languages and their dialects. It was Greek intellectuals of the diaspora, for the most part, that formulated ideas of revolution, statecraft, and nationhood.

Piero Vereni notes that political, historical, and territorial arguments of intellectuals define the frontier and make it legible for politicians to turn it into a political-administrative border (Vereni, 1996: 80). In the Greek case, significant intermingling of intellectual and political communities in the nineteenth-century Hellenic State, elaborated and continuously adjusted ideas about the nation-state, its inhabitants, and its borders with a view to territorial expansionism. As the extent of the state increased in stages between 1864 and 1947, the annexation of new lands required adjustments to the territorial-identity-driven dogma. New facts on the ground – notably, the incorporation of non-Hellene minorities to the polity – required interpreting and instrumentalizing. The management of difference emerged as a geopolitical concern.[4] Importantly, expanded bordering included new sets of "brother-Hellenes" and non-Hellene foes. As the case may have been, their assimilation or their countering created new kinds of borders and borderlands, no less significant than the interstate ones.

Countering Jakob Philipp Fallmerayer's thesis

Foremost among the critics of Hellenic "racial" continuity, even before Paparrigopoulos articulated his "continuity" thesis, was Austro-Bavarian historian Jakob Philipp Fallmerayer (1790–1861). In fact, Fallmerayer's works on Hellenism predate the entirety of Paparrigopoulos' scholarly opus and served as a catalyst for Paparrigopoulos's strident support of the cultural and "racial" continuity of the Hellenes (Dimaras, 2006: 94). Strongly anti-clerical, liberal, and at times openly anti-monarchist, Fallmerayer adeptly navigated the shifting politics of Bavaria to gain significant academic appointments (University of Landshut, 1826; Ludwig Maximilian University of Munich, 1848). Notably, he gained access to the Bavarian court through his mentorship of King Maximillian II (Leeb, 1996: 161–73). His work on the Balkans and Asia Minor needs to be understood in terms of German culture politics and nineteenth-century European geopolitics. Concerning the former, Fallmerayer was quite explicit about his distaste for the Roman Catholic Church, conservative politics, and the German establishment's "hellenolatria" (the adoration of all things Hellenic), which he considered the German world's "Greek sickness" (Skopetea 1999: 18). Elli Skopetea suggests that difficulties in mounting any systematic criticism of and opposition to the Church and state compelled Fallmerayer to direct the bulk of his energy to theorizing about the Greek question and producing his vituperative thesis on the Hellenes of his age. Concerning nineteenth century geopolitics, Fallmerayer emerged as a staunch opponent of Russian imperialism and pan-Slavism, and in that sense, his positions were consistent with and supportive of Austria's Balkans strategy. Where his views parted ways with Austrian diplomacy was in his strong support of the Ottoman Empire. He considered it a ready and useful bulwark against Russian expansionism in Europe's southeastern littoral, while Austria perceived it as a weakening state actor ripe for dismemberment. Fallmerayer's scholarly works on Byzantium, the Empire of Trapezous, and the Morea region, and a broad range of journalistic publications on the new Hellenic State and the Ottoman Empire became the outlets for his historical and geographic theses.

Fallmerayer's Greek thesis was stark in its absolute negation of the Hellenicity of the inhabitants of the new Hellenic State and the Hellenicity of populations in the broader region. In his *Geschichte der Halbinsel Morea während des Mittelalters* (1830), he posits that "[t]he race of the Hellenes has been wiped out in Europe. Physical beauty, intellectual brilliance, innate harmony and simplicity, art, competition, city, village, the splendor of column and temple – indeed, even the name has disappeared from the surface of the Greek continent ... Not the slightest drop of undiluted Hellenic blood flows in the veins of the Christian population of present-day Greece"[5] (Fallmerayer, 1830: III; Leeb, 1996: 55). For Fallmerayer, Ancient Greece was dead and without successors. Yiorgos Veloudis claims that Fallmerayer's objective was less to demolish the early nineteenth century national project of fledgling Hellas and more to deal blows to his domestic foes: the Church and the conservative politics of Bavaria, whose unabashed philhellenism was a keystone of its culture politics and geopolitics. Fallmerayer claimed that German philhellenism was a fraud since the political, scholarly focus was on studying ancient monuments and literature and not on the lives of contemporary Hellenes (Fallmerayer, 1830: XI, Veloudis, 1982: 30).

Fallmerayer lays out a new argument for an Orientalist reading of the Hellenic World (Skopetea, 1999: 35).[6] As such, Rome is the progenitor of Europe, while Greece takes one to the East. He lauds the openness and cosmopolitanism of the Roman experience, and the fineness of its laws, and contrasts it to the closed and brutal character of an, admittedly, otherwise brilliant and inventive Greek society (Fallmerayer, 1830: 8-9). The openness and cosmopolitanism of the Hellenistic Era – what Paparrigopoulos introduced as Macedonian and "Asianic" Hellenism – does not subvert Fallmerayer's argument, as he dismisses the Hellenicity of ancient Macedon entirely. He epigrammatically states that the long spears of Philip II's phalanx put Hellas to death in the Battle of Chaeronea (338 BCE) (Ibid: 12). In this claim, as we argued earlier, he was assisted by Greek scholars of the Enlightenment. For Fallmerayer, Greece was further compromised following the defeat of the Achaean League by Rome in the Battle of Corinth (146 BCE), and ultimately pulverized by Alaric I's Visigoth invasions (396 CE). According to him, the physical devastation wrought by the invasion was instrumental in annihilating the ancient world and creating the ground conditions for the implanting of Christianity. Hellas, therefore, is subjected to a state of prolonged death, or specifically, per Fallmerayer, to a series of lethal blows: Sparta's defeat of Athens in the Peloponnesian War prepares the ground for Macedonian dominance; the Roman conquest of Greece; Alaric's invasion; and the corrosive agency of Christianity and Byzantium upon a reduced Hellenism to follow. In later writings, he recognized that there might be some vestiges of the Hellenic "race," although only in the Peloponnese (Fallmerayer in Feigl and Molden, 1913: 139–40). Accordingly, he conceded that the Hellenes maintained their racial identity, their names, and their toponyms for some time and until they were demographically and culturally supplanted in the Slavized parts of the historic Hellenic territory (Skopetea 1999: 40–41).

Paparrigopoulos emerged as Fallmerayer's most forceful opponent. His earliest and politically significant work was a published response to the Fallmerayer thesis. In 1843, he published *Concerning the Settlement of Certain Slavic Populations in the Peloponnese* [*Περὶ τῆς ἐποικήσεως σλαβικῶν τινῶν φυλῶν εἰς τὴν Πελοπόννησον*]. In 1846, he popularized his approach by offering three of his introductory courses on Greek history in French (*Lė Moniteur Grec*, 1846). In 1850, he published in Greek (*Pandora* Journal), and then in 1854, he republished in French "L'opinion Grecque sur le Système de Fallmerayer" in the Athens-based "Spectateur de l'Orient" (Gotsi, 2006: 40). He laid out his counterarguments and, at the same time, articulated important elements of his own "continuity" thesis that he was soon to unveil. In his manifesto-style narrative, he claimed for the Hellenes their national consciousness:

> They tell us that our independence is the work of a foreign conspiracy[.] [L]et us demonstrate that it has the deepest roots in history; let us raise the most illustrious monument to our Revolution in the panorama of the centuries; let us study that awesome and bountiful drama of our past, and let us show the world that we know well who we are, where we come from, where we are headed. The zeal for the study of the history of the forefathers has always been proof of the greatness of the course of the nation! (Paparrigopoulos in Dimaras, 2006: 151).[7]

Paparrigopoulos then undercuts Fallmerayer's racial continuity imperative (unless the alleged Greeks of the nineteenth century are blood descendants of the ancient Hellenes, then they are not Hellenes) by pointing to the facts on the ground. Historical mixing of nations and races has been inevitable and denying it is futile: "Many [Hellenes] among us consider this mixing of blood [ἐπιμειξία] as a great misfortune, desiring to prove it historically unfounded (which is impossible), considering any other state of affairs as beneath Hellenic honor." He continues, "That may be a noble sentiment; however, the deception is obvious." He then entirely discounts it as an issue and claims it to be an advantage and a characteristic of the nations that accomplish magisterial acts: "The greatest achievements of nations have been byproducts of such or even greater mixing. The English are a brew of Celts, Saxons, and Normans..." and so on (Ibid: 152). He writes, "intercourse among nations is, from time to time, necessary for the development of the genius of people" (Ibid). The issue is not that mixing of nations occurred but the nature and characteristics of the residual nation. He continues by making a forceful claim that "amalgamation" of nations [συγχώνευση, which, based on context can also be understood as "acculturation"] has allowed a vital and spirited Hellenism to acculturate and, in some instances incorporate (and genealogically and culturally absorb) both the Albanian and Slav populations in the region. In fact, he claimed that the Albanian nation could be subsumed under the banner of Hellenism, given Albanians' apparent alignment with the aims of the Greek War of Independence. Even more controversially, he claimed that the Albanian "race" [φυλη] had only one ethnic characteristic, language, which was gradually receding in the face of a conquering Hellenism (Ibid: 153). Paparrigopoulos demonstrated significant pragmatism in his treatment of Fallmerayer's charge of racial discontinuity only to validate the national exceptionalism that would drive the Hellenic State's foreign and security policies, at least until the end of the century and until the embers of the Great Idea are entirely put out by the disastrous Greek-Turkish War (1920–1922). Importantly, exceptionalism has served as an important driver of Turkish foreign and security policy from the beginnings of the Republic, and especially during the Cold war (Yanik, 2011: 80-82).

Negotiating Ottoman demography, travel narratives, and zero-sum ethnographic cartography

Even if the "racial" continuity question were addressed to the satisfaction of audiences in Greece and abroad, the size and geographical distribution of the Hellenic, the Hellenized, and the claimed as belonging to the Hellenic collective populations presented significant challenges and opportunities for the State: on the one hand, the Hellenic element appeared in great demographic force in the south of the Balkan peninsula, the Aegean archipelago, the Ionian Islands (the Heptanesus), and in the Aegean Sea littoral of Asia Minor. Further, Hellenic heritage diasporic communities – thinner on the ground – spanned from the Black Sea littoral to the north to Egypt's metropoles of Alexandria and Cairo to the south. Of massive symbolic, if not strategic, significance was the demographic strength of the Eastern Orthodox *Millet* in the imperial capital, Constantinople. In contrast, the demographic profiles in aspirational territories directly to the north of the new state – Thessaly, Epirus, Macedonia, and Thrace – were much contested among other emerging Ottoman-successor states (notably, Albania, Bulgaria, Montenegro, Romania, and Serbia). To the extent that Ottoman-era demographic data could reveal it, and based on observer reports, Hellenes were thinner on the ground in these regions, and geographically clustered.

This gross demography described the Ottoman Empire, while, through the *Tanzimat* reforms, it attempted to change long-standing regulatory religious community governance. Structural entities, such as the long-established *Millet* system that regulated non-Muslim subject communities, had to be substantially reformed under pressure from abroad, and especially by the Great Powers, which wanted the rights of their subjects or citizens protected and, by extension, the safety of their economic activities sheltered.[8] Importantly, for the Hellenic cause, geopolitical convulsions related to the creation of Bulgaria and the fleshing out of Bulgarian nationalism based on both ethnolinguistic and ethnoreligious distinctiveness precipitated the break-up of the Orthodox Christian *Millet*: Following a decade of agitation by Bulgarian nationalists, in part under the patronage of the Russian Empire and over the objections of the Orthodox Patriarchate of Constantinople, Sultan Abdülaziz granted the Bulgarian community *Millet* status in the imperial *firman* of 12 March 1870. The action rendered the Bulgarian Orthodox Church independent (and free of Patriarchal-*cum*-Hellenic-Phanariot influence) and, by extension, undercut a key Hellenic structural advantage – the claim of Orthodox religious unity under Hellenophone leadership. Crucially, Ottoman law forbade community bodies, like the *Millets*, to engage in associational politics, either through the creation of corporate entities or through sponsorship. Such limitation meant to control governmentality, had the unexpected outcome of encouraging the proliferation of nationalist expression through associations inside the Empire and the newly created successor states.

Agents of Hellenism: the society for the diffusion of Greek letters

In the thrall of expansionist and liberatory imaginings, a Hellenic State served as an incubator of national and nationalist intellectual activism. A vital association of that time was the *Society for the Diffusion of Greek Letters* (Σύλλογος για την Διάδωσην των Ελληνικών Γραμμάτων).[9] The roster of 505 members reveals the prominent agency of a prosperous, patriarchal, urban class in pursuit of Hellenic geopolitical ends. Only nine women were included – the plurality in Istanbul – although the use of only first initials of first names obscures gender. A review of the names suggests that several members were related. Some bear Vlach names, which suggests that, at least to a degree, the Hellenic national project successfully co-opted the community's leading members into the ranks of Hellenism. The provenance of the membership is remarkable in its diversity and the breadth of the diaspora (Figure 3.1).

It is not surprising that as many as 63 members list Athens (48) or nearby Pireaus (15) – the capital region – as their residence. Several places across the new state feature much smaller memberships (Corfu with 11 members is the most significant). However, it is striking that sites with a significant membership, sometimes superior to Athens', abound outside the State: in the "unredeemed lands," Smyrne/Izmir counted 50 members and

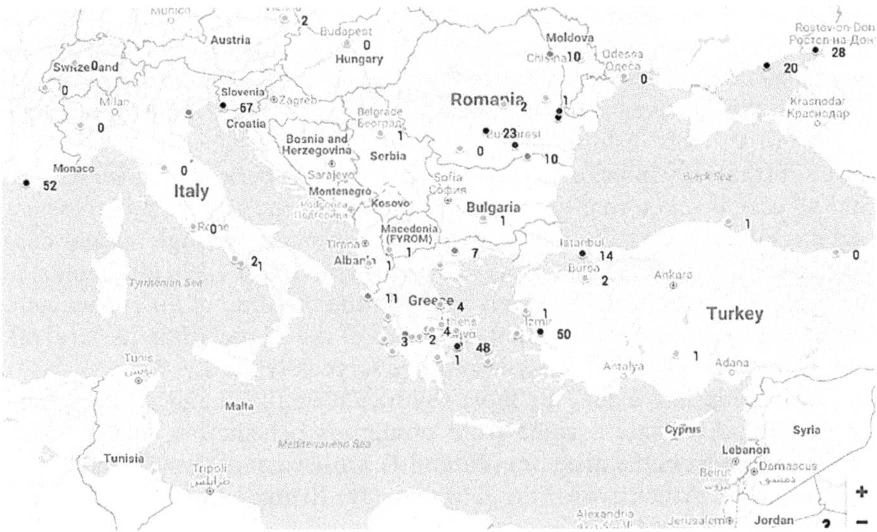

Figure 3.1 Distribution of *Society for the Diffusion of Greek Letters* members.

Constantinople/Istanbul 14, with many other sites, including Thessaloniki, Prussa/Bursa, Ikonion/Konia, Trapezous/Trabizond having much smaller memberships. It is the extra-regional diaspora, however, that features the most impressive communities of members, with Trieste (67), Marseille (52), and far-away Taganrog on the northern coast of the Sea of Azov (28) topping the. The highest concentration of members of the diaspora (115) falls within a narrow northern Black sea zone that spans from the Bucharest region in the West to the Sea of Azov in the East (Figure 3.2).

The annual report covers in considerable detail the financial particulars of the Society, placing great emphasis on the sustainability of support through member contributions. Membership fees, gifts, and bequests supported its activities. In 1877, it had accrued the considerable reserve of

Figure 3.2 Cluster of Society members in the north Black Sea region.

213,052 drachmas (Society Annual Report, 1878: 129). Concerning the fees, there is little question that the annual fee of 25 drachmas would have constituted a structural barrier to most Greeks of the time. It appears, however, that even the moneyed members did not always meet their obligations to the Society in a timely fashion: it laments that the funds harvested in 1876–77 were substantially inferior to those in the previous period: "[I]t is essential that all care is taken that expenses are limited to the absolutely necessary, and it is [also] essential to increase income, especially through regular contributions, which rose to only the negligible amount of 5,296 [drachmas] in the year that just ended. It is only through the building of an endowment, supported by annual income that the needs of the Association, its survival, and the meting out of its objectives can be served" (Ibid: 129-30).

Membership fees aside, the most significant financial and political support to the Association came from numerous benefactors, the majority coming again from the diaspora (Table 3.1). Only eight of the 68 benefactors were based in Athens, with the plurality (36) listing cities in the Russian Empire as their residence.

The geography of financial support for this entity and others like it raises the question of agenda and the allegiances of the members to different Great Powers, and Russia in particular. By the 1870s, intellectuals of statecraft like Paparrigopoulos had realized that the Russian Empire's patronage of Slavic revolutions and its territorial interests trumped the Hellenic State's traditional leadership position of Christian Orthodoxy – more symbolic at that point than actual or practicable. On the one hand, the geographies of membership and financial support reflected the importance of international circuits of capital. Foreign capital was essential to economic growth

Table 3.1 Provenance of Society benefactors

City	*State (in 2021)*	*State (in 1878)*	*"Unredeemed lands"*	*Benefactors*
Taganrog	Russia	Russian Empire	No	23
Rostov	Russia	Russian Empire	No	10
Trieste	Italy	Austria-Hungary	No	9
Athens	Greece	Kingdom of Hellas	No	8
Bucharest	Romania	Ottoman E/Moldova-Wallachia (auton.)	No	4
Istanbul	Turkey	Ottoman Empire	yes	2
Vienna	Austria	Austria-Hungary	No	2
Alexandria	Egypt	Egypt	No	2
Saint-Petersburg	Russia	Russian Empire	No	2
Marseille	France	France	No	1
Iași	Romania	Romania	No	1
Brăila	Romania	Romania	No	1
Galați	Romania	Romania	No	1
Craiova	Romania	Romania	No	1
Odessa	Ukraine	Russian Empire	No	1

and the state's survival (capital for military procurement was essential to a forward-thrusting power). On the other hand, crises in the core of the world industrial economy made investing in the periphery compelling. The Panic of 1873 (the "Long Depression" of 1873–79) precipitated by inflationary policies in the United States after the Civil War (1861–65), economic convulsions following the Franco-Prussian War (1871), insurance runs on Chicago real estate equities because of the Great Fire of that same year, the New York Stock Exchange crash of 1873 and shifts in German monetary policy that same year, made investments in the Balkans and the Middle East. Members of the Hellenic diaspora invested in Athens real estate and supported infrastructure projects through bonds-buying, making their deeper involvement into Hellenic politics a good gamble.

An expression of elite cultural-political will, such societies and associations extended, and at times supplanted, national partisan politics. They provided prominent intellectuals, legal professionals, ranking members of the military and the Church, and politicians with a *quasi*-institutional, parastatal environment to engineer ideas, strategies, and policies. Indeed, there was much to understand, counter and oppose, supplant and advocate in the Balkan geopolitical scene in the service of Hellenic foreign and security agenda: the ambiguity of Ottoman statistical data; the discourses and counter-discourses on the cultural profile of the region by domestic and international correspondents; the struggle to create, promote, and crystallize in the public's mind cartographic and territorial imagining of one's version of historical and population geographies.

Ottoman censuses and surveys: ambiguity in the service of nationalism

Partisan historiographers of the Balkans, Greeks and others, took advantage of the opacity of Ottoman census and tax records to promote self-serving national imaginaries of historic demographic presence or even primacy.[10] Fuat Dundar notes that Ottoman censuses (1831, 1844, 1881–93 and 1906–07) and the *tahrir defters* (tax/land surveys) were used in political exercises establishing the contours of ethnic and religious Ottoman society and its transformation through successive conversions, or Islamization in target provinces (Dundar 2015: 138). To complicate things further, Muslims were not classified in these counts by ancestry or any other characteristic, such as language. Justin McCarthy notes that although the Balkan Turkish population was mainly located in Thrace and Bulgaria, there were also sizeable Turkish-speaking communities in Macedonia and elsewhere (including the new Greek State). Further, he notes, "the majority of Albanian speakers were Muslim" (which puts into question Paparrigopoulos' sweeping statement that the Albanian nation was on its way to being amalgamated into Hellenism). "Among Slavic-speaking Muslims, Bosnians spoke Serbo-Croatian, and Pomaks spoke

Bulgarian ... Muslims located in Central and Southern Greece and Crete, who self-identified as Turkish, spoke Greek. For many Turkish, Albanian, Slavic, and Greek-speaking Muslims, the language they spoke was not that of their ancestors" (McCarthy, 2000: 31). The rurality of the Empire contributed to the censuses' unreliability. İpek Yosmaoğlu notes that "when modernizing states came in contact with rural communities and attempted to integrate them into new sociopolitical structures, they often found out that these communities did not give up their long-established references for social differentiation on-demand" (Yosmaoğlu, 2006:70). Among markers of identity, religion was the most profoundly ingrained, and census subjects could be co-opted, coerced, even terrorized into classifying themselves according to modern national categories. He remarks that in the violent contestation over Macedonia between such tactics were commonplace among both Greeks and Bulgarians:

> Don't have yourselves registered as Rum [Greek Orthodox and by extension Greek], because then no good will come to your children, your goods, and your animals, you will [have to] go up the mountains. Until today we somehow forgave your mistakes, but you could lose your well-being. Know that it is thanks to me that you have stayed alive...
>
> (Letter in Bulgarian by Vlad Vasil, a guerrilla leader, read to shepherd Taşo Petre, outside his village, Negorçe (Negortzi), on 20 August 1905) (Ibid: 62).[11]

Whether serving Ottoman Turkish, non-Muslim Ottoman, successor Ottoman, or Western interests, statistical information was classified in a manner that allowed manipulation, rearrangement, and interpretation that undermined their exegetic potential. Dundar confirms that the religious character of the imperial regime privileged taxonomies based on religious identity: "The main reference point of taxonomy in the *tahrirs* was religion: "Muslims" (muslim/islam/*islamiyye*) and "infidels" (*gebr/gebran, kefere, zimmiyan*). In the regions where the Armenians (*Ermeniyan/Eramine*) and Jews (*Yahudiyan*) were numerous, the *tahrirs* emphasized these communities, thus the term *gebran* specified mainly the Roman Orthodox [Rum/Greek Orthodox in religion]" (Ibid: 142). Official population records published in Turkey during the first decades of the Republican era still listed more than 30 mother tongues (*ana lisanları*; *langues maternelles*) for Istanbul (Strauss, 2011: 127).[12] On the one hand, one would have to reconcile with the fog of the informational landscape, and on the other, accept the apparent diversity of the region's population.

The most complete and successful censuses were undertaken during the reign of Abdulhamid II (1876–1909), following the *Tanzimat* reforms (1839–76) and the State's related bureaucratization. It was a period of extraordinary geopolitical flux, of decline in Ottoman power, and increased population mobility. Between 1831 and 1870, bowing to pressure from the

Great Powers, and, according to Dundar, because of the Empire's increased governmentality, Catholics, Protestants, and Bulgarians were recognized as *Millets* – religious communities with negotiated rights and obligations: "The main aim of the 1881–93 census was to determine the numbers and proportions of the Christian population to reorganize the provincial councils, because the San Stefano and Berlin treatises had been conditioned on the determination of the numbers of Christians in the problematic regions, that is, in Macedonia and Armenia. Thus, the 1881–93 census (and also 1906–07) was the most politicized and "ethnicized" Ottoman census carried out under the pressure and direct involvement of the Great Powers, the Ottoman Empire, and minorities (Ibid: 150). Concerning the 1903 census of Ottoman Macedonia, Yosmaoğlu notes that "establishing demographic supremacy-through whatever means that it might take-had become part of the struggle for Macedonia" (Yosmaoğlu 2006: 62).

Paparrigopoulos and the politicians and diplomats of the Hellenic State had to negotiate this unsteady ground of demographic information and its various interpretations. It was futile to proclaim either the "racial"/ethnic purity of the Hellenic nation or the ethnically clean character of the Balkan territories, regardless of whether they had become incorporated into the new state or claimed as "unredeemed." In practical rather than intellectual terms, Papariggopoulos was effective in countering the Fallmerayer's charges because the competing successor states – Bulgaria, Serbia, Romania, Montenegro, and later also Albania – had also had to finesse the quilting, layering, and intersectionality of identity signifiers as part of their own national projects. Moreover, the historic equation of *Rum,* or Roman Orthodoxy with the Ottoman Empire's Hellenic element, lost its strategic significance for the Hellenic cause once the Bulgarian Church, under the Exarchate, declared its independence from the Patriarchate and started counting its own faithful in the contested lands. It appeared that no successor state had the statistical advantage.

Traveler and correspondents' narratives: subjectivity and the construction of "the Balkan"

As censuses and surveys could not fix the region's cultural profile and could not fully support nationalist projects, other sources were instrumentalized by the Ottoman successor states to support claims to regional cultural primacy (Panaretou, 2005; Hammond, 2010; Krivokapić, 2015). Their inherent subjectivity notwithstanding, descriptions of landscapes and ground conditions in travel narratives filled some gaps. Works by Western, Russian, and Balkan writers imagined the Balkans as "undiscovered country," as if Ottoman dominion and their legacy cast a veil upon the cultural landscape.[13] The narratives communicate the thrill of the exotic, disappointment in the apparent backwardness of the Balkans, disapproval of everyday practices that deviated from national folkloric expectations, or all of the above. Russian travelers looked for the influence of Russia upon their Slavic brethren (McArthur, 2010: 5-6). Bulgarian, Greek, Serbian, and Romanian writers were often promoting nationalist positions, although, at times, they could also be brutally critical of practices of their co-ethnics. Travel narratives without overt nationalist ideological leanings, such as Jelena Dimitrijevic's travelogue (1918) centered on Salonica and its environs during the Young Turk Revolution, provide unexpected and precious perspectives; in this case a proto-feminist geography of the region. Written in the national language with primary readership their national community, their geopolitical influence would have been limited if not translated into widely spoken European languages and disseminated in the West and Russia.

Some travel narratives claimed to be systematic in their description of the region, as in the case of Dinicu Golescu's *Account of my Journey* (1826). He reports that he had compiled a statistical map of Wallachia, which included the following:

> [A]ll the counties and their demarcations, the towns, districts and ports both in the uplands and on the Danube, the quarantines, the livestock exchange markets, the paths, the sub-lieutenancies, the villages, families, priests, deacons, boyars, sons of boyars, company men, pandour soldiers, foreigners, gypsies, Armenians, and Jews, monasteries, metochs, domestic and dedicated sketes, lakes, fisheries, sawmills, wine presses, rivers, streams, brooks, fairs, weekly markets, minerals and the productions of each county (*BARP* 1: 567 in Bracewell and Drace-Francis, 2009: 60).

However, travel narratives written in widely read languages, especially if serialized in popular magazines, effectively shaped public opinion. Such is the 1867 travel narrative on the Balkans, by Georgina Mackenzie and Adeline Ibry. They describe with apparent certainty Thessaloniki, Ottoman Macedonia's largest city and a major territorial objective of the Hellenic State, as follows:

> The Therma of ancient history, and the Thessalonica of Saint Paul's Epistles, yields at present the curious instance of a city historically Greek, politically Turkish, geographically Bulgarian, and ethnographically Jewish. *They* [our emphasis] have said that Salonica is geographically Bulgarian; in other words, it is one of the ports of that country, with a Slavonic-speaking population that stretches from the Aegean to the Danube. Indeed, Salonica itself forms a point on the ethnographical boundary, which, in this part of Turkey-in-Europe, divides the Slavonic population from the Greek… [N]evertheless some miles of country inhabited by Bulgarians stretch south of the Via Egnatia, Greek colonies lie to the north of it; and in the towns, the population is mixed,

> in part consisting of Osmanlee Turks. The other boundary cities are Monastir, Vodena [Edessa], and Yenidje [Yianitsa], in all of which dwell few or no Greeks, whereas in Salonica itself there are only about 500 families of Slavs (Mackenzie and Ibry, 1867: 14-15).

"They" referred to local informants who likely had a specific, if not partisan, perspective on the region's human geography. In a later part of the narrative, they describe Vodena/Edessa as "The Bulgarian 'city of waters,' once Macedonian Edessa" (Ibid: 65).

Travel narratives by Europeans, Russians, and Americans on the Grand Tour may have had a dedicated popular following, but it was dispatches and reports by newspaper correspondents and diplomats that influenced government policy. The description of the ethnic composition of Balkan states and regions of Turkey-in-Europe, read at the House of Lords on Monday, 16 March 1863, and reported on the next day by "The Evening Standard," concerned "The Affairs of Greece" and the question of royal succession in the country. The ethnography of the Balkans was clearly a consideration for how the British Government weighed its policy alternatives. The balancing involved an array of "Sclavonic"/Slavic peoples," the Greeks, the Albanians, and the Turks, the term "Mohamedan" usually but not exclusively referring to Turks:

> What composed the Turkish dominions in Europe? ... Servia was inhabited by a pure Sclavonic race of the Greek faith, but not dependent upon the Greek Church; on the contrary, one of the first struggles of Servia was to release herself from the Greek Patriarch government at Constantinople, and have an independent patriarch at Belgrade. Then came Bulgaria, inhabited by a race speaking the Sclavonic language, and of the Greek faith; but not of a Sclavonic but Tartar origin. For the most part, the landowners were Turks, and the working population were Christians, though they refused to [recognize the dominance] of the Greek Patriarch at Constantinople... In Roumelia there was a Sclavonic substratum, but the landowners were Turks, descendants of conquerors. They came to Albania, and except in the South the population there was Mohamedan; they were not Sclaves, and had nothing to do with the Greeks. Thessaly was a very small province, and there was a Greek race substratum, but the landowners were, for the most part, descendants of the Mohamedan conquerors. In Macedonia, there was the same state of things, except a few seaport towns inhabited by Greeks (*The Evening Standard*, 17 March, 1863: 2).

Similar correspondence appears in French political journals, such as the influential *La Lanterne*. An important part of the reportage on the mobilization of Greek forces on the Thessalian-Macedonian border is the ethnographic profiling of Macedonia: "I am certain," the reporter writes, "that Bulgaria and Serbia have confidentially informed the Royal Government of Greece that these two Slavic states will do nothing as long as the struggle centers on Crete, but their own direct interests would compel them to intervene at the border at the first sign of invasion of Macedonia by Greek troops. It is natural, after all; Bulgaria like Serbia having a very great number of its nationals [ethnics] in Macedonia, where the North-West is inhabited exclusively by Serbs, the North-East and the East by Bulgarians, while Greece can only have pretensions on the Southern part of this great province (*La Lanterne*, 19 March 1897: 2).

Until the end of the First World War, new knowledge about the Balkans, however inaccurate, was being produced and diffused through foreign media from the middle of the nineteenth century. Paparrigopoulos and Hellenic diplomacy were, clearly, not in a position to control the ethnographic discourse that had become source knowledge for the conduct of Balkan geopolitics of statecraft and difference. That remains true to this day, at least with respect to Greece's desire to retain exclusive rights on the place-name "Macedonia." John Agnew notes that since the dissolution of Yugoslavia, efforts to stop the Former Yugoslav Republic of Macedonia from adopting the name "Republic of Macedonia" could be aptly described as "territorial hysteria" (Agnew, 2007: 403-08). The Treaty of Prespa that settled the name issue in 2018–2019 by establishing "North Macedonia" as the name of the Former Yugoslav Republic of Macedonia is a rare example of pragmatic Balkan geopolitics of difference.

Morbus ethnographicus: ethnographic and "ethnocratic" maps in the service of nationalism

A significant tool used in building Balkan geographical imaginaries was ethnographic cartography, a discipline the importance of which did not escape Paparrigopoulos. Cartographically imagining homelands and, in the case of Balkan states, "greater homelands," served territorial claim, as long as either critical Great Power support could be secured for it, or domestic or collective military forces could be applied to it. Mapping the state and the nation in both their actual and aspirational expressions produced both the politics and pathways to their realization (Peckham, 2000: 77–78; Batuman, 2010: 221–22).

The classic guide to understanding Balkan cartography is Henry Robert Wilkinson's work on Macedonia's ethnographic cartography (Wilkinson, 1951). He was responsible for two key developments: first, a greater understanding of the political character of such thematic mapping in the conduct of European geopolitics. Second, he pioneered the map-deconstructionist strategy that was later theorized, elaborated, and established by John Brian Harley (Harley, 1988, 1989a, 1989b). "All powers," Wilkinson writes, "both great and small, imperialistic and nationalistic, discerned the importance of putting exactly that interpretation on the ethnography of Macedonia which might best extend their influence in the area and so prepare the way for the establishment of local hegemony [an in the case of the Ottoman successor states], or Near Eastern ascendancy [as in the case of the British and French thalassocracies]" (Ibid: 6). Concerning Wilkinson's groundbreaking method, he demonstrated the importance of a new approach to the analysis of the more than 200 maps in his Macedonia study: "The maps are treated in chronological order because so many of them are derivative, and each may be studied in its historical context. A comparative method of approach has also been attempted which has necessitated a policy of standardization and simplification of maps, and the adoption of a common formula of analysis for the content of each. A survey of the ethnic classifications used by various authorities since 1821 reveals more than 150 diverse groups" (Ibid: 6–7). Indeed, maps by French, British, Austrian, German, Russian, Bulgarian, Serbian, and later also Greek cartographers, some in their official capacity as cartographers of state, but in most cases, as scholarly or commercial cartographers, vied to construct persuasive and visually striking visualizations of the Balkans' ethnic geographies, almost invariably to influence public opinion and diplomatic outcomes.

The last half of the nineteenth century, and especially the 1870s, epitomized nothing less than a *morbus ethnographicus* ("ou de la manie de résurrection des nationalités")–"map-mania," per Peckham (2000: 77). The term was coined by Father H. Lammens of the *Society of Jesus* in his travelogue about the Levant (Lammens, 1897: 86): noting that nowhere else in the world was that sickness more pronounced than in the "gréco-balkanique" peninsula, he remarks with some sarcasm that "to support their rights, Slavs, Greeks, and Latins invoke again and again religion, history, archaeology, statistics, philology. In their books, their pamphlets, their proclamations, cluster lists, figures, and extremely patriotic diagrams." A little further he mirthfully addresses the cartographic profession:

> They have above all a pronounced vocation for maps. They make magnificent ones with beautiful red, violet, blue tints, etc., for the different nationalities. Only, in each case, the maps differ as much from each other as a modern atlas from that of Ptolemy. Instead of tracing with an impartial brush the motley color of the races, it is always the case that the geographer, an excellent patriot, lets a large blot of national color splatter on his drawing. The blot, stained with oil, spreads step by step, becomes enormous, and ends up invading the peninsula. If you visit a Greek high school, you'll find there a map of a Macedonia and an Epirus that is entirely populated with Greeks (Ibid: 153).

If that were true in 1896 and nationalist ethnographic maps were, indeed, hanging in Greek classrooms when Father Lammens disembarked from the cruise ship *Sindh* of the *Messageries Maritimes* in Pireaus and Samos, this was not the case 20 years earlier. The most dramatic expression of Greece's strategic deficit in the Balkan cartography wars was its participation in the Congress of Berlin without its own official positions enshrined in maps disseminated in the public domain.

There were, however, two unofficial maps that publicized Hellenic territorial claims: one was published in 1877 by the *Edward Stanford Ltd* cartographic establishment (often referred to in the literature as "the Stanford map"), which Yosmaoğlu reveals as the unsigned work of Ioannis Gennadius, the Greek *chargé d'affairs* in London (Yosmaoğlu, 2013: *Kindle* Locations 3028–3029) (Figure 3.3). The Gennadius map was dismissed as

Figure 3.3 The "Stanford map" also published in 1877 by A. Synvet, a philhellene French educator at the Galatasaray Lisesi, generously supported Hellenic positions, albeit using a different mode of data collection based on the rosters of the Orthodox Patriarchate.

outrageous (by cartographers Kiepert and Sax, among others), although it was influential in two senses: it spoke to Greek elites, emboldened them, animated educational imaginaries, but it also expressed a high bargaining position that, though unattainable, may have moved the needle (or the compass, as it were) to some degree in favor of Greece. Gennadius was clearly exploring ways to capitalize upon what was understood by ethnonationalists as the exceptional character of Hellenism: its potential to catalyze difference into the Hellenic.

The map's legend is critical: the "Greeks" category is asterisked. In the lower margin of the legend, the qualification is explained as "including kindred races identified by ties of language and traditions." Accordingly, "kin" to the core Hellenic community are speakers of the Greek language, though unspecified as to how they use their fluency in the language. The notion of common traditions was even murkier. It pointed to another gap in the informational landscape. However, it was inevitable that such allegations would be subject to systematic study again for nationalist reasons. Folk and popular traditions (λαογραφια), were to become instruments of nation- and statecraft in the period 1878–1922. The effort to build institutional support for ethnology in the Hellenic academy was spearheaded by Nikolaos Politis and the *Society for the History and Ethnology of Greece* (Ιστορική και Εθνολογική Εταιρεία της Ελλάδος) he helped found in 1882.

Circles in the Hellenic academy were aware and concerned about the informational gap. By 1876, Paparrigopoulos was well established as the country's "national historiographer," and was also well known in academic circles abroad, even before the publication of his History in French.[14] His transformation from scholar to national political agent was complete by 1876 when he delivered a passionate speech on the Pnyx, the ancient parliament site across the Acropolis, about the need for Greece to mobilize in every sense – popular, military, diplomatic – to gain in the region what was its territorial legacy. Accusing the Great Powers of rewarding the revolt of Serbs and Montenegrins against the Ottomans with favor and concessions, and neglecting their moral obligation to the Hellenes (the morality of the claim based on his own academic thesis of "continuity"), he suggests a territorial partition of Turkey-in-Europe: "[t]the Slavs in the North, the Hellenes in the South [of the Balkan peninsula]" (Dimaras, 2006: 340–41). Paparrigopoulos found cause for the Ottomans' exit from Europe: A heavily indebted Turkey should want to rid itself of these troublesome and expensive lands (Ibid: 341–42). The operative question, of course, was which geographic parallel was this partition to follow. Ergo, maps were needed to establish the "rightful" division of Europe's Southeast in scientific terms.

Paparrigopoulos played a critical, though indirect, role in the Congress. Starting in July 1877, he struggled to carve out a cartographic niche for Hellenic positions on Southern Balkan ethnography in what was otherwise a rather dismal terrain of alternative ethnographic visualizations benefitting the Bulgarians and Serbian.[15] Following up on his correspondence with respected cartographer Heinrich Kiepert,[16] Paparrigopoulos visited him in Germany to lobby for changes to his 1876 ethnographic map of the Balkans.

Paparrigopoulos' objective was to have Kiepert and his publisher, Dietrich Reimer Verlag (Berlin), blunt some of the most injurious, to Hellenic positions, claims made in that critical map: the vast footprint of Bulgarians deeply set into the "unredeemed lands" of Macedonia and Thrace. He did have some leverage, but it was not sufficient to change the widely distributed map: Paparrigopoulos' reputation in academic circles was outstanding, and he could count on it to persuade Kiepert to edit his original map, or create another one for a Hellenic audience. Moreover, less discussed in the literature, the Association for the Diffusion of Greek Letters, which sponsored Paparrigopoulos' German travel, was a significant customer of the *Reimer* cartographic company – Kiepert's publisher. The 1876–77 annual report lists the purchase of 100 large, Kiepert-designed hanging maps for Greek elementary school classrooms, at least half of which were destined for the "unredeemed lands" of Macedonia and Thrace (ADGL, 1878: 64–66).

Not to oversell the notion of "customer is always right," Kiepert did, ultimately, consent to the drawing of a new Balkans map, translated into Greek, titled Map of Hellenic Lands including the Nearby Albanian, Slavic, and Romanian (*Πίναξ των ελληνικών χωρών μετά των παρακειμένων αλβανικών, σλαυϊκών και ρουμανικών*). The innovation involved how identities were classified and symbolized cartographically: building on the primacy of Christian Orthodoxy as a Hellenic institution and signal trait, and asserting that the dominance of the Hellenic language in commerce was constructive of state identity, Kiepert and Paparrigopoulos articulated the concept of "ethnocratic cartography," drawing, accordingly, on the Classical Greek meaning of "κρατος" as "power." Henceforth, a profoundly Hellenized Orthodoxy (clearly under assault at the time by Bulgarian nationalists) and the *lingua franca* status of Greek in everyday life in market towns and cities (again, a contestable notion) were alleged as the most consequential delimiters of "race," "ethnos," and "culture."

Diamandouros, Dragonas, and Ceyder suggest that the map had a dual purpose. Concerning the sharp and absolute delimitation of "ethnocratic" territories on Europe, Paparrigopoulos was signaling how a territorial partition of Turkey-in-Europe could be made (Diamandouros et al., 2010, 26). In that sense, his policy inclinations are made manifest, no matter how wildly optimistic. That Hellenic diplomacy made such a scientifically questionable map – and other such pro-Greek maps like the "Stanford" and Synvet maps – available to negotiators at the Congress of Berlin in 1878 was a failure of the Hellenic State to grasp the importance of the information war that featured the weaponization of statistical, journalistic, and cartographic knowledge. Diamandouros correctly interprets the cartographic design's ambiguity concerning the drawing of Asia Minor's "ethnocratic" character as an expression of the massive geographic boundedness of the Great Idea (Μεγαλη Ιδεα). The blue tint fades eastward to the vast interior allowing the map's reader to set the limit (to the

Black Sea coastal cities of Trapezous/Sinope, or even further to encompass the entirety of the peninsula) (Ibid: 25–26) (figure x: "ethnocratic map").

At the time, Paparrigopoulos's exertions in the area of ethnographic cartography were not deemed essential to the national mission. Ambassador Raghavis remarked in a letter to the historian that action much more critical than map-making was necessary, such as rousing and assisting their subjugated compatriots in Ottoman lands (Dimaras, 2006: 344).

The Russo-Turkish War of 1877–78 signaled the weakness of the Ottoman Empire and its inability to serve as a bulwark against an expansionist Russian Empire. The Congress of Berlin, held in June and July of 1878, sought to reverse the most deleterious for the Ottomans aspects of the Treaty of San Stefano signed earlier in the year. The Treaty envisioned a Greater Bulgaria, very much in Russia's interest, which would incorporate vast swaths of Thracian and Macedonian territory, thus seemingly putting an end to Greece's expansionist aspirations. Greece's participation in the conference was of existential significance. It does, ultimately, produce enough diplomatic élan to see the Ottoman Empire's concession of Thessaly to Greece in 1881.

National toponyms, national topographies, and the establishment of the geographic discipline in Greece

From 1865 to 1909, the development of literary, scientific, historical, and ethnological institutions and societies of national scope served as intellectual connective tissue between the academy and the political establishment. They served the national interest in various ways: by supporting the state's modernization goals through the promotion of the natural sciences in education, by countering international scholarship on Greece perceived as damaging, by revealing the historical geographies of places through topographical research (this being very important to the properly Hellenic naming of places), and by developing taxonomies of customs and material culture, as well as theories of how they fit in Hellenism.[17]

New and restored toponyms called into life a Hellenic world of historical events and personages, myths, faith, beasts, nature, and "cityness" (as in the toponyms for Traianoupolis or Alexandroupolis). As such, they were (and are) the output of political production of "place." Reestablishing the pre-Ottoman Hellenic nomenclature of enduring places was an expression of politics of memory (or the erasure of memory, for the altered Ottoman-era toponyms). Such practice of recall conforms fully to Paparrigopoulos' dogma of Hellenism's "continuity." The Ottomanization and Slavization of place-names were treated as parenthetical to an unbroken Hellenic toponymic tradition that locks down Greece's and, after 1920, also Macedonia's and Thrace's space-time as Hellenism's space-time.

Concerning the significant number of new settlements built after the catastrophic Greek-Turkish War (1920–1922), the State encouraged naming them after lost places of origin in Eastern Thrace and Asia Minor, by attaching "New" to the name (as in "Nea Moudania," "Nea Silivria," and "Neo Sofiko"). Individually and collectively, refugees attempt to rebuild their version of the sociality they enjoyed before and create a new place identity from the debris of memory, custom, and salvaged material culture. Semiotically, these appellations worked as metonyms by standing in for the original. Refugee settlements as material and social constructs also operated synecdochically, each one an island representing an archipelago of new villages, towns, and urban quarters signified by the collective experience of forced migration.[18] In more literal terms, they represented new beginnings, albeit in an unfamiliar social and material environment. They were built, governed, and regulated by a putatively centralized, poorly resourced and managed state (and a distant one, in the case of Western Thrace). Importantly, Hellenic Statism featured leadership and a bureaucracy that were quite adept at exercises of authoritarianism, even at the granular level of imposing Hellenic or Hellenized toponyms.

In 1834, the State launched the re-toponymy of places bearing non-Hellenic names.[19] How such policies were articulated and implemented over the next 180 years reveals the variable control Athens' political establishment and the Hellenic academy exercised over the symbolic geographies of regions and municipalities. Efforts by the center to shepherd the Hellenizing turn in toponymy in the periphery reveal waxing and waning state centralization and governmentality deficits. Old Turkish, Slavic, Albanian, Vlach, and Pomak toponyms persisted in common usage even years after the official redesignation of the name. Moreover, when the spirit of patriotism did not move civil servants to action, there were penalties: "obdurate" or "indolent" local officials could be subject to fines and imprisonment. Such practices persisted into the twentieth century.

As much a policy as an emerging political paradigm, during the short-term Bavarian regency and in King Othon's name, toponymy was dictated by the Palace. In the thrall of philhellenism and aware of its domestic and international political value, the Bavarian Regency commanded, "to resurrect, with this measure, antiquity, to link Greece's present with its glorious past, to reflect an image of Classical Greece abroad, and, finally, to enrich the Modern Greek language with high-sounding names" (Royal Edict of 23 March 1834).

Importantly, that enterprise was more than nationcraft. It signaled the thickening of state capacity. Starting in 1827, the Kapodistria regime, the Bavarian Regency during King Othon's minority that followed it, and subsequently, King Othon's governments, carried out censuses at irregular intervals of one- to three years. A review of these early censuses reveals the political elites' efforts to modernize and Europeanize (also de-Ottomanize) the new state by accounting for land resources, taxable activities, and populations, and by recording, albeit in gross scales and to an incomplete degree, the properties of the Church and large Ottoman-era agricultural estates (Çiflik).[20] A significant challenge to instrumentalizing census information stemmed from the state's territorial expansion in stages between 1864 and 1947. The incorporation of the Ionian Islands (1864), Thessaly and Arta (1881), Macedonia, Epirus, and Crete (1913), Western Thrace (1920), and the Dodecanese Islands (1947), required the incorporation, standardization, and operationalizing of population, settlement, and resource statistics accumulated under different jurisdictions and on different bases. Each of these national territorial enlargements introduced its toponymic heritage that had to be litigated into the national toponymic paradigm.

As a prerequisite to Athens' effective administration of the periphery, the Bavarian Regency instituted by Royal Edict (10/22 January 1834), a nested, hierarchical administrative geography of prefectures constituted of municipalities, and municipalities constituted of town and village communities. The establishment and staffing of these regional and local governments and bureaucracies, however poorly resourced, was an elementary but essential exercise of statecraft. The renaming of old and new settlements, as root exercise of statecraft, linked territorial governance to Hellenization.

Credible scholarship had to support the alterations. The academy and the state worked in concert in furthering the enterprise. Created in 1865 as a private institution dedicated to "the advancement and the common good of the people," the Philological Society "Parnassos" quickly emerged as a meeting place for Greece's powerbrokers in the worlds of politics, the academy, justice, and the military. National historian Konstantinos Paparrigopoulos' election in 1872 as its honorary President further signaled the society's parastatal character. Enjoying the Palace's support since its foundation, "Parnassos" was recognized by Royal Edict on 17 March 1875 as a nonprofit private institution in the public interest.[21] Notably, the society was on the vanguard of promoting the natural sciences (including geography), which were considered to be lagging behind in Greece (Mergoupi-Savaidou, 2013: 233–234). In 1882, the newly created Historical and Ethnological Society of Greece, which included several "Parnassians," founded the National Historical Museum. Its mission was to collect and curate material about Hellenism and the new state's historical role in the region. Historian and geographer Antonios Miliarakis was appointed director of its important library.

Miliarakis was one of the founders of the geographic discipline in Greece and a staunch advocate for its elevation from a primary school subject to a field of higher learning. To that end, and at the encouragement of German Classical archaeologist and mentor, Gustav Hirschfeld, Miliarakis published an annotated bibliography of scholarly works on geography written by Greeks between 1800–1889. It covered, to some degree, all regions of the world and several European and non-European states in some detail (Miliarakis, 1889). The introductory section provides a thumbnail of Greek geography's subdisciplinary structure: it included works in "general," historical, "philological" (substantially, cultural), military, ethnological, economic and statistical, geological, botanical, and zoological geography, which were followed by travel and itinerary accounts and works on the teaching of geography. The bulk of the work is dedicated to "forensic topography" and travel accounts organized according to Greece's prefectural departments. In his survey, Miliarakis includes sections on Constantinople and Thrace (at that time under Ottoman rule) and the newly formed states of Bulgaria and Romania. Notably, Miliarakis omits any direct reference to the Ottoman Empire, although the bibliography list works on Ottoman regions of special historical interests to Greeks. Works on the Aegean and Ionian islands and Crete are listed in a separate section. Works on the topography of the non-Greek world follow.[22]

Miliarakis, as a science booster, saw the discipline of geography as inextricably linked to Greece's entry into a forward-thinking European world. He also ascribed to it a role in fostering a national esprit de corps and good territorial governance. In 1877, he wrote the following in the notable journal *Estia*:

> [T]he reading of specialist topographical and chorological essays, of travel accounts, and descriptions of the civil and ethnographic character of the country's provinces, [and] the ceaseless study of their particular geographic state, would kindle among all Greeks ever warmer, vigilant, and present the spirit of nationalism; the spirit of brotherhood. Seeking to demonstrate the extent of Hellenic lands and making bonds between Hellenes of the diaspora all the stronger and close, such literature greatly heightens the passion and amplifies the energy necessary to realize the sacred goal (Miliarakis, 1877¨425).[23]

However, categorizing Miliarakis as an uncompromising nationalist would be wrong. He enjoyed recognition in German and French scholarly circles as a skilled, meticulous, and credible scholar. Therefore, at issue was whether Miliarakis' geographies and topographies of Greek places could satisfy both Greece's nationalist elites and international scholars. The preamble to his study on the toponym "Messaria" (commonly found in the Greek islands, Crete, and Epirus) provides clues. Speaking generally to the academic subject of toponymy, he insists on exhaustive historical research and abhors the thoughtlessness of officials in changing toponyms, even in the service of Hellenization.

> It would be right and just to consider any substitution of contemporary toponyms by ancient ones, and any changes attempted by the geographer or the authorities to the root term or word ending, as equivalent to the destruction of living monuments of Hellenic history and language. [The act would be] devoid of any practical and moral benefit other than the short-term demonstration of the destroyers' half-knowledge or complete ignorance of the history of places and their names (Miliarakis, 1893: 423-424).

In a footnote to that passage, Miliarakis provides the comparative European context of his critique:

> Once they created sovereign states, the Italians, the French, the English, the Spanish, [and] the Germans neither in their geographical treatises nor in their practice of administration did they replace standing geographical names of places with ancient ones. Instead, they preserved the later ones in the exact form they found them, while in our case, and for no good reason, there was a reversal (Ibid: 424, fn. 1).

Miliarakis' position suggests that Greece's establishment was not unified in its approach and goals of Hellenization. The annihilation of the non-Hellenic past in language and landscape was not necessary for Hellenism to flourish. "Ground-truthing" Hellenism was enough.[24] On the one hand, Miliarakis appears to be primarily concerned with aporias in the historical record and the provenance and meaning of toponyms, which, he claims, "are sculpted on the land" (Ibid: 423). On the other hand, his forensic topographies addressed elements at the center of folkloric studies (customs, material culture, linguistic idioms), the raison d'être of which, under the leadership of "Parnassian" Nikolaos Politis, was to prove the unbroken continuity of popular Hellenic culture from antiquity to the present.

Co-founders of the Historical and Ethnological Society of Greece, Politis and Miliarakis were axiomatically committed to new scientific methods to reveal the historical, geographical, and ethnological bases of Hellenism. The ethnographic modus established under Politis' leadership was to compare ancient and contemporary practices and material culture for evidence of continuity (even when that evidence was vestigial). It also compared Hellenic folkloric practices and objects with those in other regions to situate them in a grand, racialized chronology and topography of culture.[25]

Concerned about the quality of work on toponymy, the state reaffirmed its control over the enterprise by creating a blue-ribbon committee (the Curatorship of Toponyms of Hellas) charged with overseeing it (Royal Edict of 31 March 1909). As committee chair, Nikolaos Politis was unequivocal in his criticism of past practices. He writes, "Although noble and, no less, most useful and necessary the [toponymy] enterprise [Greece's] first governments set upon; we cannot consider it to have been successful" (N. Politis, 1899: 1-6; N. Politis, 1920: 3). He noted several failures in the proper replacement of "barbarous" (in the Classical sense of "non-Hellenic") and "ill-sounding" toponyms: misattributions of ancient toponyms to unrelated places; the fabrication of parodic, Hellenic-sounding toponyms; the invention of toponyms lacking any connection to the locale; ungrammatical articulations of toponyms (inflicting violence upon the national language was considered anathema); selection of toponyms that obscured or erased vestiges of the historical record scribed upon the land; and replications of the same toponym in a municipality, contributing to confusion (Ibid: 3–11).[26] Politis' and

Miliarakis' collaboration is made bare by Politis' finding that "historical evidence, encoded in toponyms, is important and precious when it illuminates lesser-known periods of our national history. One of our geographers compared the evidence with information 'sculpted on the land'" [an expression, which was first used by Miliarakis in his Messaria study] (Ibid: 6).

Lexical dysphoria aside, Politis considered the decision-making process for the replacement of toponyms broken. His first concern was to assess and harness past and on-going scholarly research on toponyms. It was to that end, and at his recommendation, that the State created the Magistracy over the Toponyms of Hellas and placed Politis at its head. It was, essentially, a steering committee that sat in Athens on matters related to toponymy. The initial 14-member committee was staffed by university professors of geography, archaeology, history and letters, and experienced administrators, such as ephors (State overseers of antiquities), and it was open to the counsel of other scholars, as needed. A Member of the Hellenic Parliament also served (Ibid). It reported its findings to the Ministry of the Interior, the latter having ultimate jurisdiction over any changes.

The committee had the power to establish prefecture-level subcommittees along the same lines of subject-matter expertise and political adjacency. By law, the subcommittees were charged to meet monthly to adjudicate field information harvested at the local level. They reported their findings to the steering committee, which added its recommendation before sending it up the ladder to the Ministry of the Interior (Ibid: 7). Therefore, the new, non-adversarial process was designed to scrutinize locally sourced information about places and filter it through layers of decision-making bent to national orthodoxy. Closest to the ground, however, were village committees charged with ferreting out information about the locality. Paradoxically, they were also given a priori license to select and use their chosen toponym without a clear and easy path of appeal to higher levels of oversight. Politis reports on two adverse outcomes of the overall procedure: first, the steering committee's findings and recommendations were routinely ignored up and down the decision-making ladder. Second, the law, he claimed, gave village committees too great a say over the name. The village's de facto adoption of toponyms negated the presumed oversight of Athens. Lastly, requiring the prefecture's involvement in the appeal process made the work of the Athens-based committee all the more complex and laborious (Ibid: 8–9).

This ostensibly domestic and academic, if not antiquarian, enterprise needs to be understood in its geopolitical context. In the time between the creation of the steering committee in 1909 and Politis' report in 1920, Greece and the region had been greatly transformed by seismic political events: the Young Turk Revolution had broken out in 1908 ushering in the final life-stage of the Ottoman polity; the Balkan Wars changed the balance of power in Turkey-in-Europe; and Greece gained in the 1913 settlement of these wars significant territory at the Empire's and Bulgaria's expense, in partial fulfillment of the Great Idea. Greece would benefit further by siding with the

Entente in the First World War. The 1918–1919 Treaty of Versailles gave rise to a new geopolitical paradigm in which national self-determination became an operative dimension of international politics and geopolitics only to be instrumentalized by the Great Powers to service of their interests. In 1920, Greece, as beneficiary of the Treaty of Sèvres, would come the closest to realizing the Great Idea's ambition. Geographic inquiry and the toponymy enterprise as handmaids of territorial Hellenization had to adjust to new configurations of borders and mind the Hellenicity of the new borderlands.

Following the 1923 Treaty of Lausanne and the ensuing population exchange, diplomatically isolated Greece found itself menaced by enemies to its East and North. Part mythological, part doctrinal, the expressions "εξ ανατολών κίνδυνος" and "εκ Βορρά κίνδυνος" ("the peril from the East" and "the North," respectively) were introduced to the national vocabulary. New geopolitics emerged whereby international security concerns found resonance in domestic intercommunal relations and the management of difference. A stronger perception of non-Hellenic ethnics as disloyal and suspect would gradually crystalize and drive Greece's domestic and international policies. A qualitative shift occurred concerning the meaning of re-toponymy: it no longer served the building a national home alone; it also served as a means of containing external foes. The tone became strident, and the scientific legitimacy of the enterprise sidelined.

In 1926, there were further refinements of and corrections to the state's policy on toponymy. Issued by the Ministry of the Interior's Directorate of Local Self-Government, the manual titled *Instructions for the Renaming of Communities and Settlements with Turkish and Slavic Names* is quite explicit on the subject of Hellenism's superiority over other civilizations (πολιτισμοί) and the importance of re-toponymy.

> [D]ifferent raiders and conquerors did not leave [our land] without a trace, which, regardless of their inconsequence, pollute the face of our beautiful fatherland. Most importantly, they need to be gradually expunged, to go away, because the enemies of our nation and race use them oftentimes as an excuse to contest our lands. Such traces are various Turkish and Slavic names of villages. They must be replaced by Hellenic names, especially now, since, with the influx of refugee Hellenes, the population [of the villages] is also Hellenes (Kyriakidis, 1926: 3).

Its author, Stilpon Kyriakidis, exhorted local officials to select, with local teachers' assistance, toponyms that were properly "neo-Hellenic," – beautiful, easy to pronounce, and referencing the qualities of the region. He writes, "in this manner [the new name] would be readily accepted and easy to adopt by the neighbors, and will erase and silence the current, barbarous one" (Kyriakidis, 1926: 3).

Village communities were allowed to select toponyms, but had to follow strict guidelines from that point on. If not directly resuscitating ancient

Greek or Byzantine toponyms, a new one would need to be derived from one of the following categories listed in the following order: either the name or the official title of a notable Byzantine landowner; a military land-use; the geology or geomorphology of the site; the hue of the soil; water features; flora; fauna; attributes of sunlight; vestiges of the built environment (especially churches); agricultural land-use; occupational profile of inhabitants; name of a notable local family; local history; local events of note; the name of Revolutionary hero/martyr (only when killed in action on-site) and folk-tales (Ibid: 4–6). Lists of appropriate toponyms in each category make up the bulk of the manual (Ibid: 8–39).

The state's preoccupation with toponyms will continue into the 2000s nearly unchanged by vast political change in the intervening period: cycles of democratic and authoritarian rule, the Second World War followed by the Civil War, Greece's entanglement in the Cold War ushered in by the Truman Doctrine, two military coups (most notably in 1967), seven years of Junta rule followed by a return to democratic rule, the Cyprus crisis that resulted in the partition of the island, the abolition of the monarchy, and Greece's accession to the European Communities in 1981 and, later, the European Union.

The policy itself proved to be extraordinarily durable, although gradually and only after the return to democratic rule did the tone moderate and the grip of the state loosen. A comparison of the law on toponymy promulgated by the Junta[27] to those who followed during democratic rule illustrates this progressive change: the national blue-ribbon committee of academics and administrators now included representatives of local self-government. It considered new toponyms proposed by local authorities before, but there was a renewed emphasis on scholarly accountability. The Ministry of the Interior was still the final arbiter. There were, however, vestiges of old geopolitical anxieties in the description of aims. The 1992 law stated that a new toponym should replace one if it "is derived from another language, is ill-sounding, or is inappropriate," although, in this case, inappropriateness may as much refer to technicalities as to red lines concerning the national sentiment.[28] By 2006, the admonitions disappeared. Legal language became substantially less prescriptive and only provided examples of how the toponym may be grounded: local history, mythology, local attributes of the site, etc. – a very political "etc." since, in theory at least, it opened the way to the consideration of less nationally orthodox names.[29]

Conclusions

Great Powers were ambivalent about supporting the 1821 Greek Revolution, and, later, Greece's territorial ambitions. No power, including Russia, responsible for dismantling the Ottoman Empire, had an interest in resurrecting some version of a Hellenic one. By the mid-Nineteenth century, it was clear that the balance of power in the broader region, which included the Black Sea and the Caucasus, was favoring the Russian Empire. It was

Russia that needed to be balanced – not Turkey – and unless Greece was going to contribute to that grand Anglo-French strategy of, at least, retarding and territorially fencing in Russia, then Greece's foreign policy and security aspirations would find their disapproval. After all, this was still a conservative Europe remade in Vienna in 1814–1815.

The Crimean War (1853–1856) was a test case of how far Greece could deviate from the geopolitical course set by Great Power priorities. Britain and France counted on Turkey's survival to contain Russian expansionism. Greece's aspiration to wrest Thessaly, Epirus, and Crete from the Ottoman Empire by instigating uprisings in these regions clashed with Anglo-French strategy in support of the Ottomans. Russia's gambit to retrench itself southward, deeper into the Balkans and the Black Sea regions, did not include a role for Greece. For Russia, the putative defender of all Orthodox subjects, Greece's adventurism was a distraction. In 1854, Franco-British forces first blockaded and then seized the Port of Piraeus to prevent further actions by Greece, thus dooming the Epirus revolt. This was not an even playing field and Greece was a Small Power. Its historical territorial claims had little currency in the conduct of power geopolitics, unless they could be leveraged in support of one or another Great Power's grand strategy.

In stark contrast stood Greece's geopolitical conduct during the Russo-Turkish War of 1877–1878. Reassured by other Great Powers that its claim of Thessaly would be given serious consideration should Greece stay out of the war, Greece did just that. At the Congress of Berlin (1878), the Great Powers assented to Thessaly's and Arta's and much of Epirus' incorporation into Greece. The Convention of Constantinople formalized the region's union to Greece in 1881.[30]

Greece's geopolitical dependency on Great Power assent did not preclude it from pursuing an independent foreign and security policy, albeit, at times at a great cost, as the disastrous Greek-Turkish Wars of 1897 and 1920–1922 proved. It is on such a scaffold of power geopolitics that Greece's political, military, and intellectual elites negotiated the enduring nineteenth century contours of Hellenic history and the racial, and territorial signifiers of its Helleno-Orthodox identity. Greece earnestly mobilized the international philhellenic milieu and its extensive diaspora to support the idea of a resurrected Hellenism deserving of the restoration of its historic lands. In the diplomatic milieu, it instrumentalized Ottoman censuses (and the censuses it carried out in its jurisdiction), ethnographies, cartographies, topographies, and toponymy as bases for counterarguments against Bulgarian and Turkish statecraft's territorial and population claims.

In the chapters that follow, we adopt a more granular approach to questions of difference, territory, and geopolitics, with studies on the Hellenization of Thessaloniki, the formation of majority-minority geopolitics in Thrace, the settlement-as-geopolitics of Macedonia's and Thrace's borderlands, and the contested building of the first new mosque in Athens since Greece's independence in 1830–1832.

Notes

1. Politis (2008) and Konstantinos Dimaras (2006) extensively cover the views of Enlightenment-driven purists of Hellenism in their works on Hellenic historiography.
2. "Εξησθένισεν οχι ολίγον το έθνικόν τών Ελλήνων φρόνημα" [άφου οι Βυζαντινοί] "είχον άπομακρυνθή τής υγιούς φιλοσοφίας, έμπεσόντες εις σχολαστικότητα." [Το] "έλληνικόν στοιχεΐον δεν ήγέρθη" [σέ ολη τή διάρκεια τής βυζαντινής περιόδου], "διότι ούκ ήν ό έγείρων, ή ελευθερία και ή παιδεία." Στεφάνου Α. Κουμανούδη Ανέκδοτα. 'Ιστορία του 'Ελληνικού "Εθνους, ο.π. σ. 48, 51.
3. Stefanos A Koumanoudis was one of the most vocal polemicists of the view that religion – and Orthodox Christianity in particular – was instrumental in defining Hellenism. He wrote on the ills of Christianity: "Many praise Christianity in many ways. When it was at its peak and was draining all potency from its contemporaries, it was not the best of times; tyranny and murder were commonplace, as was injustice. The courts were weak. Such were, undoubtedly, conditions in the middle ages, profoundly benighted, ignorant, and superstitious. The natural sciences hardly progressed, and despite that, Papism shone [«'Επαινούν πολλοί τον Χριστιανισμόν πολλαχώς. 'Αλλ' εγώ βλέπω δτι δταν ήτον αυτός εαυτού εγκρατέστατος και άπερρόφει ούτως ειπείν ολην τήν ικμάδα τών συγχρόνων του, οι καιροί εκείνοι δεν ήσαν οι κάλλιστοι* τυραννίαι, φόνοι συχνοί, άδικίαι έπράττοντο. Τα δικαστήρια ήσαν αδύνατα. Ταύτα ήσαν εις τον μεσαίωνα αναντιρρήτως, προς δε και σκότος πολύ και αμάθεια και δεισιδαιμονία. Αι φυσικαί έπιστήμαι εκειντο άδρανέστατα, και ομως τότε ελαμψεν ό Παπισμός."] (Ματθαίου, 2007: 264–65).
4. In 1864, Britain ceded the Ionian Islands to Greece in the Treaty of London, and in 1947, Italy ceded the Dodecanese Islands in the Treaty of Peace as partial compensation for the 1940 war of aggression. In the intervening period, Greece annexed Thessaly, Epirus, and substantial portions of multicultural Macedonia and Thrace. Following its catastrophic defeat in the Greek-Turkish War (1920–1922), Greece became significantly less hawkish, except for the 1974 attempted forced union of Cyprus to Greece by the Colonels' Junta.
5. "*Das Geschlecht der Hellenen ist in Europa ausgerottet. Schönheit der Körper, Sonnenflug des Geistes, Ebenmaß und Einfalt der Sitte, Kunst, Rennbahn, Stadt, Dorf, Säulenpracht und Tempel, ja sogar der Name ist von der Oberfläche des griechischen Kontinents verschwunden.... auch nicht ein Tropfen echten und ungemischten Hellenenblutes in den Adern der christlichen Bevölkerung des heutigen Griechenlands fließet*".
6. He demonstrated a clear preference for Rome over Greece in his comparison of Sparta to Rome as political-military regimes.
7. Dimaras reproduces the *Pandora* version of the article in its entirety in his 2006 comprehensive study of Paparrigopoulos' life and works.
8. Most notable among foreign intrusions on Ottoman sovereignty was the 1838 Anglo-Ottoman (commercial) Treaty of Balta-Liman, which followed a series of critical geopolitical setbacks for the Empire: the Anglo-Turkish War of 1807–09 (culminating in the Treaty of Dardanelles or *Kale-i Sultaniye Antlaşması*, 1809), the Great Power naval action at Navarino in 1827 that saw the near-annihilation of the Ottoman navy, and Britain's patronage of the Hellenic State in-the-making between 1827 and 1830, that led to statehood in the Treaty of Constantinople, again substantially under British aegis (21 July 1832). While *Article 1* of Balta-Liman confirmed that privileges and immunities of British subjects were to be enjoyed in perpetuity, *Article 2* was the denouement of the Treaty's unequal character: "The subject of Her Britannic

Majesty, or their agents, shall be permitted to purchase, at all places in the Ottoman Dominions (whether for the purposes of internal trade or exportation), all articles, without any exception whatsoever, the produce, growth, or manufacture of the said Dominions; and the Sublime Porte formally engages to abolish all monopolies of agricultural produce, or of any other articles whatsoever, as well as all Permits from the local Governors, either for the purchase of any article, or for its removal from one place to another when purchased; and any attempt to compel the subjects of Her Britannic Majesty to receive such Permits from the local Governors shall be considered as an infraction of Treaties, and the Sublime Porte shall immediately punish with severity and Viziers and other Officers who shall have been guilty of such misconduct, and render full justice to British subject for all injuries or losses which they may duly prove themselves to have suffered." It left Ottoman sovereignty in tatters, only to see it assaulted again, legally and territorially, in successive Great Power actions throughout the nineteenth century.

9. The name of the association also appears as *Association for the Dissemination of the Greek Culture.*
10. E. Ruzhdavichka (2004), Kirzioglu (1979), Todorova (1995), in Dundar.
11. Kaymakamlik of Gevgili to Rumeli Vilayeti, Balbakanlik Devlet Arlivleri, Osmanh Argivi, Rumeli Miifettigligi Selinik Evrtki (henceforth TFR.I.SL) 81/8053, 8 Agustos 1321/21 August 1905.
12. "According to [early Republican] censuses, the following languages were spoken in the city of Istanbul as mother tongues: Turkish, Abkhazian, Persian, German, Arabic, Albanian, Bosnian, Bulgarian, Czekoslovakian (!), Circassian, Armenian, Flemish (Dutch), French, Georgian, Croatian, English, Spanish (most of the speakers being Jews probably Judaeo-Spanish), Swedish, Italian, 'Coptic' (= Gypsy), Kurdish, Laz (a Caucasian language spoken on the North-Eastern part of the Black Sea coast), Polish, Hungarian, Pomak (a variety of Bulgarian spoken by Muslims), Romanian, Greek, Russian, Serbian, Tatar, and 'Jewish' (may refer to Judaeo-Spanish or Yiddish, or both)" (Strauss, 2011: 127).
13. Thornton, Thomas (1807), Dimitrii Bantysh-Kamenskii (1810), Dinicu Golescu (1826), Iurii Gutsa-Venelin (1829), Robert Walsh (1836), John Lloyd Stephens (1838), Georgina Mary Muir Mackenzie and Adeline Paulina Ibry (1867), G. F. Abbott (1903), Mary Edith Durham (1905), Henry Noel Brailsford (1906), Carnegie Endowment for Peace (1913-14), and Lucy Mary Jane Garnett (1917) among others.
14. In 1870 the *Association pour l'encouragement des études greques en France* had recognized the *History* with its award. In 1875 Paparrigopoulos was made an honorary member of the faculty of the University of Odessa; and in 1881 he was made a member of the Serbian Academy (Dimaras 2006: 329).
15. s. Spiridion Gopčević's map "Ethnographical map of Macedonia and Old Serbia" (Gotha: Justus Perthes, 1889) was probably the boldest pro-Serbian cartographic statement. Better know is Jovan Cvijic's later map that identifies and names a Slavic Macedonian ethnicity, as distinct from the Bulgarian, as a demographic and political bulwark against Bulgarian territorial expansionism.
16. Kiepert was a collaborator of Carl Ritter, one of the founders of modern geography
17. For Classically-trained Greek elites, the foundation of cities and towns would have resonated as an act of poiesis in the Platonic sense of the *Symposium*: a form of immortality and most certainly rebirth. In Heideggerian terms, the eventfulness of 'bringing forth' new identities of place and then establishing

them *in* the world and as *part* of the world would critically contribute to the nation-state's meaning. The Hellenizing of place-names in the service of ethnonationalism reflected the zeal of nationalist scholars and politicians to lend Hellenic meaning to the country's topography.

18. There is significant geographic scholarship on toponymy's politics and semiotics, referenced by Rose-Redwood, Alderman, and Azaryahu as "critical place-name studies." See Masalha, 2015; Rose-Redwood et al., 2010; Georgiou, 2010; Kadmon, 2004 and 2000; Papapavlou, 2003). A significant part of that literature focuses on memory, and the socially and politically constructed toponymy of streets in different settings and scales, including case studies of cities in the American South, Palestine, and Francophone Canada.
19. An excellent review of the Royal Edicts and legislation regarding the history of toponymy in Greece appears in Giorgos Koumaridis's, "Hellenising the Country: Changing Place-Names in Greece," unpublished Master's essay, Birkbeck College, University of London, 2013.
20. Censuses would improve in later decades, the census of 1861 is recognized as paradigmatic of that evolution. The road to institutional maturity has been long. The national land registry is still unfinished in the 21st century.
21. The history of the Society is profiled in its official site: http://www.lsparnas.gr/index.php.
22. For an extensive discussion of Antonios Miliarakis' work, see Robert Shannan Peckham's *National Histories, Natural States: Nationalism and the Politics of Place in Greece* (2001), pp. 79-82.
23. Translation by authors. Miliarakis indirectly references the "Great Idea," which at the time was all but official doctrine.
24. His critique anticipates much later scholarly discourse about "epistemicide": the extermination of epistemes of absent or weaker Others by groups in power.
25. Theories connecting Hellenism to "Indo-European" civilizations were in the mainstream in Politis' and Miliarakis' time. Today, they are in the domain of Greece's extreme Right (Bistis, 2013).
26. Concerns over the proliferation of popular toponyms would persist for years. Stilpon Kyriakidis, Politis' successor to the deanship of the enterprise, leveled new admonishments to village committees in a 1926 official manual on toponymy.
27. Legislative decree 133/25 February to 03 March 1969 (ΦΕΚ Α′ 35).
28. Law 45942/6-17 August 1992 (ΦΕΚ Β′529), article 1, paragraph 4.
29. Law 3463/2006 (ΦΕΚ Α 114/30.6.2006). Changes to the legal environment have allowed in some cases the return to non-Hellenic toponyms. In a recent example, on 25 January 2019, the national committee on toponymy approved the request of the village community Platania ("Plane trees") in Epirus to reestablish the original toponym, Tseritsana. In this case, the rationale was based on two historical events: the assistance that Tseritsana women gave to the besieged women of the village of Soulio during the Greek War of Independence, and the massacre of Tseritsans by the German army during the Nazi occupation of Greece in the Second World War.
30. This was already the second instance of geopolitical largess on the part of Great Britain since their disciplining of Greece during the Crimean War. In 1862 Britain ceded the Ionian Islands to Greece on the occasion of the assumption of Greece's throne by King Georgios I, the brother of Alexandra, Princess of Wales. It was a hugely popular gesture, and the British could count on favorable sentiments in both the Palace and Hellenic Parliament.

Sources

Abbott, G. F. *The Tale of a Tour in Macedonia*. London: Kessinger, 1903.
Agnew, John. "No Borders, No Nations: Making Greece in Macedonia." *Annals of the Association of American Geographers Vol.* 97, No. 2 (2007): 398–422.
Bantysh-Kamenskii, Dmitrii. *Puteshestvie v Moldaviiu, Valakhiiu, i Serbiiu (Travels in Moldavia, Wallachia and Serbia)*. Moscow, 1810.
Batuman, Bülent. "The Shape of the Nation: Visual Production of Nationalism Through Maps in Turkey." *Political Geography* Vol. 29 (2010): 220–34.
Bistis, George. "Golden Dawn or Democratic Sunset? The Rise of the Far Right in Greece." *Mediterranean Quarterly* Vol. 24, No. 3 (2013): 35–55.
Bracewell, Wendy; Drace-Francis, Alex, eds. *Balkan Departures: Travel Writing from Southeastern Europe*. New York & Oxford: Berghahn Books, 2009.
Brailsford, Henry Noel. *Macedonia. Its Races and Their Future*. London, 1906.
Carnegie Endowment. *Report of the International Commission to Inquire into the Causes and Conduct of the Balkan Wars*. Washington, DC, 1913-14.
Diamandouros, Nikiforos; Dragonas, Thalia; Keyder, Çaglar, eds. *Spatial Conceptions of the Nation: Modernizing Geographies in Greece and Turkey (International Library of Historical Studies, Book 66)*. London: I.B. Tauris Publishers, 2010
Dimitrijevic, Jelena. *Letters from Salonica on Young Turk Revolution*, 1918.
Dundar, Fuat. "Empire of Taxonomy: Ethnic and Religious Identities in the Ottoman Surveys and Censuses." *Middle Eastern Studies* Vol. 51, No. 1 (2015): 136–58.
Durham, Mary Edith. *The Burden of the Balkans*, 1905. Reprint. Ann Arbor, MI: The University of Michigan Press, 2006.
Feigl, Hans; Molden, Ernst, eds. *Schriften und Tagebücher: Fragmente aus dem Orient. Neue Fragmente. Politisch-historische Aufsätze – Tagebücher (in Auswahl) (Writings and diaries: Fragments from the Orient, New fragments, Political-historical essays – Selections from the diaries)*. Munich and Leipzig: Georg Müller, 1913.
Garnett, Lucy Mary Jane. *Balkan Home Life*. London, 1917.
[Gennadius, Ioannis]. *An Ethnological Map of European Turkey and Greece: With Introductory Remarks of the Distribution of Races in the Illyrian Peninsula and Statistical Tables of Population*. London: Edward Stanford Ltd., 1874.
Georgiou, Vasiliki. "Competing Discourses in the Debate on Place Names in Cyprus: Issues of (Symbolic) Inclusion/Exclusion in Orthographic Choices." *Journal of Language and Politics* Vol. 9, No. 1 (2010): 140–164.
Golescu, Dinicu. *Account of my Journey*. 1826.
Golescu, Dinicu. *Account of my Journey*, trans. M. Kovacs. In B. Trencsényi and M. Kopećek, eds., *Discourses of Collective Identity in Central and Southeast Europe (1770-1945). Texts and Commentaries, Volume 1: Late Enlightenment – Emergence of the Modern 'National Idea.'* Budapest: Central European University Press, 2006.
Gotsi, Georgia. "Empire and Exoticism in the Short Fiction of Alexandros Rizos Rangavis [Raghavis]" *Journal of Modern Greek Studies* Vol. 24, No. 1 (2006): 23–55.
Gutsa-Venelin, Luri. *Uchenoe puteshestvie Iu. I. Venelina v Bolgariiu (The scientific expedition of Y. I. Venelin to Bulgaria)*. Moscow: Rossiiskaia Akademiia Nauk, 1829/2005.
Hammond, Andrew. *British Literature and the Balkans: Themes and Contexts*. Amsterdam and New York: Rodopi Press, 2010.
Harley, J. B. *Maps, Knowledge, and Power*. In D. Cosgrove, & S. Daniels, eds., *The Iconography of Landscape: Essays on the Symbolic Representation, Design, and Use of Past Environments*. Cambridge, UK: Cambridge University Press, 1988: 277–312.

Harley, J. B. "Deconstructing the Map." *Cartographica* Vol. 26, No. 2 (1989a): 1–20.

Harley, J. B. "Cartography, Ethics and Social Theory." *Cartographica* Vol. 27, No. 2 (1989b): 1–23.

Kadmon, Naftali. "Toponymy and Geopolitics: The Political Use — and Misuse — of Geographical Names." *The Cartographic Journal* Vol. 41, No. 2 (2004): 85–87.

Kadmon, Naftali. *Toponymy: the Lore, Laws, and Language of Geographical Names.* New York: Vantage Press, 2000.

Kallivretakis, Leonidas. "Historical Study of Settlements in Greece (15th-20th centuries)": Research objectives and problems of the sources." In *Πληθυσμοί και οικισμοί του ελληνικού χώρου: ιστορικά μελετήματα*. Athens: Institute for Neohellenic Research/National Hellenic Research Foundation, 2003: 7–24.

Kirzioglu, M.F. "1593 yılı Osmanlı vilayet tahrir defterlerinde anılan gence-Karabag sancakları ulus ve oymakları" *(Nations and tribes mentioned in the Ottoman provincial registers of 1593)*. Ataturk Universitesi, Edebiyat Fakultesi Ara, stırma Dergisi, Ataturk University (1979): 199–222.

Kitromilides, P.M. "On the Intellectual Content of Greek Nationalism. Paparrigopoulos, Byzantium and the Great Idea." In D. Ricks, D. and P. Magdalino, eds., *Byzantium and the Modern Greek Identity.* London: Publications of the Centre for Hellenic Studies, King's College, London, 1998.

Krivokapić, Marija, ed. *The Balkans in Travel Writing.* Newcastle upon Tyne: Cambridge Scholars Publishing, 2015.

Leeb, Thomas. *Jakob Philipp Fallmerayer: Publizist und Politiker zwischen Revolution und Reaktion.* Munich: C.H. Beck Verlag, 1996.

Mackenzie, Georgina Mary Muir; Irby, Adeline Paulina. *Travels in the Slavonic Provinces of Turkey-in-Europe.* London: Bell and Daldy, 1867.

Masalha, Nur. "Settler-Colonialism, Memoricide and Indigenous Toponymic Memory: The Appropriation of Palestinian Place Names by the Israeli State." *Journal of Holy Land and Palestine Studies* Vol. 14, No. 1 (2015): 3–57.

McArthur, Sarah. *Being European: Russian Travel Writing and the Balkans, 1804-1877,* Doctoral dissertation. London: University College London School of Slavonic and East European Studies, 2010.

McCarthy, Justin. "Muslims in Ottoman Europe: Population from 1800 to 1912." *Nationalities Papers* Vol. 28, No. 1 (2000): 29–43.

Panaretou, A. "Particularities of Greek Travel Writing in a Balkan and European Context." *Journeys: The International Journal of Travel and Travel Writing* Vol. 6, No. 1–2 (2005): 117–22.

Papapavlou, A N. "Implementing Language Policies: the Standardisation and Transliteration of Toponyms in Cyprus." In G. Katsimali, A. Kalokairinos, E. Anagnostopoulou, and I. Kappa eds., *Proceedings of the 6th international conference of Greek linguistics.* Rethymnon: University of Crete, 2003.

Paparrigopoulos, C. "Histoire de la Régénération du Peuple Grec.Introduction." *Le Moniteur Grec* Γ′ (1846): 70–71.

Peckham, Robert Shannan. *National Histories, Natural States: Nationalism and the Politics of Place in Greece.* London: I.B. Tauris, 2001.

Peckham, Robert Shannan. "Map Mania: Nationalism and the Politics of Place in Greece, 1870-1922." *Political Geography* Vol. 19 (2000): 77–95.

Ruzhdavichka, E. "Situation Ethnodemographique et Processus Religieux à Nikopol au Début du 16e s." *Etudes balkaniques* Vol. 2 (2004): 88–108.

Stephens, John Lloyd. *Incidents of Travel in Greece, Turkey, Russia and Poland*. New York: Harper & Bros, 1838.

Strauss, Johann. "Linguistic Diversity and Everyday Life in the Ottoman Cities of the Eastern Mediterranean and the Balkans (late 19th–early 20th century)." *History of the Family* Vol. 16 (2011): 126–41.

Synvet, A. *La Carte Ethnographique de la Turquie d'Europe et Dénombrement de l'Empire Ottoman*. Paris, 1877.

Synvet, A. *Les Grecs de L'Empire Ottoman: Étude Statistique*. Constantinople, 1878.

Thornton, Thomas. *Present State of Turkey*. 1807.

Todorova, Maria. "Bulgarian Historical Writing on the Ottoman Empire." *New Perspectives on Turkey* Vol. 12 (1995): 97–118.

Vereni, Piero. "Boundaries, Frontiers, Persons, Individuals: Questioning 'Identity' at National Borders." *Europea* Vol. 2, No. 1 (1996): 77–89.

Walsh, Rev. Robert. *A Residence in Constantinople During a Period Including the Commencement, Progress and Termination of the Greek and Turkish Revolutions*. London: Frederick Westley and A.H. Davis, 1836.

Yanik, Lerna K. "Constructing Turkish "Exceptionalism": Discourses of Liminality and Hybridity in Post-Cold War Turkish Foreign Policy," *Political Geography* Vol. 30, No. 2 (2011): 80–89.

Yosmaoğlu, İpek K. "Counting Bodies, Shaping Souls: The 1903 Census and National Identity in Ottoman Macedonia." *International Journal of Middle East Studies* Vol. 38, No. 1 (2006): 55–77.

Zervas, Theodore G. *Formal and Informal Education During the Rise of Greek Nationalism*. New York: Palgrave Macmillan US, 2017.

Βελουδής (Veloudis), Γιώργος. *Ο Jakob Philipp Fallmerayer και η Γένεση του Ελληνικού Ιστορισμού* (Jakob Philipp Fallmerayer and the birth of Hellenic historicism). Αθήνα: Εταιρεία Μελέτης Νέου Ελληνισμού – Μνήμων, 1982.

Δημαράς (Dimaras), Κ Θ. *Κ. Παπαρρηγόπουλος* (Konstantinos Paparrigopoulos). Αθήνα: Μορφωτικό Ίδρυμα Εθνικής Τράπεζας, 2006

"Ελλάδα (τέλη 19ου αιώνα) (Greece (end of the nineteenth century))." In *Επιστήμη και Τεχνολογία. Ιστορικές και ιστοριογραφικές μελέτες*. Εταιρεία Μελέτης και Διάδοσης της Ιστορίας των Επιστημών και της Τεχνολογίας τ. 1. Αθήνα: Εκδοτική Αθηνών, 2013: 227–249.

Καραμανωλάκης (Karamanolakis), Βαγγέλης Δ.. *Η Συγκρότηση της Ιστορικής Επιστήμης και η Διδασκαλία της Ιστορίας στο Πανεπιστήμιο Αθηνών, 1837-1932* (The emergence of historical episteme and the teaching of history at the University of Athens, 1837-1932). Αθήνα: Γενική Γραμματεία Νέας Γενιάς – Ιστορικό Αρχείο Ελληνικής Νεολαίας, 2006.

Κιτρομηλίδης (Kitromilidis), Πασχάλης. «Ιδεολογικά Ρεύματα και Πολιτικά Αιτήματα: Προοπτικές από τον Ελληνικό 19ο Αιώνα» (Ideological movements and political questions: Nineteenth century perspectives). In Δ. Γ. Τσαούσης, ed., *Όψεις της ελληνικής κοινωνίας του 19ου αιώνα*. Αθήνα: Βιβλιοπωλείον της «Εστίας» Ι. Δ. Κολλάρου και Σιας ΑΕ, 1984: 23–38.

Κιτρομηλίδης (Kitromilidis), Πασχάλης. «Το Ιστοριογραφικό Εκκρεμές και ο Κωνστ. Παπαρρηγόπουλος» (The historiographic pendulum and Konstantinos Paparrigopoulos). *Νέα Εστία – Αφιέρωμα στον Κωνσταντίνο Παπαρρηγόπουλο. Εκατό χρόνια από του θάνατό του*, 1546 (12/1991): 1571–1583.

Κουμανούδης (Koumanoudis), Στέφανος Ν [ed]. Ανέκδοτα. Ιστορία του Ελληνικού Εθνους (Unedited history of the Hellenic nation). Αθήνα: Ύψιλον, 1983.

Κουμπουρλής (Koubourlis), Γιάννης. «Εννοιολογικές Πολυσημίες και Πολιτικό Πρόταγμα: Ενα Παράδειγμα από τον Κ. Παπαρρηγόπουλο» (Conceptual polysemies and the political project: An example drawn from K. Paparrigopoulos' work). *Τα Ιστορικά* Vols. 28-29 (June-December 1998): 31–58.

Κυριακιδης (Kyriakidis), Στιλπων. *Οδηγίαι δια την Μετονομασίαν Κοινοτήτων και Συνοικισμών Εχόντων Τουρκικόν η Σλαβικόν Όνομα* (Directions for the renaming of towns and settlements, which have either a Turkish or a Slavic [place] name). Εν Αθήναις: Υπουργείον Εσωτερικών, Διεύθυνσης Τοπικής Αυτοδιοικήσεως. Εκ του Εθνικού Τυπογραφείου, 1926.

Λάιος (Laios), Γεώργιος. «Κωνσταντίνος Παπαρρηγόπουλος, Α′ (1814-1851)» (Konstantinos Paparrigopoulos, A' (1814-1851). *Μνημοσύνη* Vol. 5 (1974-1975): 284–324.

Λιάκος (Liakos), Αντώνης. «Εθνική Ιδεολογία και Εθνικισμός: Προβλήματα Ανάγνωσης» (National ideology and nationalism: Issues in reading them). In *Σεμινάριο 17. Εθνική συνείδηση και ιστορική παιδεία.* Αθήνα: Εκδ. Πανελλήνια Ένωση Φιλολόγων, 1994: 28–37.

Λιάκος (Liakos), Αντώνης. «Προς Επισκευήν Ολομελείας και Ενότητος. Η Δόμηση του Εθνικού Χρόνου» (Restoring plenum and unity. The structure of national time). In *Επιστημονική Συνάντηση στη Μνήμη του Κ. Θ. Δημαρά.* Αθήνα: Εθνικό Ίδρυμα Ερευνών/Κέντρο Νεοελληνικών Ερευνών, 1994: 171–195.

Ματθαίου [Matthaiou], Σοφία, «Ο Στέφανος Α. Κουμανούδης, "οι διδάσκαλοι του έθνους" και ο επικήδειος στον Νεόφυτο Βάμβα» (Stafanos Koumanoudis, "the teachers of the nation" and the funeral oration for Neofytos Vamvas). *Ερανιστής* 26 (2007): 255–80.

Μεργούπη-Σαβαΐδου (Mergoupi-Savaidou), Ειρήνη. «Ο Πόλος του "Φιλολογικού Συλλόγου Παρνασσός" στη Συγκρότηση της Επιστημονικής Κοινότητας στην Ελλάδα» (The contribution of the "Parnassos" philological society in the emergence of Greece's scientific community). In Χρήστος Τριανταφύλλου (Triantafyllou), Μαρία Κουρσή (Koursi), eds., *Επιστήμη και Τεχνολογία. Ιστορικές και Ιστοριογραφικές μελέτες.* Αθήνα: Εκδοτική Αθηνών, 2013.

Μηλιαράκης (Miliarakis), Αντώνιος. «Περί της Ωφελείας των Γεωγραφικών Επιστημών» (Concerning the development of geographical sciences). *Εστία* Vol. 4 (1877): 422–426.

Μηλιαράκης (Miliarakis), Αντώνιος. *Νεοελληνική Γεωγραφική Φιλολογία, ήτοι Κατάλογος των από του 1800-1889 Γεωγραφηθέντων υπό Ελλήνων* (Modern Greek geographic scholarship, including a catalogue of geographic studies by Greeks, 1880-1889). Αθήνα: Βιβλιοπωλείον της Εστίας, 1889.

Μηλιαράκης (Miliarakis), Αντώνιος. *Οδηγός των Απλών Τοπογραφικών Περιγραφών* (Guide to simple topographical descriptions). Αθήνα: Εκδόσεις Κολλάρος, 1901.

Παπαρρηγόπουλος (Paparrigopoulos), Κωνσταντίνος. «Εισαγωγή εις την Ιστορίαν της Αναγεννήσεως του Ελληνικού Εθνους» (Introduction to the history of the Hellenic nation's renaissance). *Πανδώρα* Vol. 1 (1850): 9–10.

Πολίτης (Politis, A), Αλέξης. *Ρομαντικά Χρόνια. Ιδεολογίες και Νοοτροπίες στην Ελλάδα του 1830-1880* (Romantic years. Ideologies and mentalities in Greece of 1830-1880). Αθήνα: Μνημων, 1993.

Πολίτης (Politis, A), Αλέξης. «...Αν ήρχιζε μετά είκοσι χρόνους... Ο Κοραής, οι Κοινωνικές Ιδέες του Διαφωτισμού και η Ελληνική Επανάσταση» (...If it had started twenty years later... Koraes, the Enlightenment's social ideas and the Greek Revolution). *Διάλογος.* Ένωση Φιλολόγων Νομού Ηρακλείου, τχ. 18-19 (2008) 21-30.

Πολίτης (Politis, N), Νικόλαος Γ. *Τα Ονόματα των Δήμων* ([Place] names of municipalities). Εν Αθήναις: Τυπογραφείον της Εστίας, 1899.
Πολίτης (Politis, N), Νικόλαος Γ. *Γνωμοδοτήσεις Περί Μετονομασίας Συνοικισμών και Κοινοτήτων* (Opinions concerning the renaming of settlements and towns). Εν Αθήναις: Τυπογραφείον Δημ. Μ. Δελή, 1920.
Σκοπετέα (Skopetea), Έλλη. *Φαλμεράυερ. Τεχνάσματα του Αντίπαλου Δέους* (Fallmerayer: Adversarial artifices). Αθήνα: Θεμέλιο, 1997.

4 The Hellenization of Salonica

From Jewish metropolis to Greece's "co-capital" (1912–1932)

> "...I do not know the extent to which a city with a motley society, nine-tenths of which consists of Jews, can be liked by others; it has nothing Greek about it, it has nothing European. It has nothing" (Triha, 1993: 309–310).

It was in these words that the Greek officer Hippocrates Papavasileiou described Thessaloniki in a letter to his wife on 19 May 1913, six months after the city's occupation by the Hellenic Army.[1] Although Thessaloniki's conquest may be considered one of Greece's most significant gains from the Balkan Wars (1912–1913), the image of the city did not correspond to the Hellenic Kingdom's imaginary of it. On the contrary, it presented the image of a typical Ottoman city, with numerous minarets and, in this case, an equal or greater number of synagogues, which confirmed that the dominant population group was the Jews. For the Hellenic State, Thessaloniki had to be Hellenized so that it belonged to Greece and was also "Hellenic" or did not appear Ottoman.

This chapter's subject is the Hellenic State's policies towards the Jews of Thessaloniki between 1912 and 1932. The devastating fate of the community during the Nazi occupation of Greece, although massively historically significant, falls outside our thematic scope. Throughout the period 1912–32, Eleftherios Venizelos and the Liberal Party dominated Hellenic politics. Their main project was the urban modernization of the country as a springboard for implementing their irredentist and territorially expansionist visions. The Balkan Wars (1912–1913) resulted in Greece annexing the "New Lands," as Ottoman territories taken by the Greek army were called. The geographical extent of Greece doubled, and its population rose by 77 percent. There followed Greece's delayed entry into the First World War on the side of the Entente (1917), two years of intense conflicts, complications and interventions called the "National Schism." During this period, Greece's King Konstantinos I, supported the country's neutrality (a boon to the Central Powers). He opposed his Prime Minister, Venizelos, who sought Greece's participation in the war as a pretext for, and geopolitical frame in which, the State could pursue its irredentist and expansionist policies against Bulgaria and the collapsing Ottoman Empire.

The Greek-Turkish War (1919–1922) was Greece's final attempt at expansionist geopolitics drawing on the aspiration of a Greater Greece – the Great Idea (Mavrogordatos 2016: 33–155). The Greek defeat in 1922 signaled an end to this ideology as it had hitherto been expressed. The following period to 1932 (the last year in which Venizelos served as Prime Minister) was a decade of significant government instability. The loss of the war and the massive forced migration to Greece of approximately one and a half million Greeks from Asia Minor and Eastern Thrace made economic restructuring an imperative. The crisis necessitated the search for new meaning to the Hellenic national narrative and territorial project. The rise of the Left further complicated domestic politics and regional geopolitics.

Greece's geopolitics of difference enters a new phase after the Balkan Wars and the Greek-Turkish War. We study its underpinnings in the Greek legislative agenda between 1912 and 1932 concerning the Jewish community of Thessaloniki and the city as a multicultural place. Hellenization as strategy and practice redefined difference in cosmopolitan Salonica by reshaping its demography, urban space, and the profile of everyday life for its constituent communities. Hellenic placemaking was built on creeping assimilationist policies. The Jewish community's reaction to such legislation describes the asymmetric contestatory politics between the growing, forward-thrusting national state and a demographically diminished religious minority. This urban contestation, in turn, shaped the state as much as it shaped the city. Therefore, the "national," "ethnic," and "transnational" are challenged and transformed in a framework of complex social interaction (Papamichos Chronakis 2011: 19–26).

The Jewish community of Thessaloniki on the eve of the Balkan Wars

Despite their differences, every population census of Ottoman Thessaloniki in the early twentieth century agrees that the Sephardic Jews formed the majority of the city's population (Melios 1997: 300–302, Molho 2006: 29–52, Karavas 2014: 346–347, Kavala 2015: 17). The Jews outnumbered other groups demographically, controlled most of the city's economic life, and, despite their substantial social differences, demonstrated a cohesive social organization, which included a multitude of schools, philanthropic foundations, associations, and hospitals (Avdela 1993: 171–175, Molho 2006: 93-109, 139-240). For the Jews, Thessaloniki was the "New Jerusalem" far removed from Palestine, a city that formed part of the Ottoman state but which was, in effect, dominated by their community.

Regarding the ideological profile of the city's Jewish population, we discern three tendencies: assimilationist, Zionist, and socialist. The growth of assimilationist ideas is largely associated with the Ottoman Empire's Tanzimat reform policies implemented in the second half of the nineteenth century. The Alliance Israelite Universelle promoted modernization, in part, through the development of a group of schools in Thessaloniki after 1873.

The absence of a Jewish territorial nation-state meant that Jewish communities in the New Lands, as in other Balkan countries, could not rely on a homeland state's geopolitical agency to protect and champion their interests. Statelessness proved to be a significant disadvantage in an age of extreme national conflicts (Molho 2006: 147–180, Mazower 2006: 286–287). The growth of assimilationist ideas was the first significant breakthrough in the ideological orientation(s) of Salonican Judaism. During this modernization period, assimilationists were called upon to redefine their place in the ideological and political milieu of Ottomanism and the new framework created by the Empire's organization into *Millets*, while contributing to secularizing the city's Jewish community. A result of these views was the active participation and representation of Thessaloniki's Jewish element in the Young Turks movement.

The second significant breakthrough was the spread of Zionist ideas in support of a Jewish territorial nation-state. In this era, most of the Thessalonican Jewish Zionists wished to see it founded in Palestine. This view, however, could not be openly expressed by Ottoman Jews, given that Palestine was an Ottoman province and the Ottoman State could take any public expression of such a position as sedition (Molho 2006: 229–240). Zionism developed out of the necessity to modernize the Jewish society of Thessaloniki, and functioned as an opposing force to assimilationist ideas, strengthened by the insecurity created by the policies of the Young Turks as well as by the subsequent arrogance of Hellenic governments (Molho 2006: 286–287, Tzioumakis 2009: 78–79).

The third breakthrough relates to the spread, starting from the early twentieth century, of socialist ideas among the working classes, as these were expressed by the Federación Socialista Laboradera (better known as the "Federación"), founded in 1909. The Federación, created by the Jewish working classes of Ottoman Thessaloniki, appealed to all popular strata in the city independent of ethnic or national identity. The objective was to organize them on a class basis linked by socialist ideas. The typically multiethnic but essentially Jewish Federación emerged as a vital political force. Its activity influenced not only the city of Thessaloniki but also the Hellenic State's political life in the early decades of the twentieth century. In 1918, it participated in the founding of the General Confederation of Greek Workers (ΓΣΕΕ), the country's top workers' union, and the Socialist Labor Party of Greece (ΣΕΚΕ), which in 1924 evolved into the Communist Party of Greece (ΚΚΕ) (Avdela 1993: 180–182).

The Hellenic state and the Jewish community of Thessaloniki (1912–1922)

Greece's victories in the Balkan Wars created new circumstances both for the Hellenic State as well as for Thessaloniki's Jewish community. Greece doubled its territory, but now had to manage a broad spectrum of nationally and ethnically diverse groups, typically recorded in sources of the era as "aliens" (αλλογενείς), meaning, persons of different origin than the Greeks'. Hellenic nationalism perceived them with various degrees of hostility (Marketos 1994: 53–54). Therefore, all communities excluded from Hellenic national identity were treated as discrete entities requiring special administration. There were undoubtedly some alternative, inclusive views raised (mainly by the Patriarchate of Constantinople with its stance against "racism," and by socialists proposing radical transnationalism as an alternative to nationalism), but these were marginal to the prevailing national ethos (Marketos 1998: 68–73).

The primary objective of the Hellenic State was the direct and extensive control of regions it had incorporated and the reinforcement of the Greek presence there. Hellenic statecraft was territorialized in both state and civil ways: Through the projection to the New Lands of Athens' military-administrative complex, supported by the robust presence of armed forces and a range of refurbished Ottoman military bases and several new ones. It was also grounded by exchanging Ottoman for Hellenic governmentality, which entailed state centralization and was supported by administrative cadre relocating to Thessaloniki from Athens. Non-Hellenic native elites, therefore, experienced diminishing regional and local political control.

The Hellenization of the New Lands' economy, Thessaloniki's included, meant the severing of regional and local economies from their formerly imperial hinterland that encompassed the undivided (now transnational) lands of Macedonia and Thrace. Before the Balkan Wars, Thessaloniki's economy flourished under the auspices of Ottoman and Mediterranean cosmopolitanisms much like Smyrne's (Izmir) and Istanbul's. The loosening geopolitical grip of the Empire and the opening of its economy to European industrial capitalism enhanced their status as entrepots.

Moreover, Hellenization in the New Lands was cemented, post-1923, by the resettlement, mainly in Macedonia and Western Thrace of massive numbers of refugees of the Greek-Turkish War. Thus, the New Lands were demographically reprofiled as substantially Hellenic and in line with the State's national ideology of a territorially and historically coherent and continuous Hellenism.

The policy the State applied towards "aliens" was not a unified one. It varied depending on the national/ethnic group to which it was addressed, on who was articulating and implementing it, on domestic pressure of politically powerful interest groups (such as the Orthodox Church, the military, and the Greek capitalist class), and on historical circumstances and the particular needs and maneuvers the latter imposed. The political elite, state actors (including, again, the army), regional representatives of Athens, local societies, and different social classes did not always view the State's New Lands geopolitics in the same light, nor did they rank priorities in the same way. Differences also existed among members of the same group, and in the way different policies were applied (Abatzopoulou 1998: 22, Margaritis 2005: 38, Tzioumakis 2009: 67).

Regional power geopolitics and geopolitics of difference were linked. Opposition to, and enmity toward Turkey and Bulgaria were structural elements of the Greek national ideology. For this reason, as recorded in contemporary sources, the prevalent policy of the Greek State towards Muslim and Bulgarian populations was either planned or unplanned expulsion (*xenelasia*, *ξενηλασία*) or exchange of these populations. Awarding political rights was the exception to the rule. Even in the liberal Venizelist camp would not accept the awarding in 1915 of political rights to "aliens" without objections (Maurogordatos 2016: 189–233).

Within this framework, the Jews of the New Lands, specifically those of Thessaloniki, were atypical. Unlike the Muslim community of Western Thrace,, which could, respectively, and in principle, look to Ankara for support, Jewish communities lacked a clear regional champion. Generally speaking, Hellenic political elites considered the Jewish community pro-Ottoman but less dangerous than the Muslims and Slavophone populations. The prevalent perception was that in the absence of foreign state advocacy, the Jewish community could not be mobilized against Greek interests. Accordingly, they were less threatened by "*xenelasia*," given that there was no state willing to receive them or to which they would have wanted to flee. Furthermore, the overwhelming majority lived in Thessaloniki, where control by the Hellenic State was greater and more direct than in smaller cities and in the countryside.

The presence of foreign consulates presented an additional constraint for those in charge, thereby preventing and limiting any sort of outrages against the city's Jewish population. Moreover, in the estimation of Hellenic elites, Jews in the New Lands were considered good candidates for Hellenization. There was a clear expectation that they would ultimately adopt Hellenic national identity more easily than other national/ethnic groups in the city, since there was the favorable precedent of the successful integration of the Jews of "Old Greece" (Marketos 1998: 77–78). However, the Liberal governments, which dominated Greek politics during the period 1912–1922, did not articulate a comprehensive plan for integrating the Jewish populations of the New Lands into Greek society. The policies they implemented were largely dictated by national and international circumstances and the extent to which these allowed them to advance their vision for a more Hellenized Thessaloniki.

For the Jewish population in Thessaloniki, the Balkan Wars signified a violent rupture with their Ottoman past, both remote and recent. With the occupation of the city by the Hellenic Army and its definitive annexation to the Hellenic State, the resident Jewish population found itself facing a new reality, which was confronted with skepticism, suspicion, and reservations about the future and its place in what was now an assertive, if not aggressively nationalist, state. On the one hand, Thessaloniki Jews expected the Hellenic State to strengthen the Greek element in the city where, until recently, they had been predominant. Under such circumstances, they would be socially and financially marginalized. They questioned how they could retain all they had achieved under the Ottomans and, above all, sustain economic growth, which had been severely impacted by the Balkan Wars. The drawing of new borders created among Thessaloniki's commercial class the fear that the city would be cut off from its Macedonian hinterland. From imperial heartland city, it would become a border city where they would have to confront increased Greek competition and new economic policies imposed by Athens and serving Athens' priorities (Molho 1994: 25–32). On the other hand, by its numerical superiority vis-à-vis other national/ethnic groups that made up the city's population and its financial robustness, the Jewish community of Thessaloniki felt sufficiently powerful to stake its claim. Towards this end, it organized at the national and international levels to influence developments to its advantage.

The Hellenic Army's entrance into the city precipitated unrest that involved the local population and caught international diplomacy's attention. However, they were quickly brought under control following Prime Minister Eleftherios Venizelos' direct intervention (Kavala 2015: 20–22). During the ensuing period, many members of the Jewish community sought and successfully attained citizenship of a western European country to enjoy the benefits of its protection, or resorted to outright immigration (Molho 2006: 254–271, Pierron 2001: 102–104). However, the Hellenic political leadership soon realized the geopolitical benefit of affiliation with Thessaloniki's sizable Jewish community. Thus, the State attempted to rein in anti-Semitic speech in the public realm and contain like episodes on the ground. It simultaneously launched a "friendship offensive," chiefly towards the city's Jewish elite. Many different reasons were imposing such a turn on the Liberal government. One assumption was that Thessaloniki's Jews, through their international networks, could be another diplomatic bulwark. Their "good witnessing" on behalf of the Greeks would strengthen Greek positions during interstate negotiations concerning the final territorial settlement on Thessaloniki's status and its ultimate and full incorporation to the Hellenic State. At the same time, Hellenic governments wanted to leverage the Thessalonican Jews' international connections to the finance community to obtain loans critical to statecraft, including regional development and the resupply of the armed forces. Furthermore, the peaceful incorporation of Thessaloniki's Jewish population into the Hellenic State would advance a new policy for including minorities within the nation-state framework. In this paradigm, "the dominant *ethnos* would not be threatened by existing ethnic minorities, but rather would embrace them as equal members (of the nation-state)" (Molho 2006: 290). In this way, the Hellenic State aspired to serve as a model and dictate equal respect for Greek minorities in the other Balkan states, above all, in the Ottoman Empire of the Young Turks. Hellas would emerge as a model nation-state that could succeed the Ottomans, occupying their territories and controlling their populations in the Aegean and Eastern Mediterranean. Such a state would embody the geopolitics of difference subsumed under "the Great Idea."

Finally, the Jews of Thessaloniki represented a significant electoral body. As potential voters, every accommodation made to them could be redeemed in the future with votes.

The Liberal government satisfied many of the community's demands to those ends, through a series of laws passed in 1914 and 1915. Among others, the Jews were exempted for three years from military service obligations with the added option to buy out their period of service (N 231/1914). They had the option not to work on Saturday and could maintain commercial and financial accounts in their community language (Ladino) (N 180/1914). They could participate in public administration and were guaranteed freedom of the press for the newspapers and print materials they published. The state undertook to fund the philanthropic activities of the Jewish community; marriages contracted under Mosaic Law were recognized (N 147/1914); unleavened bread for Passover was exempted from import duties; and the position of interpreter was established in the courts of Thessaloniki (N 148/1914), although the teaching of Greek became obligatory in Jewish schools (N 568/1915). While these measures affirmed privileges already established under the Ottomans, Liberal governments were unable to dispel the Jewish community's suspicion and skepticism of them. In the elections of 31 May 1915, the Jews voted the Liberals down, while the majority of Jews abstained in the following elections of 6 December 1915. Eleftherios Venizelos's insistence on Greece taking part in the First World War on the side of the Entente revived fears among the city's Jews from the Balkan Wars, and rallied them to vote en masse for the followers of King Konstantinos I and for neutrality (Dodos 2005: 106–127).

The Liberals were the primary constituency supporting national policies and geopolitics whose main objective was to fulfill Hellenic society's irredentist and territorially expansionist visions. The modernization and democratization they professed and claimed to practice aimed at strengthening the Hellenic State in ways that would have made possible the acquisition and Hellenization of new territories in the Balkans and Asia Minor. Therefore, affiliation with the Jewish community of Thessaloniki and satisfying its demands were dictated by a broader policy of promoting national positions and geopolitical claims by Greece. It did not allay the Jewish community's insecurity vis-à-vis the more long-ranging objectives of Venizelist policy. Moreover, due to near-continuous warfare in the region, significant numbers of Greek refugees (mainly from the Balkans and the Ottoman Empire) made their way to Thessaloniki, altering the demographic balance in favor of the Greeks: Already by 1915, Greeks comprised 41.2 percent and Jews 37.1 percent of Thessaloniki's population (Dodos 2005: 99–100).

The Great Fire of 1917 in Thessaloniki and all that followed confirmed many of the Jews' fears, and led to a rupture in relations with Liberal governments. Nevertheless, in 1920 the Jewish community of Thessaloniki won passage of Law 2456 "On the Jewish Communities," with which the rights of Jewish communities were guaranteed.

A problem of representation: picture postcards and army photographs of Thessaloniki (1915–18)

The extensive corpus of picture postcards of Thessaloniki dated from 1915 to 1919 – the period Entente expeditionary forces were resident in the city – provides abundant evidence on its urban morphology and cultural geographies. The pictorialism of foreignness and war reveals the significant control foreign photographers had over urban narratives. The strong preferences of some themes over others and some places in the city over others determined the degree to which Thessaloniki's difference would be made manifest to audiences abroad. The subjectivity of favored themes – historical monuments, picturesque street scenes, vernacular organic urbanism, human subjects of different cultures with an emphasis on their non-Western character (on the street and in staged, photography studio settings), urban panoramas revealing the city's antiquity and exoticism (the castellated Old Town, synagogues, mosques and minarets) – reproduced and reinforced the Balkanness of the place (Figures 4.1 and 4.2).

Figure 4.1 Picture postcards provided nominal explanations to international audiences, such as that of the 1520-1530 Ottoman conversion of the Church of the Twelve Apostles into the Soğuk su Mosque. The absence of human subjects accents the monument's air of antiquity. The consumer is empowered to construct different and contrary narratives about the city. (Source: ELIA, with permission).

Figure 4.2 Members of the Jewish community in Salonica walking along the waterfront road on the Sabbath, May 1916. Jewish-themed picture postcards, as in Ariel Varges' photograph captioned "Type of old Jew," highlight traditional roles, dress, and occupations, thus exoticizing and typecasting the members of the community. The community's extensive Europeanized entrepreneurial and intellectual classes, material culture, and built environment are documented incidentally, at least in this particularly pervasive and influential medium (Source: Imperial War Museum, Q 31948, Ariel Vargas, photographer for the US International Film Service, with permission).

The docility of portrayed urban scenes contrasted with the subgenre of postcards and official army photographs referencing the war effort (navy ships on the waterfront; Allied commanders; Entent troops in barracks, encampments, on parade, and posing in the street often with natives; and, famously, a downed enemy lighter-than-air Zeppelin craft) (Figure 4.3). They shared a great deal with the images captured by official army photographers or the US International Film Service.[2]

Ultimately, there was less interest on the part of the flourishing picture postcard industry in recording the emerging modernization of the town plan and built forms, than 'the picturesque' and the martial qualities of the urban landscape (the soldiery in the streets and shops, the steel carcass of a downed airship, the devastation of the Great Fire). All the same, collectively, the great abundance of postcards captured visual and material expressions for other subject matter. Importantly, in these mass-produced images

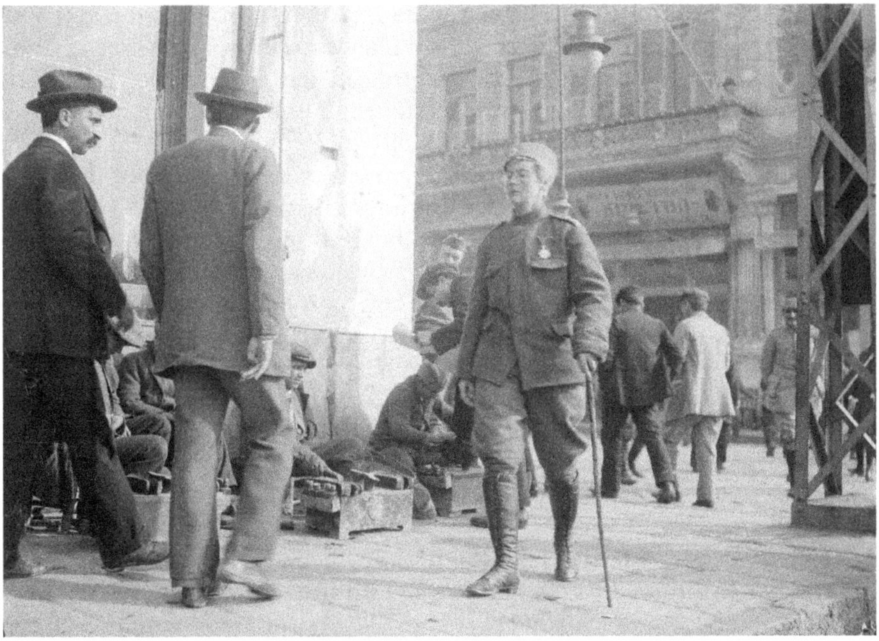

Figure 4.3 Sergeant Major Flora Sandes walks through Thessaloniki in January 1917 during her rehabilitation from wounds inflicted by a Bulgarian hand grenade." The waterfront streetscape reveals intersections of class and occupational structure and planned urbanism. The Sargent Major's military dress and confident stride queers the street (Source: Imperial War Museum, Q 32702, Ariel Varges, photographer for the US international Film Service, with permission).

posted from Thessaloniki to the families of soldiers in places like Yorkshire, Grenoble, and Perth, the Jewish community was vastly underrepresented in the West's imaginary of Greece, again, with political consequences.

Urban modernization, the Great Fire of 1917, and the laws concerning the rebuilding of the city

Fires in Ottoman cities were not an unusual phenomenon. Organic town plans, dense built fabric, and pervasive use of wood in the construction of ordinary buildings contributed to the rapid spread of, and extensive destruction by fire. Thessaloniki was no exception. In the nineteenth century alone, the city endured through sixteen major fires, the fires of 1839, 1846, 1854, 1856, and 1890 considered the most destructive (Papastathis and Hekimoglou 2010: 17). The Tanzimat Reforms set in motion urban modernization intended to reduce hazards and promote economic development. The fire of 1890 devastated a large portion of the urban zone bound to the north by Egnatia Avenue and the waterfront to the south, and the area surrounding the Church of Hagia Sophia and the Metropolitan Church of Saint Gregorios Palamas. The new town plan for the impacted areas exhibited morphological regularities that expressed the urbanism of European cities at the turn of the century: wide rectilinear avenues and built fabric open to the street (Karadimou-Gerolympou 2007: 4). By the end of the nineteenth century, urban fires had stimulated the growth of a fire insurance industry, which was significantly capitalized by international investors. The 1890s modernization notwithstanding, generally speaking, Ottoman state policy left the rebuilding of areas affected to private initiative. As a result, the already-existing urban plan was more or less reproduced, thus depriving the city of the chance to upgrade its infrastructure across the urban area.

With the notable exception of the street blocks that were modernized after the fire of 1890, pre-1917 descriptions of Thessaloniki conjure a city center of narrow and unkempt streets, wretched sanitary conditions, an absence of public spaces, and no visible urban planning (Pierron 2004: 108–109, Mazower 2006: 291–306, Molho 2006: 120–121, Karadimou-Gerolympou 2007: 2–5).

The Great Fire of Thessaloniki broke out on 5 August 1917. When it subsided two days later, it had burned an area of 12,000 "stremmata" (1,200 hectares) and some 8,000 buildings. The burnt area encompassed the entire commercial center and the neighborhoods of the lower city, inhabited mainly by Jews, leaving 73,447 people homeless (64,000 according to other sources), the overwhelming majority of whom were Jews (Pierron 2004: 114–115, Mazower 2006: 380–386, Molho 2006: 120–121). However, most important for the city's Jewish community was the total "destruction of its religious, philanthropic, educational, social, and cultural foundations" (Molho 2007: 121) (Figure 4.4).

The Fire of 1917, although proven accidental, opened the way to new geopolitics of difference. Urban rebuilding would operate as an exercise of "landscape cleansing" of much of the built environment that was not explicitly Hellenistic, Roman, Byzantine, or otherwise evocative of the city's Hellenicity and, more specifically, Helleno-Christianity. In his mid-sixteenth century survey of Thessaloniki, cartographer Nicolas de Nicolay counted eighty synagogues (de Nicolay 1576: 197), the names of which – Aragon, Kastilya, Catalan, Geruş Sefarad, Italya, Mayorka, Puglia, and Sisilya, among others – conjured the geography of origin of mainly Sephardic Jewish populations that arrived in the city from different locales at different times during the Fifteenth Century, to join the "Griego" Romaniote Jewish community that lived in the city since its foundation in the Hellenic era (Dimitriades, 1983: 154–55). By the end of the nineteenth century, there were approximately 42 functioning synagogues in Thessaloniki (Moraitopoulos, 1883: 33). Importantly, as Joseph Nehama notes, synagogues served both religious and civic functions, and, as built fabric, they were historical and cultural landmarks and significant features in a "sentimental topography" experienced and shared by all members of the community. When a

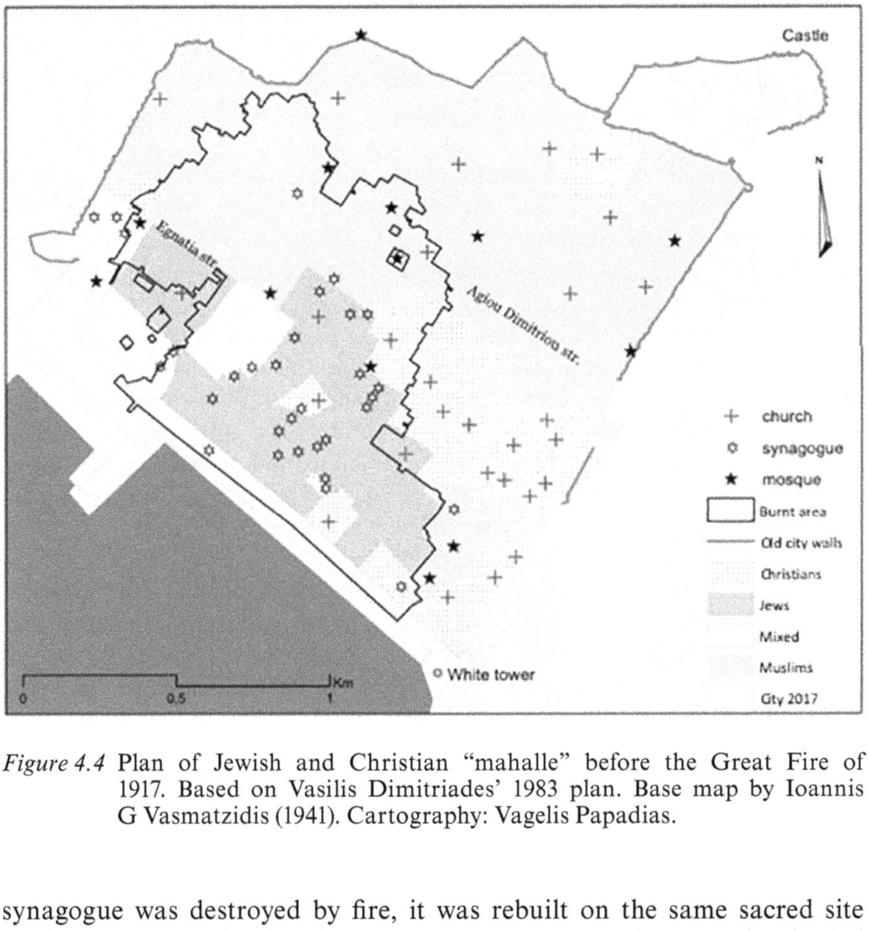

Figure 4.4 Plan of Jewish and Christian "mahalle" before the Great Fire of 1917. Based on Vasilis Dimitriades' 1983 plan. Base map by Ioannis G Vasmatzidis (1941). Cartography: Vagelis Papadias.

synagogue was destroyed by fire, it was rebuilt on the same sacred site (Nehama, 1978: VII, 768). Thus, community preservation was closely tied to the preservation of the built fabric of Thessaloniki's 16 Jewish neighborhoods.[3] The fire destroyed the vast proportion of ordinary buildings that made up the Jewish quarter, alongside more than 30 synagogues and many rabbinical libraries, schools, and community archives. Whether the community would be able to rebuild on the same sacred sites, or rebuild at all in the historic Jewish quarter, was a question that rested on the community's financial resources, its ability to influence the urban planning and design processes initiated post-Fire immediately, and, most importantly, on the legal-regulatory frame for rebuilding that was articulated by the Liberal government. The community had lost many synagogues and rabbinical schools and libraries before, notably in the fires of 1620 and 1890, but it never faced near complete material annihilation until 1917. Anything less than mapping a rebuilt Jewish sentimental and functional topography upon the destroyed quarter would have been an unfavorable outcome and a blow to the community's sustainability.

In contrast, the Turkish and Greek Orthodox quarters of Thessaloniki were significantly less impacted by the fire, although a few years later, following Greece's 1923 defeat in the Greek-Turkish War, and in a spirit of revanchisme, Thessaloniki's most visible evidence of the Ottoman past – the 26 remaining minarets of significant mosques – were demolished (Stefanidou 2009: 28). Once occupying both the lowlands and highlands of the castellated city as of the nineteenth century, the Turkish population was substantially settled "uptown." Yet, in 1913, Ottoman Turkish urbanism was still legible throughout the city (Epaminondas 2014).

Elviya Çelebi, who visited Thessaloniki in 1668, remarked on the city's great garrison of 33,000 and the city's 150 mosques (Elviya Çelebi, 1668: Volume VIII). Although most likely an exaggeration, there is little doubt of the city's importance to the Empire, which would have justified both a significant presence of a Muslim demographic and a substantial Ottoman and, specifically, Islamic Ottoman built environment. Land register of the religious endowments (vakıf) no. 336 (1835) accounts for 34 mosques and 49 smaller "madjid" in the city (Dimitriades, 1983: 282–85). The earliest photograph of Thessaloniki on record, captured by photographer Josef Székely in 1863, while in the employ of ethnographer Johann Georg con Hahn (Epameinondas 2014: 7), clearly undercounts them as it reveals no more than thirty mosques and tékes. By 1883, the number had dwindled to 59 (Moraitopoulos 1883: 32), while 1906 tax records list only 54 (38 significant structures and the rest small "mahalle"-level establishments) (Dimitriades, 1983: 282–85). As the Muslim population declined in numbers and the remaining Muslim population migrated uptown and closer to the Byzantine-era Acropolis, large and small mosques were abandoned, falling into disrepair and, ultimately, ruin. Importantly, the accounting of Islamic and Christian Orthodox sacred spaces was further complicated by the conversion of Christian churches into mosques (or to other types of building use), and their subsequent reconversion to churches following the incorporation of Thessaloniki into Greece and, especially, after the populations exchange of 1922–23 in the context of the Treaty of Lausanne. The actual number of Islamic religious buildings was surely substantially greater, not counting other built fabric vestiges of Ottoman urbanism, such as covered markets, Koranic schools, baths, caravan "palaces," inns, and government infrastructure.

The preservation of Ottoman Turkish built fabric – especially buildings of religious character – was subject to political debate since the foundation of the Hellenic State. In the early years of statehood, proponents of preservation of Ottoman-era buildings noted their picturesque qualities. They advocated their preservation either as exotic "curiosities" or "useful" (as in the case of *bezesten*-covered markets), when they did not encroach upon 'higher value' built fabric from the Hellenic past (Stefanidou 2009: 33). By not prioritizing its preservation, vernacular Ottoman-era built fabric was quickly demolished in the new Hellenic State. Legal protections were extended to elite Ottoman urbanism only in 1899 (Law ΒΧΜΣΤ'/1899 "Concerning antiquities" και Royal Edict 11.8.1899 "Concerning the constitution and function of the Archaeological Committee").[4] Protections were expanded in scope in 1920 during Eleftherios Venizelos' premiership (Ibid: 35). Thessaloniki's Greek conservative elites were generally opposed to the preservation of Ottoman urbanism, which they derided as "foul," "unattractive," and "an obstacle to modernization" (Ibid: 33). Venizelos' political instincts, however, pointed to the diplomatic and strategic advantages of preserving the rights of religious minorities and nurturing inter-communal relations. Preservation of the built environment of cultural Others was part of that political paradigm. That political stance also benefitted – at least until 1917 – Thessaloniki's Jewish community, which feared living under nationalist Hellenic rule.

With Ottoman Islamic and Jewish urbanisms in retreat, if not ruined, the Liberal government treated the 1917 fire as an opportunity to reshape Thessaloniki and modernize it along the lines of French beaux-arts urbanism. It desired to give it a very different material character from what it previously had through the eradication of its Ottoman-era aspect (part of which in this case was also Jewish). Less than a month after the fire, Law 823/1917 was published, which forbade any building activity in the burned-out area, stipulating that a committee would prepare a new city plan. The committee, formed a month later, was composed of seven people: four architects (one British, one French, two Greek), two civil engineers (one French, one Greek), and the mayor. None of its members were from the Jewish committee of Thessaloniki (Pierron 2004: 123). French architect Ernest Hébrard was appointed as the committee supervisor (Figure 4.5). A few months later, a new law determined how the private properties which had burned would be registered and how their value would be assessed (N 1122/1918). However, the property registration process did not proceed smoothly since many owners systematically obstructed it, refusing to cooperate with the demarcation committees, which had undertaken the task of recording private properties with their liens/encumbrances and an assessment of their value deemed unfavorable.

Completion of property registration in the burnt area was followed by the passage on 3 May 1918 of Law 1394, "On the implementation of the city of Thessaloniki's new city plan." This was an essential piece of legislation that determined how the city's new town plan would be implemented. The new plan drawn by Ernest Hébrard's atelier, with important contributions by Greek architects Aristotelis Zachos and Konstantinos Kitsikis, aspired to reform the city, placing its rebuilding in a new regulatory, functional, and aesthetic context.[5] The plan foresaw the complete expropriation of the burnt area and a new system of land parceling through the creation of new, larger, and more formally shaped lots and street blocks. The new cadaster

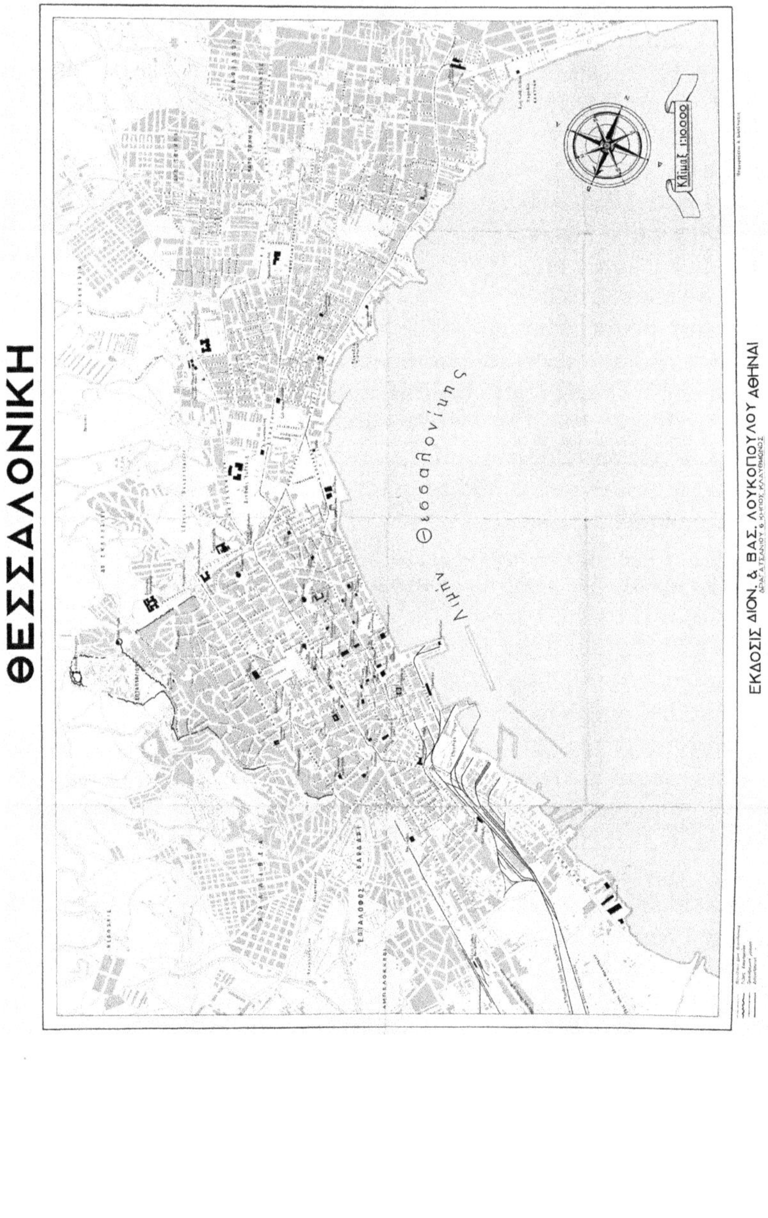

Figure 4.5 Thessaloniki plan after the Great Fire of 1917, architects Ernest Hébrard, Aristotelis Zachos and Konstantinos Kitsikis (inset) The 1935-1938 plan includes regular fringe belt developments built to accommodate refugee populations post-1922 (Savvaidis Collection) (Ethini Hartothiki, with permission).

reduced the total number of lots and auctioned them off, thereby reducing the number of potential landowners. The town plan's wider, straight streets and the provision made for sites to accommodate "public, municipal, religious, and other public service buildings," further reduced the historic core's buildable area, directly impacting the financial position of Jewish landowners. Regulations on the built fabric type, building height, and "architectural character of buildings" (article 18) aimed to refashion the urban landscape according to prevailing European planning norms.

The Jews of Thessaloniki did not embrace Law 1394/1918 and reacted by forming protest committees. They mobilized their international support network and engaged in systematic obstruction by adopting delay and abstention tactics from the land parcel auctions. Finally, in the elections of 1 November 1920, they voted against the Liberals. The contribution of Thessaloniki's Jews to the outcome was small compared to the size of the country's electorate. The shift against the Liberals was due to more complex reasons. However, the Liberals' defeat in these elections brought a monarchist government that was substantially responsible for the Greeks' defeat in the Greek-Turkish War and the "Asia Minor Catastrophe" of 1922. Importantly, it was thought that anti-Venizelist votes cast by minorities living in Greece (including the Thessalonican Jews) contributed to the overall disaster. Thus was created the ideology of the "unbearable oxymoron," of "alien" Greek citizens whose electoral preferences were dictated by their desire to, allegedly, harm the country where they lived. Such behavior was deemed both anti-nationalist and anti-Greek. (Hassid 1999: 363–372, Dodos 2005: 134–149, Mavrogordatos 2016: 191–200).

In July 1921, the government of Dimitrios Gounaris, which had succeeded that of Venizelos, passed Law 2633/1921, "On the allocation of real estate within the new city plan of the burned area of Thessaloniki." This new law, while not annulling the preceding Law 1394/1918, brought about a series of modifications, the most important of which involved the increase in the number of available building lots in the burned-out area. Accordingly, the new rule oversaw the division of planned lots in the Hébrard town plan into smaller ones, thereby decreasing their unit price in proportion to their reduced size. This revaluation of land made lots more affordable to more of their old owners – many of whom were members of the Jewish community. The new framework by which properties were auctioned off foresaw the participation of representatives of the Jewish community on the Town Planning Committee of Thessaloniki.

An assessment of the laws mentioned above shows that the 1917 Fire allowed Hellenic governments to promote the Hellenization of Thessaloniki, bringing changes to its urban morphology and ethnoreligious organization to the terms of urban planning, regulation, and economic development. The plan, which was approved and implemented starting in 1918, created an entirely different city from that of the past: more functional, and with new, larger building blocks, a formal, geometrized town plan (at least in the part of the urban core that was annihilated by the fire), with wider, straighter streets, and numerous public spaces. Observing Western standards of building- and land-use spatial sorting, the plan visioned a Central Business District, defined by government administration, commercial businesses, and a new service sector that included insurance and transportation companies, distributed across the center-west and southwest parts of the historic core. New residential neighborhoods included commercial retail ribbons, replacing the old sectarian structure of Ottoman "mahalle" (neighborhoods). The vise of the Byzantine-era circuit of walls was released with their near-complete demolition, allowing the extension of suburban fringe belt developments to the east and west of the historic city (Karadimou-Gerolympou 2005: 10).

The choice by the Liberal government to pass on the cost of reconstruction to previous owners (a choice which was not retracted by the Gounaris government) contributed to the ethnic and socio-economic reordering of Thessaloniki's core (Papamichos Chronakis 2011: 292–296, Kavala 2015: 25). The Jews, who were the burnt region's primary residents, maintained their professional activities, were not excluded from the auctions, but only the wealthiest among them could afford the new lots. Competition for urban land foresaw the emergence of a new investor class. On the one hand, Jews of fewer means were forced to leave the historic core where they had been the dominant population group. On the other, they resettled into new, fringe belt, planned refugee settlements, where, in new housing, their quality of life improved (Pierron 2004: 133–135, Mazower 2006: 386–391). Furthermore, the new street plan and the imposition of unified architectural standards for buildings contributed to a homogeneous whole, which was constructed top-down with state intervention. The emphasis was on uniformity rather than diversity; on modernization rather than cosmopolitanism.

In the new plan, the destroyed mosques and synagogues in the burnt-out area were not rebuilt. In contrast, the Byzantine and post-Byzantine churches and other Roman and Byzantine monuments were highlighted as centerpieces of new public spaces or monumental termini of new avenues. Thus, the memory of the city became associated with "genteel" building. The urban renouveau rejected all past except the city's Hellenic past. Its historic landscapes, now integrated into a new morphology, served as a storybook of Hellenism's continuity from antiquity through Byzantine times and into the modern era (Karadimou-Gerolympou 1995: 127). Thessalonican morphology's Ottoman-era features were erased through state intervention, and Jewish culture became marginalized or disappeared (Hastaoglou 1994: 42).

Furthermore, the failure to rebuild the destroyed synagogues contributed to an attenuation of Jewish urban memory. Most of these synagogues were named after their founding community's origins and served as the focus of the congregation. Synagogues cultivated a broad range of Jewish diasporic memory, and each one was the depository of distinct, historic cultural experience. Together they composed Thessaloniki as a Jewish homeland.

Over time, Thessaloniki's Jewish history ceased to be visible in the public sphere. It could not be easily materially evoked, and so for the majority (apart from experts), it did not exist; they dealt with the city's Jewish culture as an epiphenomenon, "whose denial is even easier today due to the Holocaust" (Pierron 2004: 135).

The Jews of Thessaloniki, feeling threatened by the implementation of a Hellenic national policy expressed through legislation meant to address the Fire's consequences, sought to obtain legal representation as a religious minority subject to legal protections (Hristopoulos 2002: 129–130). It was a favorable juncture for the community. At the Paris conference following the end of the First World War, the principle of legal guarantees of minority rights prevailed, and most states in Eastern Europe were obliged to accept it and sign the related conventions. Law 2456/1920 can be credited to Thessaloniki's Jewish community's initiatives, whose members worked on the preparation of the final legal text. For this reason, its structure reflected to some extent the place-specific positions and practices adopted by the Sephardic Jews of Thessaloniki.

Law 2456/1920, "concerning the Jewish communities"

The law codified in a single text earlier legislation, which had regulated various issues concerning the Jews in Greece. Its first section concerns the composition, role, and responsibilities of Israelite Communities in the country. According to it, Israelite Communities in Greece were Legal Entities of Public Law (Public Entities). As such and in accordance with a negotiated legal accommodation, they were recognized as subjects with rights and obligations, which undertook to perform some state functions and services while exempt from others. Israelite Communities could be created in any city or village in the country when 20 Jewish families lived there and operated a synagogue. However, only one Jewish community could be founded in any city or village, and the entire movable and real property of the community belonged to it. Jews living in Greece belonged in religious terms to one of the country's Jewish communities. Jewish communities had the right to found and maintain religious, philanthropic, and educational foundations. The supreme body for each community was the community assembly, which was made up of male Jews 21 years and older who lived permanently in the area of jurisdiction of the community and who had not been deprived of their political rights. The community assembly elected its community council and was represented by a President and Chief Rabbi, who was also its religious leader.

The law's second section concerned the rabbinate and the religious court's competencies, the *Beth din*. In each Jewish community, there was also a rabbinical council, headed by the Chief Rabbi, which was responsible for all the community's religious affairs and appointed its *Beth din*. The religious court was responsible for Jewish family law (marriage and its dissolution, dowries, obligations and rights of spouses, divorce and issues of alimony, parental responsibility for children, rights and obligations of parents and children), whose decisions had to agree with "sacred Jewish law" (Naar 2018: 61–76). The law's other articles concerned among other things the protection of the "Israelite religion," which enjoyed similar protections to the "prevailing religion in Greece [...] the Eastern Orthodox Church of Christ," in the protection of the office of Chief Rabbi, in the possibility for Jews to maintain their accounting books in Ladino or French, in their right to the Sabbath as a day of rest (instead of Sunday), and in the administration's obligation to respect it.

Law 2456/1920 formed the framework for the recognition and protection of Greek Jews as a religious minority, since the criteria set forth were religious and associated with the organization of subjects into a community. Such a view originated in the earlier communal organization of ethnoreligious communities in the Ottoman Empire (the *Millets*) and formed a recognizable framework for relations between the Hellenic State and Thessaloniki's Jewish population. At the same time, it delimited the scope of what both sides understood as "minority rights," as well as how these could and should be guaranteed.

Elements constitutive of a complex Jewish ethnic identity were recognized as part of a distinct religious affiliation, and their protection was associated with the communal organization of the subjects included in it. Minority rights were attributed to Jews in a mediated fashion through the Jewish community to which they belonged and not to Jews as individual subjects of a modern state and, therefore, holders of rights. In this manner, the law prevented the formation of socially-defined collective bodies based on other criteria (such as gender or class). For example, on the subject of marriage, the powers assumed by the religious court, the *Beth din* (corresponding to those held by the Orthodox Church in Greece) ensured supervision by community members, preventing – or at least, limiting to the greatest extent possible – the possibility of mixed marriages, since these were impossible unless one of the members of the couple changed their religion. Ergo, the marriage could not be considered "mixed." This measure proved exceptionally robust since civil marriage in Greece only became established in 1982.

For both sides, the Hellenic and the Jewish, this sectarian interpretation of identity guaranteed the requisite lines of separation and was equally acceptable to both, though for different reasons. For the Greeks, it was acceptable because it generally left untouched the existing national ideology of Helleno-Christianity, protecting it from any revisions that could have been called for based on contact with religiously, linguistically, and/or nationally/ethnically "alien" elements. For the Jews, the measure was acceptable because it appeared to ensure continuity with an earlier form of organization (the *Millet* system) to which security was attributed, cultural efflorescence, and a robust economy. The Jews believed these factors were at risk following their inclusion in the Hellenic nation-state. In short, this was a cultural-legal regulatory framework shaped under conditions peculiar to the majority and minority interlocutors. It nevertheless was given new meaning in the context of the nation-state and instrumentalized by the two sides for different reasons, irrespective of the contradictions it contained and the limits it set – or, perhaps, because of them.

After 1922 – in the "refugee capital"

The Hellenic State's defeat in the Greek-Turkish War (1919–1922) was a geopolitical watershed, since it signaled the end of Hellenic irredentist aspirations. The arrival of one and a half million refugees in Greece, together with the simultaneous departure of most of its Muslim inhabitants, entirely altered Greece's population geography. In Northern Greece and Thessaloniki, where the majority of refugees were primarily resettled, Greeks became the majority. The main issue for the Hellenic State throughout the interwar period became the resettlement and integration of refugees. The end of the Great Idea caused political and diplomatic introversion. Long-cherished visions and hopes were dashed, highlighting the need to invest the faltering national ideology with new meaning. For many people, this need became more pressing due to the influence of the Left and socialist and communist doctrines. "Perils from the North" and "East" conditioned how the state came to perceive difference as a national security liability. The enemy was now inside. It would be identified with either ethnic or religious minority groups, which continued to live in the Hellenic State, with the Left, or both.

In this context, relations between the Hellenic State and the Jewish community of Thessaloniki, the country's most visible and prominent Jewish community, entered a new phase. In 1922 to 1932, relations between them were put to the test on numerous occasions. Reasons included the Hellenic government's backsliding and the abolition of some rights the state had established for its Jewish citizens. There were four main points which exacerbated relations between the Jews of Thessaloniki and the State: the question of Saturday as a day of rest, the obligatory maintenance of accounts in Greek for businesses owned by Jews, the conclusive arrangement of the issue of military service for Jews in the Hellenic armed forces, and the establishment of a separate constituency exclusively for Thessaloniki's Jews.

According to Law 2456/1920 (article 27), the Jews had the right to a day off on Saturday (instead of Sunday). However, on 17 July 1924, the Greek Parliament, accepting a decision by the Municipality of Thessaloniki, passed Law 3103/1924, making Sunday the obligatory day of rest for the entire city.[6] The Jews could also take Saturday off if they so wished. The Jewish community, rightly believed that this law trampled on and limited its members' religious freedom. Should they choose to exercise their religious duty to maintain the Sabbath, and they would have to shoulder the economic cost of their choice. The community reacted and mobilized in numerous ways (demarches to the government, activation of international support networks, public protests, resolutions condemning the measure, and targeting specific politicians to obtain their promise to change the law), but in vain. Despite a plethora of legislative adjustments concerning the Sunday holiday issued in 1924 and 1925, the change they sought proved impossible.

The Thessaloniki Municipal Council's decision and its adoption by the Hellenic State suggested that significant changes were in the works, in which the balance of power would shift in favor of the Helleno-Orthodox element. Thessaloniki's Greek population had grown with the refugees' arrival, and now that it formed the majority, it both wanted and was able to impose its views. Law 3103/1924 has all the "characteristics of anti-Semitic legislation," since it denied the religious identity of the Jews, equating them in the public sphere with everyone else (Pierron 2004: 190, Molho 2005: 56, Molho 2003: 53-54)). The fact that it referred to "a Hellenic citizen" was nothing more than a democratic pretext, as it essentially demonstrated an intolerance for difference, denying the Jews the right to have their difference respected and acknowledged (Pierron 2004: 193–194). Just as different urbanisms of the past gave way to western-Hellenic urbanism, cultural difference had to be eradicated from Thessaloniki and yield to homogeneity. In parallel, it satisfied the financial interests of the refugees. It showed the apparent willingness of Hellenic governments to support refugee populations in competition with native ones, even more so when the latter were not viewed as Greeks from the standpoint of Hellenic national ideology (Καβάλα 2015: 30–31). During the years that followed, the Jews of Thessaloniki conformed to the law, and for most of them, Saturday became another workday, while the Jewish Shabbat was confined to the private sphere.

There were similar developments on the question of maintaining accounts. Most of the Jews of Thessaloniki did not know Greek (Mazower 2006: 475). Thus, Law 2456/1920 (article 25) had recognized their right to maintain their accounts in French or Ladino. In 1925, a ministerial decision revoked this right and made it compulsory for accounts to be maintained in Greek. This decision falls within the broader framework of Hellenic governments' policy to maximally homogenize-as-Hellenic the image of Thessaloniki. This law, too, sought to erase any differences and render Jewish culture invisible in the public sphere while making it even more difficult for small Jewish businesses, which now had to hire accountants for this job, to compete in the marketplace (Pierron 2004: 194–195).

The third issue, concerning Thessalonican Jews' service in the Hellenic armed forces, emerged after 1922. It brought relations between the Hellenic State and Thessaloniki's Jewish community to a new crisis level. As noted earlier, in 1914, the community was exempted from obligatory military service. This facility was available until 1921 (Laws 2400/1920 and 2728/1921), with the majority of Jews making use of the relevant provisions to buy out their military service. The related legislation was primarily adopted for Muslims in Greece and was indicative of how they were perceived as a Trojan

Horse. Nevertheless, it was expanded to include the Jews of Thessaloniki. A year after the Hellenic State's defeat in Asia Minor, and following the final arrangements on the withdrawal of the Muslims of Greece to Turkey, new legislation (Decree of 17.8.1923 and Law 3066/1924) did away with the earlier provisions, and the Jews of Thessaloniki were called up to serve in the Hellenic Army. All those born between 1898 and 1901 were obligated to serve three months and buy off the remaining period of service for 3,000 drachmas. Those born in 1901–1902 were obligated to serve for two years without the possibility of a buy-out. Subsequent legislation (Law 3108/1924) brought back the possibility of purchasing "army service with a lump-sum payment of 5,000 drachmas," an exceptionally high amount for the period, and one which the overwhelming majority of Thessaloniki's Jews could not pay.

For the Hellenic State, following the withdrawal of the alleged enemy (the Muslim Turks) to Turkey, the question of the exemption of Thessaloniki's Jews from military service appears to have become a non-issue in the belief that, since they remained in Greece, they were obliged to serve in the ethnically-homogenized Hellenic army. For their part, the community's concern and reactions, ultimately, had less to do with exemption from military obligations, and more with the Hellenic State's reneging on its previous arrangements and commitments. The State's retrenchment posed a serious issue of credibility, eroded the Jews' trust, and raised questions about their treatment, henceforth. It represented more evidence of a slippery slope of dwindling minority rights. The city's Jewish community mobilized to resolve the issue, and a new law satisfied the main demands of the Thessalonican Jews as regarded any outstanding issues they may have had with service in the Hellenic Army (Law 2718/1921). However, from 1927 onward, the city's Jews would serve in the Hellenic Army just as their fellow citizens did.

The fourth issue that led to a crisis in relations between the Hellenic State and the Jewish community after 1922 had to do with creating special electoral districts exclusively for the Jews of Thessaloniki (as for the Muslim minority in Thrace). In previous elections during the decade 1912 to 1922, Thessalonican Jews had voted alongside their fellow citizens. Nevertheless, the role and weight of the "alien" vote was an issue, which kept returning to political discussions, especially among Liberals. The 1920 election results and the crushing defeat of the Liberals, strengthened the ideology of the "unbearable oxymoron" and the need to control the vote of "aliens," both Muslims and Jews. After the Greeks' defeat in Asia Minor and the arrival of the refugees, this myth was reinforced and led to an overt effort if not to limit voting rights, then certainly to control the electoral will of one part of the Greek citizenry. In the elections of 16 December 1923, it was specified that the Greek Jews of Thessaloniki (like the minority population in Thrace) would comprise a separate electoral district formed by ethnic rather than geographic criteria. They would elect their own Members of Parliament in proportion to the number of citizens of Thessaloniki of the Hebrew faith, with the total number allocated as four (Legislative Decree of 12 September 1923 and Revolutionary Decision 31446/5411 of 12 November 1923). The government justified the electoral law by invoking the need to secure the Jewish MPs election, claiming that the Hellenic State was obligated to conform to the provisions for religious minorities in the Special Convention of Sèvres. This Convention, however, was only later ratified by the Hellenic State (on 2 August 1924), and the implementation of any such measure nowhere derives from the Convention itself (Divani 1999: 61–63, Dodos 2005: 151). Despite objecting to their obvious separation from the rest of the electorate, the Government did not rescind the measure. In a show of protest the Jews abstained from the 1923 elections. The measure was withdrawn a few months later, but returned in altered form in 1925 with the provision that the Jews of Thessaloniki would not comprise a separate electoral district, but that they would vote in separate precincts (Law 3355/1925), as happened in the parliamentary elections of 1926. However, in the next elections for Parliament in 1928, the measure mandating different electoral districts for Thessaloniki's Jews and the minorities in Thrace returned (Legislative Decree of 11 July 1928). The Liberals retained the measure despite the reactions of the Jewish community of Thessaloniki, whose members were, at that point, compelled to vote in this manner (Laws 3824/1929, 5493/1932). The measure was finally abolished in 1934 by the People's Party, the Liberals' opponents (Law 6312/1934).

The legislation above concerning the manner and place where legally represented minority populations would exercise their electoral rights is indicative of the intense insecurity, real or imaginary, domestic and international, which dominated Hellenic state policy on nationally-suspect behavior of those of different nationality, religion, or ethnicity. Insecurity led the Hellenic State to establish mechanisms for control and oversight of these populations' electoral behavior. Surveillance was intended to enforce their timely conformity on all matters the State considered sensitive. The treatment of the Jewish community of Thessaloniki on the election issue, at the time, reinforced the latent nationalism-fed anti-Semitism of the Hellenic community and the Hellenic State. Niceties of Parliamentary procedure could not conceal its perniciousness. Through Hellenic governmentality as of 1912, the Hellenization of Thessaloniki's urbanism following the Great Fire of 1917, the demographic reprofiling of the city as majority Hellenic after the Greek-Turkish War, or through sustained legislative action, albeit without violence or outright and open hostility, Jewish Salonica was transformed, in a generation, into a re-Hellenized Thessaloniki, the Hellenic State's "co-capital" in the New Lands. A worse fate would befall the community under the dictatorial regime of Ioannis Metaxas (1936–41) and Greece's Nazi occupation that followed it. Thessaloniki's Jewish community and the Jewish communities of Corfu, Rhodes, and Ioannina, faced near-total annihilation during the German occupation. The reconstruction of the small, surviving community of Thessaloniki Jews after the Second World War was shaped by regional geopolitics – primarily, the creation of the State of Israel – heightened out-migration, and tepid efforts by Greek governments to address the human, material, and symbolic losses of the past.

Notes

1. Hippocrates Papavasileiou (1868–1951) was an officer in the Greek Army and a declared royalist. He took part in the Greek-Turkish War of 1897 and the Balkan Wars. During King Konstantinos I's exile (1917–1920), he was dismissed from the army but returned to service upon the King's return. He was demobilized in 1922. He served as Minister of the Navy during the dictatorship of Ioannis Metaxas (1936–1941).
2. Official army photographers and the international press had substantial access to the Macedonian front. The *Imperial War Museum* holds extensive collections of photographs captured during the Macedonian Campaign of 1915–1919. They include photographs by Ariel Varges, Ernest Brooks, William Pascoe, Major MC Bentley, and Stanley Spencer, as well as a great number of photographs without attribution.
3. There is some discussion among scholars over the exact number of Jewish neighborhoods ("mahalle") in Thessaloniki before the Great Fire of 1917. We accept Dimitriades' count since he employs an urban morphological method in dating and classifying them (Dimitriades, 1983: 1953–77).
4. Νόμος ΒΧΜΣΤ'/1899 «Περὶ Ἀρχαιοτήτων» και Βασιλικό Διάταγμα 11.8.1899 «Περὶ Ἀρχαιολογικῆς Ἐπιτροπῆς καὶ περὶ κανονισμοῦ τῆς ὑπηρεσίας αὐτῆς».
5. Architect Aristotelis Zachos was charged by the Liberal government in early 1913 to initiate an urban modernization plan. It is indicative of the geopolitics of the pro-Entente Liberal government that, although Zachos' atelier would have been entirely capable of master-planning Thessaloniki's post-fire plan, it chose to employ Ernest Hébrard instead.
6. A total of seventeen members of the Thessaloniki Municipal Council voted in favor of Sunday as an obligatory day off, and six (all Jews) voted against.

Sources

Mazower, Mark. *Salonika City of Ghosts: Christians, Muslims and Jews 1430-1950.* New York: Vantage Books, 2007 (Greek edition: 2006).

Naar, Devin A. *Η Θεσσαλονίκη των Εβραίων. Ανάμεσα στην Οθωμανική Αυτοκρατορία και τη νεότερη Ελλάδα* (*The Thessaloniki of the Jews. Between the Ottoman Empire and Modern Greece*). Αθήνα: Αλεξάνδρεια 2018.

Nar, Alvertos. *The Synagogues of Thessaloniki. Our Songs.* Thessaloniki: 1985.

Nehama, Joseph. *Histoire des Israélites de Salonique*, 7 volumes. Thessaloniki: Librairie Molho, 1935-1978.

de Nicolay, Nicolas. *Navigazioni et Viaggi nella Turchia.* Antwerp 1576.

Papastathis, Charalampos; Hekimoglou, Evanghelos. *The Great Fire of Thessaloniki*, August 1917. Thessaloniki: E.N. Manos, Ltd., 2010.

Pierron, Bernard. *Εβραίοι και Χριστιανοί στη Νεότερη Ελλάδα. Ιστορία των διακοινοτικών σχέσεων από το 1821 ως το 1945* (*Jews and Christians in Modern Greece. History of Community Relations 1821-1945*). Αθήνα: Πόλις 2004.

Αβδελά (Avdela), Έφη. «Ο Σοσιαλισμός των "Αλλων": Ταξικοί Αγώνες, Εθνοτικές Συγκρούσεις και Ταυτότητες Φύλου στη Μετα-Οθωμανική Θεσσαλονίκη» (The socialism of 'others:' Class struggle, ethnic conflicts, and gender identities in post-Ottoman Thessaloniki). *Τα Ιστορικά* 18-19 (1993): 171–204.

Αμπατζοπούλου (Abatzoglou), Φραγκίσκη. *Ο Αλλος εν Διωγμώ. Η Εικόνα του Εβραίου στη Λογοτεχνία. Ζητήματα Ιστορίας και Μυθοπλασίας* (The other under persecution. The image of the Jew in literature. Issues of history and fiction). Αθήνα: Θεμέλιο, 1998.

Δημητριάδης (Dimitriades), Βασίλης. *Τοπογραφία της Θεσσαλονίκης κατά την Εποχή της Τουρκοκρατίας 1430-1912* (The topography of Thessaloniki during Ottoman rule 1430-1912), Θεσσαλονίκη: Εταιρεία Μακεδονικών Σπουδών, 1983.

Διβάνη (Divani), Λένα. *Ελλάδα και Μειονότητες: Το Σύστημα Διεθνούς Προστασίας της Κοινωνίας των Εθνών* (Greece and minorities: The system of international protection of the League of Nations). Αθήνα: Καστανιώτης, 1999.

Δώδος (Dodos), Δημοσθένης. *Οι Εβραίοι της Θεσσαλονίκης στις Εκλογές του Ελληνικού Κράτους 1915-1936* (Thessaloniki's Jews in Greek national elections 1915-1936). Αθήνα: Σαββάλας, 2005.

Επαμεινώνδας (Epaminondas), Γιάννης. *Θεσσαλονίκη 1863-1873. Οι Παλαιότερες Φωτογραφίες, οι Πρώτοι Χάρτες της ΠεριοχήςΣσταθμού και Λιμανιού* (Thessaloniki 18631873. The oldest photographs, the first maps [depicting] the railroad station and port). Θεσσαλονίκη: Μορφωτικό Ίδρυμα Εθνικής Τραπέζης, 2014.

Καβάλα (Kavala), Μαρία. *Η καταστροφή των Εβραίων της Ελλάδας (1941-1944)* (The destruction of the Greek Jews (1941-1944)). Αθήνα: Ελληνικά Ακαδημαϊκά Συγγράμματα και Βοηθήματα, 2015 (https://repository.kallipos.gr/bitstream/11419/4437/.../15562_02_MasterDocument.pdf, Accessed on 30 September 2018).

Καραβάς (Karavas), Σπύρος. *Μυστικά και Παραμύθια από την Ιστορία της Μακεδονίας* (Secrets and legends from the history of Macedonia). Αθήνα: Βιβλιόραμα 2014.

Καραδήμου-Γερολύμπου (Karadimou-Gerolympou), Αλεξάνδρα. «Η Θεσσαλονίκη, Πριν και Μετά τον Ερνέστ Εμπράρ» (Thessaloniki before and after Ernest Hébrard) (https://eclass.uoa.gr/modules/document/file.php/ARCH452/yerolympos_gr.pdf, Accessed on 30 September 2018).

Καραδήμου-Γερολύμπου (Karadimou-Gerolympou), Αλεξάνδρα, *Η Ανοικοδόμηση της Θεσσαλονίκης Μετά την Πυρκαγιά του 1917* (The rebuilding of Thessaloniki after the 1917 fire). Θεσσαλονίκη: University Studio Press, 2005.

Μαργαρίτης (Margaritis), Γιώργος, *Ανεπιθύμητοι συμπατριώτες. Στοιχεία για την καταστροφή των μειονοτήτων της Ελλάδας. Εβραίοι, Τσάμηδες* (Undesirable compatriots. Evidence concerning the destruction of Greece's minorities. Jews, Tsamides), Αθήνα: Βιβλιόραμα, 2005.

Μαρκέτος (Marketos), Σπύρος. «Έθνος χωρίς Εβραίους: Απόψεις της Ιστοριογραφικής Κατασκευής του Ελληνισμού» (A nation without Jews. Views on the historiographic construction of Hellenism), *Σύγχρονα Θέματα* 52-53 (Ιούλιος – Δεκέμβριος 1994): 52–69.

Μαρκέτος (Marketos), Σπύρος. «Η Ενσωμάτωση της Σεφαραδικής Θεσσαλονίκης στην Ελλάδα: Το Πλαίσιο, 1912-1914» (The integration of Sephardic Thessaloniki in the context of [the events of] 1912-1914). In *Ο Ελληνικός Εβραϊσμός* (Πρακτικά επιστημονικού συμποσίου του Ιδρύματος Πολιτισμού και Παιδείας για τον Ελληνικό Εβραϊσμό, Αθήνα 3-4 Απριλίου 1998). Αθήνα: Εταιρεία Σπουδών Νεοελληνικού Πολιτισμού και Γενικής Παιδείας Σχολής Μωραΐτη, 1998: 65-92.

Μηλιός (Melios), Γιάννης «Η Διαμόρφωση του Νεοελληνικού Εθνους και Κράτους ως Διαδικασία Οικονομικής και Πληθυσμιακής Ομογενοποίησης» (The formation of the modern Greek nation and state as a process of economic and population homogenization). In Κωνσταντίνος Τσιτσελίκης (Tsitselikis) and Δημήτρης Χριστόπουλος (Hristopoulos), eds., *Το μειονοτικό φαινόμενο στην Ελλάδα. Μια συμβολή των κοινωνικών επιστημών*. Αθήνα: Κριτική, 1997: 281–314.

Μόλχο (Molho), Ρένα. «Η Αβεβαιότητα της Ελληνικής Κυριαρχίας στη Θεσσαλονίκη Μετά το 1912: Ξένη Προπαγάνδα και Εβραϊκή κοινότητα» (Uncertainty in Greece's rule of Thessaloniki after 1912. Foreign propaganda and the Jewish community). *Σύγχρονα Θέματα* 52-53 (July-December 1994): 25–32.

Μόλχο (Molho), Ρένα. «Η Αντιεβραϊκή Νομοθεσία του Βενιζέλου στο Μεσοπόλεμο και Πώς η Δημοκρατία Μπορεί να γίνει Αρωγός του Αντισημιτισμού» (Venizelos' anti-Jewish legislation in the Interwar Period and how democracy can facilitate anti-Semitism), *Σύγχρονα Θέματα* 82 (June 2003): 53–59 (file:///C:/Users/User/AppData/Local/Temp/molho.pdf, Accessed on 8 August 2020).

Μόλχο (Molho), Ρένα. *Οι Εβραίοι της Θεσσαλονίκης 1856-1919. Μια Ιδιαίτερη Κοινότητα* (The Jews of Thessaloniki. A particular community). Αθήνα: Θεμέλιο, 2006.

Μωραϊτόπουλος (Moraïtopoulos). Γ.Κ. *Τοπογραφία Θεσσαλονίκης* (Thessaloniki's topography). Αθήνα: [unknown publisher], 1883.

Παπαμίχος Χρονάκης (Papamichos Chronakis), Πάρης. *Οι Ελληνες, Εβραίοι, Μουσουλμάνοι και Ντονμέ Εμποροι της Θεσσαλονίκης, 1882-1919. Ταξικοί και Εθνοτικοί Μετασχηματισμοί σε Τροχιά Εξελληνισμού* (Greeks, Jews, Muslims and dönme merchants of Thessaloniki, 1882-1919. Class and ethnic transformations in a trajectory of Hellenization), Διδακτορική διατριβή, Πανεπιστήμιο Κρήτης, Φιλοσοφική Σχολή, Τμήμα Ιστορίας και Αρχαιολογίας, Ρέθυμνο, 2011 (https://elocus.lib.uoc.gr/dlib/d/6/1/metadata-dlib-1331627652-282424-2822.tkl, Accessed on 30 September 2019).

Στεφανίδου (Stefanidou), Αιμιλία [ed]. *Η Συντήρηση και η Αποκατάσταση των Οθωμανικών Μνημείων στην Ελλάδα* (Conservation and restoration of Ottoman monuments in Greece). Θεσσαλονίκη: University Studio Press, 2009.

Τζιουμάκης (Tzioumakis), Γιάννης. *Ο Εβραϊκός Πληθυσμός της Θεσσαλονίκης και η Εκπαίδευσή του Πριν και Μετά την Ενσωμάτωση της Πόλης στο Ελληνικό Κράτος* (The Jewish population of Thessaloniki and its education before and after the integration of the city in the Greek state), Μεταπτυχιακή διπλωματική εργασία, Αριστοτέλειο Πανεπιστήμιο Θεσσαλονίκης, Παιδαγωγικό Τμήμα Δημοτικής Εκπαίδευσης, 2009 (http://ikee.lib.auth.gr/record/122629/files/apth.pdf, Accessed on 30 September 2018).

Τρίχα (Triha), Λύντια [ed]. *Ημερολόγια και Γράμματα από το Μέτωπο. Βαλκανικοί Πόλεμοι (1912-1913)* (Diaries and correspondence from the front. Balkan Wars (1912-1913)). Αθήνα: Ελληνικό Λογοτεχνικό και Ιστορικό Αρχείο, 1993.

Χασσίδ (Hassid), Σαμ. «Η Αλλόθρησκη Ψήφος στις Εκλογές του 1920» (The heterodox vote in the elections of 1920). In Χρ. Κωνσταντοπούλου (Konstantopoulou), Λ. Μαράτου-Αλιπράντη (Maratou-Alipranti), Δ. Γερμανός (Germanos), Θ. Οικονόμου (Economou),*eds., «Εμείς» και οι «άλλοι». Αναφορές στις τάσεις και στα σύμβολα*. Αθήνα: Εθνικό Κέντρο Κοινωνικών Ερευνών – Τυπωθήτω Γεώργιος Δάρδανος, 1999: 363–372.

Χαστάογλου (Hastaoglou), Βίλμα. «Για την Κατάσταση της Μετά την Πυρκαγιά του 1917. Ανέκδοτο Υπόμνημα και Αλλα Στοιχεία από το Αρχείο του Henry Morthenthau» (On the situation of the Jewish community of Thessaloniki after the fire of 1917. Unpublished memorandum and other data from the archive of E. Morgenthau (USA)). *Σύγχρονα Θέματα* 52-53 (July-December 1994): 33–44.

Χριστόπουλος (Hristopoulos), Δημήτρης. *Η Ετερότητα ως Σχέση Εξουσίας. Όψεις της Ελληνικής, Βαλκανικής και Ευρωπαϊκής Εμπειρίας* (Otherness as a power relationship. Aspects of the Greek, Balkan and European experience). Αθήνα: Κριτική, 2002.

5 Western Thrace's minority question

Redefining difference post-"Lausanne"

Take me back to Constantinople'
No, you can't go back to Constantinople
Now it's Istanbul, not Constantinople
(*Istanbul (Not Constantinople)* Music:
Nat Simon; lyrics: Jimmy Kennedy, 1953)

In 1920, approximately a century after the Greek War of Independence, the Greek State was in the felicitous situation of controlling, however briefly, considerable new territories in Eastern Thrace and Asia Minor's Aegean coast. These were protectorates granted to Greece by the League of Nations. At the end of the First World War, Ottoman lands in Europe, the Caucasus, and the Middle East were seized and partitioned by the victors – including Greece – their concessionary or protectorate status validated and formalized by international treaties and the newly created League of Nations. The Treaty of Versailles first articulated this new legal-territorial regime in 1919. It was subsequently extended by the 1920 Treaty of Sèvres, which was quickly abandoned in the face of massive reaction by Turkish nationalists and division among the allies.

"Versailles" recognized the ethnolinguistic and religious complexity of the populations of the shattered Austro-Hungarian, German, and Ottoman Empires and articulated new legal frames for the treatment of minorities. Its Committee on New States for the Protection of Minorities oversaw subsidiary minority rights treaties that successor states had to promulgate to receive Great Power diplomatic recognition. These "Minorities Treaties" sought to manage reterritorialized ethnic difference by "instituting a liberal minority rights regime that tried to create both 'tolerant majorities' and 'loyal minorities'" (Riga et al.,: 461). This institutional thickening of international oversight of minority affairs signified an important break with past practice: "During the [nineteenth century], the Great Powers had often recognized new states dependent on a commitment to religious freedom and toleration; such had been the case, for example, in Belgium in 1830 and Romania in 1878. What was new in 1919 was the concern for "national" rather than exclusively religious rights, for collective rights rather than individual liberties, and the provision for international deliberation» (Mazower, 1997: 51).

Notably, the fundamental objective of the Treaty of Sèvres (1920) that followed the Versailles treaty was the partition of the Ottoman Empire and, per its contested terms, the renunciation of all non-Turkish territories. A considerable part of the Middle East was to be administered by France and Britain, and a large part of Eastern Thrace (but not Istanbul) and a considerable coastal territory in Asia Minor (including the important commercial city of Izmir (Smyrne)), by Greece. It is key that "Sèvres" employed, in part, ethnonational characteristics as the parameter for how the partition and distribution of territories were to take place. In view of that, the rationale for territorial concessions to Greece was based, in part, on the ethnographic profile of Thrace and the Aegean Sea littoral. Greece's territorial overreach in Asia Minor and the spreading Greek-Turkish War that ended disastrously for Greece in 1922 were two of several reasons for scuttling the treaty. Elevating its security profile in the region only to have Greece upend the new balance of power with an expansionist war was unacceptable to the Great Powers and the United States.

A disastrous war crushes the Great Idea

Following the gains of the First World War in Thrace and Smyrne, Prime Minister Eleftherios Venizelos pressed on with an expansionist campaign in Asia Minor that tested the Allies' limits for new adventures in the unstable region. After Venizelos' loss of the national election in 1920, the recently returned King Konstantinos I continued the ill-thought campaign to expand the territorial gains of the First World War. Increasingly isolated internationally, under-resourced, and outgunned, it suffered a devastating defeat in Asia Minor that precipitated a mass exodus of ethnic Greeks and signaled the conclusive end to Greece's pursuit of the Great Idea.

The Treaty of Lausanne (1923) provided closure to the War. It addressed the future status of Western Thrace's and Istanbul's minorities, solely on religious as opposed to ethnonational grounds. The legacy of that treaty and its implementation by the signatory states, Greece and Turkey, have played an outsized role in the lives of minority subjects, Greece's democracy, the regional development of Western Thrace, and Greek-Turkish geopolitics. We do not discuss the post-1955 Turkish administrations' pernicious treatment of Istanbul's Greek Orthodox minority, which has led to its demographic decimation. Nor do we compare the two States' approaches to management of their minority populations. In this chapter, we tell the story of how "Lausanne," Thrace's ethnic difference, regional geopolitics, and Hellenic statecraft intersect. In the chapter that follows it, we ground this understanding on a study of the Pomak community, Western Thrace's Bulgarophone Muslims.

The Treaty of Lausanne and the formation of the minority

The Treaty of Lausanne and the accompanying Convention for the compulsory exchange of populations (30 January 1923) were based on, and to a significant degree reproduced, a geopolitical frame of reference with pre-national terms: the criteria established to determine which populations would be included in the exchange and which would remain in Greece and Turkey, now as minorities, reflected the organization of the Ottoman state into distinct religious communities, the *Millets* (Baltsiotis; Tsitselikis, 2008: 33–37, 61–62). For Balkan nationalisms, the religious factor was an important – if not the most important – element in determining national identity (Lekkas, 1996: 178–193, Kitromilidis, 1997: 103–117). In adopting a religious criterion for the population exchange, it appeared that the Kemalists could prevent the recognition of their own national minorities in Turkey and cater to their relations with Islamic countries (Akgönül, 1999: 26–27, Troubeta, 2001: 32–36). In the hope of preserving intact the Greek-Orthodox/Rum community of Istanbul, Greece acquiesced to the idea. Consequently, the Treaty's definition of minority status was built on the Ottoman *Millet* interpretation of community identity as religious as a world of national states was eclipsing the Ottoman world. The minorities rights framework grounded political and cultural geographies in Greek and Turkish Thrace that over time cohered less and less to the political realities and identity politics of the Greek and Turkish national states.

According to the treaty, therefore, in 1923, the Greek-Orthodox residents (Romioi or the Rum community) of Constantinople/İstanbul and the islands of Imvros (İmroz and later, Gökçeada) and Tenedos (Bozcaada) were exempted from the exchange, as were the Muslim populations of Western Thrace who remained in the Greek state. Greece's Muslims would live as a minority, enjoying a special regime of protection and autonomy of their communal, religious, and educational institutions the nature and framework of which were defined by the Treaty of Lausanne.[1]

Articles 37–45 of the Treaty refers to the protection of minorities, forming a separate section entitled "Protection of Minorities." While these articles address the protections that minorities would enjoy in Turkey, Article 45 introduced the principle of reciprocity between the signatory states on the issue of minorities, thus making clear that the Muslim minority in Greece would also be treated in the same way. More specifically, Article 38 guaranteed religious freedom to minorities in Turkey and Greece; Article 39 concerned their civil and political rights equality before the law independent of religion, and the freedom to use their mother tongue in the private and public spheres. Article 40 ensured their right to form, fund, manage their own religious and educational foundations, and be taught their native language. Article 41 foresaw the state's obligation to provide every possible financial contribution to public education so that minority students enrolled in the system would have the opportunity to be taught their mother tongue, in parallel to learning the official language of the state they belonged to. Articles 42 and 43 guaranteed protections concerning public and private life and every type of religious, philanthropic, and charitable foundation. Finally, Article 44 foresaw the framework for oversight, which would inform the implementation of the above and the procedure for reviewing their implementation.

The Muslims who were not exchanged and remained in Western Thrace formed a social-cultural set of communities that still had noticeable pre-national characteristics. This is a small minority, which in the 2011 census accounted for one percent of the national population (approximately 100,000).[2]

Crucially, however, the minority's location close to Greece's borders with Turkey and Bulgaria presented Greek governments with a decision to script it into the state's national security narrative. Muslim communities were spatially segregated in the Western Thrace administrative region (described as West Thrace in the Treaty). Their geographical clustering within that region presented political-administrative options to both majority and minority. Minority policy, therefore, would be entirely contingent on whether the State approached the minority with the intent to include or exclude it from the national identity imaginary (Hellenism).

The Muslims spoke different languages. The overwhelming majority continued to define themselves based on their religion, which was perceived as their main cohesive element. For the most part, they did not participate in processes of "national awakening" and corresponding movements in the Balkans. Their non-participation in the Balkan nationalist movements was owed to the fact that as Muslims they traditionally belonged to the Ottoman State's dominant *Millet* (millet-i hakime). Moreover, during most of the Young Turks' reforms and the dissolution of the Ottoman Empire and creation of the Turkish State that followed them – when Turkish national ideology was fully articulated – Thrace's Muslims were marginal actors in geopolitical developments: they were in succession under Bulgarian rule (1913–1919), Inter-Allied Command (1919–1920), and finally, under Greek sovereignty (1920).

Moreover, any processes addressing the Muslims in the Ottoman Empire, mainly affected urban Ottoman elites. They had a much smaller influence – if any – on agricultural populations in rural regions (Keyder, 2006: 25–34, Konortas, 2000: 188–190). With the signing of the agreements for the exchange, an important segment of the Muslim upper classes in Western Thrace immigrated to Turkey (primarily land-owners and members of the bourgeois urban classes). The overwhelming majority of Muslims who remained in Thrace were farmers in mostly montane parts of the region, who belonged to the poorer classes in local society (Troubeta, 2001: 36–37, 112). As such, their social and political positionalities in the Greek State were reduced in status and influence.

Designations for the minority and minority members

The Balkan Wars and the First World War caused significant population movement in Macedonia and Thrace. The idea of population exchange as a means of constructing nation-states (as opposed to multiethnic ones) emerged at the end of the Balkan Wars in 1913, and became one of the pillars of regional security in the Treaty of Lausanne (Dalègre, 2004: online). The brief occupation of Western Thrace by Bulgaria between 1916–1918 prompted its Bulgarization and the exodus of Greeks and Turks. Bulgaria lost the territory, first to the Entente allies, and subsequently, in 1920, to Greece, which annexed it with the consent of the Great Powers, and made every effort to reverse the impacts of Bulgarization. The loss of Western Thrace caused an exodus of significant numbers of Bulgarians (estimated at 50 or 60,000), putting in question the status of remaining Bulgarian populations and settlements in these regions, especially the Bulgarophone Muslim Pomaks (Ibid: 79–80). Once the allies assigned the region to Greek administration in 1920, and with the retreat of the Bulgarian element that followed it, the politics of population counts pivoted on the respective demographic weights of Christians and Muslims, and even more so, Greeks and Turks (Table 5.1).

Information on the size of the Muslim population (Turkish, Pomak, Roma) before the Treaty of Lausanne is unreliable, as are the ethnographic maps based on them. As we described previously, the Ottoman censuses' data's emphasis on religious affiliation, and how they were collected, provided opportunities for the Greek and Bulgarian States to claim segments of Thrace's population as their own. Bulgaria could claim the Bulgarophone Muslim Pomaks as Bulgarian, and, given the multilingual character of the region, Greece could claim Bulgarophone Orthodox Christians loyal to the Ecumenical

Table 5.1 Allied administration census of Western Thrace's population, 30 March 1920. The data is at the village level. They are aggregated and summarized in Dalègre, 1995, Vol. II, Table VII, and Aarbakke, 2000: 29

District	*Total*	*Turks*	*Pomaks*	*Bulgarians*	*Greeks*	*Others*
Orestias	27,193	5	0	10,210	15,045	1,933
Didymoticho (Demotika)	26,313	1,274	0	4,956	18,856	1,227
Soufli	21,250	2,770	0	10,998	7,435	47
Alexandroupolis (Dedeagatch)	16,317	642	0	11,543	3,355	777
Komotini (Gumuldjina)	64,951	39,601	2,341	14,794	4,773	3,442
Xanthi	48,666	30,438	9,507	1,591	6,650	480
Total	204,690	74,730	11,848	54,092	56,114	7,906
Percent of total	100	36.5	5.8	26.4	27.4	3.9

Patriarchate as Greek. Therefore, it is more likely that the population counts by the French occupation administration of Thrace (1919–1920) are more reliable, albeit animated by Great Power geopolitical concerns.

Greece's historiographic arguments about Thrace's Hellenicity notwithstanding, the 1920 population figures represented precarity for the cause of Hellenization. The population exchange of 1923 and the settlement of great numbers of Greek refugees from Eastern Thrace and Asia Minor in Macedonia and Thrace would overturn the region's demographic profile and render the Muslim/Turkish population a permanent minority.

For the Greek State, the minority was religious its designation enshrined in the Treaty of Lausanne as the "Muslim minority of Western Thrace." However, for a brief period (1954–1956/1957), the Greek State accepted the minority's characterization as Turkish.[3] But the Greek State later returned to the official designation of the minority, and any use of the national definition "Turk" or its derivatives for the name of the minority or its associations, foundations, and organizations was (and remains) impermissible. Especially since 1974, following the Turkish invasion of Cyprus and partial occupation of Cypriot territory, the Greek authorities sought a judicial ban on the public use of the terms "Turk/Turkish."[4] In contrast, there is a single minority in Western Thrace for Turkey, and it is Turkish.

The Muslim minority in Thrace manifests cultural complexities that defy a single label. It may be distinguished based on language into three ethnic groups: the first, which is the most populous, includes those who speak Turkish; the second, which is numerically significant but smaller than the first, includes Bulgarian speakers (the Pomaks), and the third, which is smaller still, includes those who speak Romani (the Roma). In practice, these ethnolinguistic communities commonly use the official Greek language in their interactions with the Greek State and in market activities above the local scale. While not stated in official texts, the Greek view accepts and promotes the division of the minority into three different ethnolingual groups so that it appears heterogeneous. By highlighting cultural differences, the minority's association with Turkish identity is diminished and the Greek viewpoint that the only shared element uniting its members is Islam is strengthened.[5] But even in this case, the use of the terms "Turk" and "Turkish" to designate that minority group which speaks Turkish is neither permissible by nor acceptable for the Greek authorities.

Designations for intra-minority languages reveal ethnic differentiation. Moreover, they indicate the identity politics and criteria that underlie the construction of this cultural differentiation. Greek scholarly and policy literatures speak among other things of "Turkish speakers" and "Turkish by origin" or "Turkogeneis" (τουρκογενείς) [(apparently by analogy referencing "Greeks by ethnic origin" (ομογενείς, omogeneis), of Agrianes, Pomaks, Pomak speakers, Slavic speakers, Bulgarian speakers, Bulgarian Muslims, and of Roma (athinganoi, tsinganoi, gyftoi, katsiveloi) (Asimakopoulou-Lionaraki, 2002: 75, Askouni, 2006: 31–42, Mavrommatis, 2007: 54–59)].

In the Turkish sources, references to intra-minority languages and ethnic modalities are limited, and the minority is generally characterized as "Turkish" or "Turkish-Muslim"; the characterization "Turkish Pomaks" is also found (Akgönül, 1999: 14; Memişoğlu, 1991: 11–40). However, these terms represent attempts to define members of the minority by others in the service of their antagonistic geopolitics. Consequently, these terms downplay what the subjects themselves believed about their origins' imaginaries. How minorities self-identify depends on the national regulatory framework within which they are called upon to express their identity and the extent to which this allows them to feel secure enough (and included in the national body-politic) to speak freely about themselves and other minorities (Plexousaki, 2009: 223–243). Subject at different times to politics of exclusion or inclusion, their identities can be variously layered and ordered. They may be Muslim, as well as Turks, Pomaks, tsinganoi, or "gypsies" (Roma).

For our purposes, we employ the term "minority" to describe Muslim populations in Western Thrace, which are otherwise differentiated by language, dialect, or origins story. Importantly, minority status also reflects the social and political marginality of Muslim Greek citizens in the region (Androusou, 2005: 73). At the intra-minority level, we employ the terms "minority Turk," when referring to members of the minority who declare Turkish identity; "Pomak," in referring to Bulgarian speaking minority members; and "minority Roma". Our use of such more granular designations was dictated by the need to differentiate these populations from the Turks of Turkey and the other Roma in Greece (Troubeta, 2001: 20–23, Frangoudaki, 2008: 492).

"Muslim" – a problematic designation

This range of identities reflects power hierarchies – social and political – that describe the asymmetric relationship between the majority and the minority. If the majority and minority interact in power hierarchical terms, so do the constituent communities of the minority. These hierarchies reveal actor frameworks; possibilities of political expression at the local, regional, and national scales; and limitations on subjects' social and geographical mobilities. Moreover, while the boundaries between the majority and the minority appear crystallized and nearly insurmountable, the same is not the case between/among the ethnic groups that constitute the minority. Minority status expressed through a shared religious affiliation does not preclude social mobility within and between the constitutive communities of the minority. How minority members choose to self-identify represents a continuous act of "becoming" and redefining their social and cultural position to others, whether they be members of the majority or other minority members. This is a political act, which is almost never performed under conditions of equality, and as such, it contains both claims to and a demarcation of boundaries (Jenkins, 2007: 156–169, 173–185; Dragona, 2008: 426–432).

In the Ottoman context, Muslim religious identity may have included all minority members in the dominant *Millet*, but in the national-Hellenic context, it differentiated them from the majority and Greek national ideology (Featherstone et al., 2011: 36). Their qualified inclusion as minority community by the Greek State resulted in political and social entrenchment by the majority community and its spatial sequestration. Spatial segregation of minority populations was expressed at the regional level in the form of separate Christian and Muslim villages and the local level in towns and villages with mixed populations, as separate sectarian neighborhoods. Importantly, they were excluded from the definition of Hellenic identity, which was shaped by the doctrine of linguistic and religious homogeneity and "historical continuity," and could neither accept nor integrate linguistic or religious difference. Inclusion in the majority national community was built on a direct relation to Christian Orthodoxy and opposition to other would-be claimants to "national space," the Muslim "Turks," and other Christian Orthodox believers in the Balkans, namely the Bulgarians and Serbs (Ventura, 2009: 106–109, Askouni, 2006: 40–42, Matalas, 2003: 311–320, Kitromilidis, 1997: 117–119).

The Greek State's definition of the Muslim minority as a homogeneous Other (despite its complex sectarian and geographic characteristics) has been ontologically based on the logic and ideology of a culturally unitary nation-state (Heraclides, 2004: 43–57, Heraclides, 2001: 312–325).[6] As long as Greek national ideology is defined in strict ethnocentric terms, it would deny to minorities access to Greek/Hellenic national identity. Exclusion simultaneously creates preconditions and a political-economic logic for their incorporation into Turkish identity (Smith, 2000: 123–126, 151–152, Lekkas, 1996: 170–173, 178–186, 210–214, Millas, 2001: 30–41, Keyder, 2006: 31–45). For Greek nationalism, differences in language, religion, origins story, historical affinity, and cultural codes bar minorities from participation in the Hellenic national imaginary and the "homogeneous" majority. On the contrary, these differences are perceived as disruptive to the nation's unity and a danger to the state. How these signifiers were specifically tailored to the Greek view of the minority is revealed the Inspector-General of Foreign and Minority Schools remarks[7] at a meeting of the Coordinating Council of Thrace in 1967:

> Any other intrusion foreign to the State (which protects the nation) constitutes a threat. When we consider the disruption of the unity of the nation [to be] the spread – even to a limited extent – of a religious movement or heresy, for example, the 'Old Calendarists,' the Evangelicals, or the Jehovah's Witnesses, and (following the disruption) the diminution of the powers of resistance; when [we witness] the spread of a foreign tongue and its literature along with concomitant sympathy for the nation, this attracts our attention; when [we witness] to a considerable degree the entry of foreign elements disfiguring the ethnographic

> image of part of the country's population, it calls forth anxiety and discontent; when [we witness] the extension of the educational system of a neighboring country over part of the national territory, one-third of which is populated by individuals for the most part of this or similar origin, for the most part [speaking] basically the same tongue, [then] it constitutes an outright and immediate national danger." [Coordinating Council of Thrace, Minutes of the 57th Meeting of 27.5.1968, Archive of the Coordinating Office of Minority Schools (ΑΣΓΜΣ), sheet 11]

However, these differences may constitute the requisite affinities that allow members of the minority group inclusion in another national imaginary – the Turkish one – and their integration into a broader Turkish "homogeneity." For Turkish nationalism, the minority populations in Thrace were considered "Turks outside (Turkey)" (diş Türkler), "enslaved Turks" (esir Türkler), "relatives" (soydaş), and (of) the "same race" (İrkdaş) (Featherstone et al., 2011: 2–3). Moreover, ethnonationalism cultivates an exclusionary discourse about the definition of citizen and their position and role in the nation-state (Birtek 2006: 92–93, 98–99, Mouzelis; Pagoulatos 2006: 75–176, 185). The newly constituted nation-state may recognize citizens as holders of individual rights, but, in the absence of legal safeguards and independent oversight, it drifts to customary discrimination of those who ill-fit the national imaginary's definition of the national subject (Hristopoulos, 2002: 31–32, 35–36, 108–109).

Two questions arise here. The first concerns the contradiction inherent in the manner the Greek nation-state recognizes the right of individual subject-citizens to self-determination but does not recognize the right of subject collectivities to self-determination. Therefore, it is acceptable for the individual to self-identify as a Turkish ethnic, but any association or collective organ that such person may create with others of the same identity is not. The second concerns the extent to which members of minorities can be considered – and consider themselves – members of the national community of the state to which they belong despite evidence of either formal or customary discrimination. Ergo, can they exercise their rights as citizens of this state, to engage in self-determination without sanction even if their expressed identity is different from that of the dominant majority? In the face of such a view of the "national," which situates them on the margins of Greek society (if not entirely outside it), members of the minority have most often chosen to self-identify as "Turks." However, their claim to a right to self-identify as Other at the individual and collective levels, and, even more, the acceptance of this right by the majority, remains an open question (Skoulariki, 2009: 69–93), especially when one considers the identities nomenclature imposed by the Treaty of Lausanne. In that context, the aims of democratic, civil society run contrary to the aims of a critical international treaty of pacification – both aims being socially valuable and politically vital.

By throwing obstacles in the path of the minority's full integration in the nation, the Greek State made Turkish identity regionally competitive. For its part, Turkey made Turkish language acquisition a centerpiece of its national identity politics. Launched by Kemal Ataturk in 1928, the "Speak Turkish Campaign" was an assimilationist strategy in the service of nation-craft. By building community around the national language, the Turkish State could downplay the multiethnic character of its population and, as articulated by Kemal Ataturk himself, make a claim that ... "[o]ne of the significant characteristics of the nation is language. One who regards himself as a member of the Turkish nation should first speak Turkish in every case. If someone who does not speak Turkish claims membership to Turkish culture and community, it would not be right to believe them" (Arar, 1981; Soner, 2004: 89). Clearly, the Speak Turkish Campaign was also intended for audiences abroad.

Turkish identity held (and holds) a dominant place among minority individuals to a significant degree because of the dissemination of the Turkish language across the groups that make up the Muslim minority. In contrast, the other non-Turkish ethnic identities of Pomak and Roma are considered by members of the Muslim minority to be socially inferior and are marginalized in different ways, with the Roma on the lowest rung of minority society.

The shift from these two lower-status ethnic identities to that of "minority Turk" was a process of adaptation and restructuring the individual identity of subjects with the adoption (at least in the public sphere) of a new language, and new codes of behavior and reference. This process was most often associated with subjects' efforts to improve the terms and prospects for their lives. Furthermore, while these shifts strengthened the Turkish character of the minority, they were not always easy to achieve, as intra-minority social mobility was not always welcomed. The following example is indicative of these attitudes: in 1962, the Greek administration, sought to "take advantage of the opportunity to show favoritism towards the Gypsies, whose influence is increasing, inciting conflicts between them and those who are Turks by ethnic origin." To that end, it requested that the central Muslim school in Komotini accept minority Roma. The Management Committee for Minority Property (ΔΕΜΠ) in Komotini initially refused, citing financial reasons. However, it was ultimately forced to adopt the Greek administration's recommendation, since it "was known that the real reason behind the exclusion was not financial, but the discomfort the Muslim minority felt because Gypsy children were overrunning their schools. In the words of the Director of the Central Muslim School in Komotini '[They] will smother us and turn us into a minority of Gypsies.'"[8] This incident, even as recorded by a representative of the Greek administration, indicates not only the state's perspective on managing the minority and the hierarchies within it, but also outlines the limitations and difficulties of social mobility by the minority's constituent groups. The Greek administration sought to create friction between the minority Turks and minority Roma, as the latter's entrance into schools also meant a series of changes that would influence everyone associated with minority education. The dominant Muslim group took these changes on behalf of the minority to contain the threat of further demotion and marginalization – a threat considered undesirable and dangerous to their social position, "homogeneity" and "purity." Their fear was externalized and projected onto the minority Roma, who were their fellow Muslims, but socially and financially even more marginalized (Dragona, 2004: 27–29; Dragona, 2008: 431–432; Dragona, 1997: 72–105).

The minority's adoption of a Turkish national affiliation was a gradual process expressed through the dispute between the "Old Muslims" and the modernizers. For most of the minority population, it had been accomplished by the mid-1950s.[9] Early post-Civil War Greece found a significant part of the minority reconciled to modernist and significantly secular and republican ideas, with an overwhelming majority abandoning the largely pre-modern Muslim identity, which had been codified by the Treaty of Lausanne. They chose instead to adopt the identity of an ethnic – as opposed to religious – minority within a Greek nation-state.

Many factors contributed to these developments, foremost being the close relationship that existed (and still exists) between the minority and Turkey, a relation which is also institutionally guaranteed by the Treaty of Lausanne.

Systematic support and protection were offered to the minority by the Turkish State, which pursued the ideological manipulation of the minority for its geopolitical ends. Pressure was exerted within the minority by their elites, who as a whole were (and are) oriented towards Turkey. A homogenizing role was played by the Turkish language between/among the various minority ethnic groups since the language was connected to a prestige identity, which could offset the multiple exclusions of the minority from majority society and ideology. Thus, Turkophone elites could (and would) leverage the majority society's disapproval of cultural difference to vitalize and valorize the minority's relation to Turkish culture and Turkey. The Greek majority obliged by nearly always considering the minority as a "foreign body" in its midst.[10]

From the mid-1950s onward, this dynamic predominated within the minority. Its consolidation can be attributed to the policies of the Greek State towards minorities, to developments in Turkey during the same period, and the entanglement of Greek-Turkish-Cypriot strategic relations in Cold War geopolitics. Catalyzed by the deteriorating Greek-Turkish international relations over the status of Cyprus and following the 6-7 September 1955 anti-Greek-Orthodox riots ("Σεπτεμβριανά"), Greek policy toward the minority gradually hardened.

The State's newly tough stance was expressed in a series of limitations and restrictions to minority rights. This policy shift contributed greatly to the minority's decision to redefine its position in relation to Greek society and move towards the adoption of Turkish identity.[11] The regime of prohibitions strengthened the minority's common perception that their current situation represented an impasse in their fraught relationship to the

Greek nation-state. The new crisis, in turn, formed the basis for conjuring a new collective memory on whose basis the minority constructed its own narrative about the past and about the histories of its communities, often contrary to the master narrative of the Hellenic nation-state (Frangoudaki, 2008: 494–498).

During the same period in Turkey, political developments also contributed to some degree to these new shifts of perception. Adnan Menderes's rise to power was associated with attempts to liberalize the Turkish State, such as "populist concessions to traditional Islam," which expressed a more moderate attitude to the role of religion in Turkish society (Güven 2006: 15). The hostile stance towards religious matters which characterized Kemalist Turkey yielded to a "Turkish Muslim" view of Turkish identity which acknowledged a place for Islam as long as it remained subordinate to the goals of Turkish nationalism (Poulton, 2000: 216–222, Zürcher, 2004: 304–306, Güven, 2006: 289–303, Kızılyürek, 2006: 111–117, Moudouros, 2012: 105–109). This new approach to Islam and Turkishness addressed the issues and dilemmas the minorities in Thrace faced since it offered a framework for the coexistence of national/ethnic and sectarian aspirations. It eased tensions between Turkey and the minority in Thrace that simmered since the foundation of the Turkish Republic in 1923.

Although it had accepted the right of self-determination for individual subjects, the Greek state disregarded this new reality of ethnization of its Thracian minority and insisted on religion as its sole defining characteristic. By 1967, the Greek authorities, even if outside their public rhetoric about the minority, appear to recognize its Turkish character. They attributed this to the minority's provenance or descent ("καταγωγή"), which also connected it to the Turkish nation.

"The subject of 'Muslim' or 'Turkish' needs to be examined in the light of the provisions of modern International Law, which ignores the provisions [of the Treaty of Lausanne] 'On Minorities.' Regardless of what a Treaty says, each Minority has a place of origin, and this is to be examined here. The term 'Muslim' has a broader meaning..." [Coordinating Council of Thrace, Minutes of its 32nd Meeting on 16-17.9.1963, p. 9, ΑΣΓΜΣ φ. 11].

> "The subject of the term[s] 'Muslim' and 'Turkish' always attracted intense attention, sometimes excessive attention. The use of the first or second term indeed characterizes a Religious or National Minority. However, it is equally true that every minority draws its "genus" ("γένος") from some Nation: that the Turks call the Greek Minority of Constantinople 'Rum,' and 'Rum' is the name of a Nation [an ethnic group appellation], that as long as the Law [ΒΔ 3065/6-10-1954] speaks of 'Turkish Schools' then Turkey's demand rests on Greek Law..." [Coordinating Council of Thrace/President to the Ministry of Foreign Affairs/A' Political Directorate, 'Διορισμός ημεδαπών μουσουλμάνων προσοντούχων δασκάλων' (Appointment of

local Muslim qualified teachers who are Greek nationals) protocol no. ΜΘ 840, 27.9.1963, ΑΣΓΜΣ, φ. 10]...

"Greece should not take it as a particularly annoying subject [that is, the presence of the minority in Thrace]; they [the Greeks] are creating issues and confrontations with a neighboring nation, where they comprise an ethnic minority..." [Coordinating Council of Thrace, Minutes of the 33rd Meeting on 6-7.4.1964].

Most notable was the airing, albeit in bureaucratic circles alone, that the majority of Western Thrace's Muslim minority had become ethnicized with Turkish as its nationality of reference.[12]

"The argument from the practical formulation of the Treaty of Lausanne is untenable. The treaty speaks about 'the non-Muslim Minorities in Turkey' because in Turkey there were and are Greeks, Armenians, French persons, etc., (all) non-Muslim Minorities. In Greece there is only a Muslim minority, which, however, is at the same time also an Ethnic Minority. Since we do not possess the courage as a State to speak on matters of lesser importance, I do not hope that we will say much about matters of great importance [such as this]. We are afraid to speak about the Pomaks and Gypsies, while we should be [speaking] of this. The Turks refuse to call the Minority 'Yunan' or 'Hellenes.' Indeed, the Rum is a National [group], but that Nation is Greek and not Roman" [Coordinating Council for Thrace, Minutes of the 44th Meeting on 6.4.1966, ΑΣΓΜΣ, φ.10]

In parallel, an effort was made by various Greek groups to highlight the different ethnic identities within the minority and their division into "those of Turkish origin," Pomaks, and Roma. However, this formulation rested on a contradiction. "Those of Turkish origin" were defined as such using the criterion of their ethnic origin (genus or γένος/φυλή), while the Pomaks and Roma were defined according to ethnic criteria. Removing this contradiction would mean using the same criterion for all ethnic groups within the minority. If the chosen criterion were "origin" or "descent," then one should be speaking of "those of Turkish origin," "those of Pomak origin," and "those of Roma origin"; if it were ethnic identity, then one should be speaking of ethnic Turks, Pomaks, and Roma (Frangoudaki, 2008: 492–494).

Insistence on the religious definition of minority status perpetuated the classification of citizens based on a pre-national *Millet*-like standard. It presupposed the presence and public confession of religious belief as the primary criterion for granting specific rights to subjects. It simultaneously raised the question of the extent to which religiously-based self-determination was possible for those members of the minority who had distanced themselves from their religion – if they had not broken away from it altogether. What is that group believed that this signifier should not define them any more than their social position, financial status, ethnic affiliation, or gender (Hristopoulos, 2002: 120–123). However, such a view coheres neither with the constitutionally established right to freedom of religion nor with citizenship and the rights deriving from citizen status (Baltsiotis, Tsitselikis, 2001: 35–36, Tsitselikis, 2007: 354–372, Baltsiotis, Tsitselikis, 2008: 82–88, Tsitselikis, 2012: 106–118). In the specific case of the Thracian minority, rejection of the Muslim identity also means the loss of the special Treaty-based rights members of the minority community have enjoyed since 1923. An example of how minority status translates into policy is the establishment of affirmative action admissions to Greek universities for Thrace's minorities: a special quota of 0.5 percent of admissions to Greek universities was established for minority graduates who formally affirm their religion-based minority status.[13] In practical terms, the policy greatly facilitates minority student access to social capital and mobility and, potentially, entry into Greek charter society. It creates the preconditions for the formation of a Greek Muslim elite that has a stake in the continuing prosperity of Greek society and polity.

This insistence on the observance of the word of the Lausanne treaty can be interpreted as a sign of the Greek State's discomfiture in managing difference and social development at the community level. Further, it communicates the State's desire to counter critical definitions of difference and defend the integrity of the ideal-typical national identity from any challenge emanating from contact with the other.

The state's constant pushback against what was (and continues to be) a painfully identity-deconstructive process taking place in minority communities, whether at the individual or collective level, perpetuates distrust. On the one hand, the minority's evolution toward ethnization and on the other, its enhanced access to Greek charter society, precipitated the displacement and existential transformation of members of the minority. It confronted them with truths and inevitabilities about their lives in Western Thrace, in a "Hellenic" Greece, and an integrating Europe. Essentially, the challenge over how Western Thrace's Muslim communities intersect with Hellenic nationhood and statehood represents a reflection on the substance of a new national identity oriented more towards the future and less the ideological re-workings of the past. Withdrawing its disapproval of ethnicized minority identities, accepting that community self-identification is consistent with liberal democracy, and recognizing that Europeanization – per Greece's acquis communautaire obligations – points the way to a civil, as opposed to a national, society and polity, describe the Hellenic State's gamble. However, a change of this magnitude would require, as a first step, the redefining of Greek national identity and a critique of its fundamental doctrine on "historical continuity" and homogeneity of the Hellenic "ethnos." It would take making place in Hellenicity/Greek history for a broader definition of a Greek political community that would be based on civil as opposed to national society criteria. The fall of the Junta in 1974, followed by the restoration of democracy and Greece's accession to the European Communities in 1981, marked points of inflection. The domestic democratic "turn," and external exigencies to Europeanize, compelled the State to reexamine its perspective on difference.

Old Muslims and modernizers or reformers

The main preoccupation of the minority during the Interwar period was the conflict between "Old Muslims" or "Old Turks" (muhafazakârlar, gericiler) – those traditionalist members of the minority who reacted to the secularist, Kemalist reforms in Turkey and did not want comparable reforms in their communities – and modernizers (inkilâpçilar, ilericiler), who embraced the reforms that underpinned the secularization of Turkish society.

Modernizers wanted similar reforms to be adopted by the minority across Western Thrace. For the Turkish State, the Thrace minorities had to abandon many former beliefs and practices, which went against the new dogmas and models it was dictating as a national center. Such a transformation was deemed a pre-requisite to the minority's integration into the Turkish people's imaginary and modernization. For the Muslim minority, this conflict simultaneously reflected a process of defining its new "minority" identity in Western Thrace and its relation to both the Greek and Turkish people. Importantly, it required the recalibration of its relation with Turkey, the successor to the Ottoman Empire and the epicenter of Turkish heritage and the main place of reference in its formation as minority (Featherstone et al., 2011: 41–45, Huseyinoglu, 2012: 125–142, Tsioumis, 2011: 577–581, Malkidis, 2004: 73–87, Mavrommatis, 2007: 41–50, Kalantzis, 2004: 11–12, Troubeta, 2001: 40–42, Aarbakke, 2000: 77–83, Alexandris, 1988: 63–68). This intra-minority conflict was chiefly expressed through the struggle for control of minority education, the Latin alphabet for written Turkish, and the elimination of many customary elements of the Muslim tradition.

Following the loss in the Greek-Turkish War (1919–23) and the signing of the Treaty of Lausanne, the Greek State, in its desire to restrict any sort of linkage between the minority and Turkey, supported the "Old Muslims" (Yioltzoglou, 2006: 555–559, Lepida, 2011: 132). However, the Greek-Turkish rapprochement in the 1930s and the creation of the Balkan Pact bolstered Turkey's position vis-à-vis the minority (Alexandris, 1988: 69–144, Lepida, 2011: 97–109). Greece's main geopolitical antagonist at the time was Bulgaria.

Until the dictatorship of General Ioannis Metaxas (1936–41), a limited number of interventions by the Greek administration in minority education occurred. The Greek State, having neither the human nor the material resources to implement a comprehensive plan for minority education, granted minorities near-complete autonomy to manage their educational affairs, provided no problems were created in Greek-Turkish relations (Divani, 1995: 68–92, Kalantzis, 2004: 9–10). The most important legislative intervention was the passage of Law 2345/1920, which established the duties and responsibilities of the Muftis in Western Thrace.[14] Simultaneously, a series of other laws secured the obligatory teaching of the Greek language in minority schools, the right of minority individuals to select the teachers, and the right to be taught their mother tongue and religion if they were enrolled in public schools. The framework for state inspections of Muslim schools was determined, and the State declared its intention to found a three-year teachers' training institute in Thrace for the education of minority teachers.[15]

The Greek administration's first attempt at intervention in minority education was during the Metaxas dictatorship. The General Inspectorate of Foreign and Minority Schools (ΓΕΞΜΣ) was created with Emergency Law/AN 248/1936. The Inspectorate, headquartered in Thessaloniki, controlled all foreign and minority schools in the Greek State as a matter of national security. Concerning minority education, its chief concern was to limit any autonomous actions so far detected. These efforts were intensified from 1939 onward: the Greek language proficiency of minority teachers was tested, there were strict recommendations that the use of the terms "Turk" and "Turkish" be avoided in schools, and there was an effort to place minority educators under the direct control of the Greek administration.[16] The minority, clearly displeased, did not manage quickly enough to react before the Second World War broke out.

The 1940s posed new dilemmas for the minorities. Bulgarian forces occupied Western Thrace except for the borderland zone of the Prefecture of Evros, which remained under direct control of German forces. Siding with Nazi Germany, Bulgaria sought to reclaim lands it lost to Greece and others in the Great War. The Bulgarization of administration, territory, and people was reinitiated under the auspices, again, of fighting a world war. On the one hand, Bulgaria's attempt to win over the Bulgarian-speaking Pomaks rallied the Muslim minority population.

On the other hand, throughout the Occupation the National Liberation Front (EAM, Εθνικό Απελευθερωτικό Μέτωπο) was neither able to organize a strong resistance movement that drew the majority and minority populations into a common action plan, nor could it create a convincing narrative which promoted – even as vision or promise – a new framework for the coexistence of these populations after the war. The minority tried to survive a difficult reality by not collaborating with the occupation forces. On the other hand, the dissolution of the Greek administration, problems involving safety, the ever-worsening economic situation and the issue of subsistence, offered the Turkish State, which remained neutral in the Second World War, political opportunities: it exercised political protection of the minority through its Consulate in Komotini while facilitating the migration of minority members to Turkey. These actions bolstered the position of the modernizers and Turkey within the minority, expanding whatever supports they already had in place.[17]

After the liberation of Greece, during EAM's (National Liberation Front) brief administration of the region (Fall 1944-Spring 1945), the strengthened position of the modernizers in the minority community became evident. The modernizers dominated the elections for the nomination of the Management Committees for Muslim Property (ΔΕΜΠ) and succeeded in having school textbooks from Turkey circulate freely. The Greek State, despite the provisions of Law 2345/1920, had not yet moved to issue a law for the organizing of community institutions, which would be responsible for minority self-governance. For the Greek State, "the rights of the Committees are limited [...] to the management of Muslim community and charitable foundation (Waqf) properties; these entities are not recognized as legal persons, nor are they granted any right of representation of the Muslim element." For members of the minority eager to secure self-governance rights, the Management Committees were the closest thing to a community institution that could be mobilized to that end.[18] Control of the Committees also allowed modernizers to oversee faculty and staff hiring in minority education and to direct resources according to their priorities. For example, they would be in a position to support those schools that accepted their worldview exclusively. Moreover, as is recorded in related sources, the most active members among the modernizers came from minority youth who had grown up after Greece annexed Western Thrace.

These persons had no memories of the pre-1920 period and the Ottoman era, a factor, which differentiated them from their older fellow Muslim Thracians. Youth activism played an essential role in the minority's search for a new identity and the expression of social mobility aspirations. Access to a modern secular education was seen as a prerequisite to social development. This paradigm was cultivated in Kemalist Turkey and readily exported to Western Thrace.[19]

Upon assumption of the administration of Western Thrace, at the onset of the Civil War, the General Administration of Thrace (ΓΔΘ) attempted to curb the modernizers' influence and undermine any potential identification of modernization with the Left. Accordingly, it "decided to regulate relations in a definitive fashion and the dependency which must exist among Muslim communities in Greece." In the summer of 1945, it intervened in minority education, seeking to restore much of what had been in effect before the world war.[20]

This time the minority reacted. Addressing the Ministries of Education and Interior, and the General Inspectorate of Foreign and Minority Schools (ΓΕΞΜΣ), minority leaders protested the State's arrogating to itself their right to oversee minority schools as a community by assigning it to the Inspectorate of Muslim Education.[21] This violation of their minority rights and the state's attempt to assimilate them were deemed comparable to what had been attempted during the Metaxas dictatorship. They also expressed that Greece's minority policy should be subject to the reciprocity clause (as established by the Treaty of Lausanne). They declared, "our fate should be coordinated with that of the Orthodox Greek Minority in Constantinople."[22] The General Administration of Thrace's initiatives were viewed as the outcome "of the political situation [the Civil War], from which we have formed the belief that this is an effect caused by chauvinistic and factional attitudes."[23] In contrast, the elections for the Management Committees for Muslim Property, carried out by the Leftist National

Liberation Front (EAM), should have been seen by the minority as affirming its right of self-determination, as guaranteed by the Treaties, and not necessarily as a tactic for questioning the Greek State's authority in the region. At the same time, minority leaders sought to disassociate themselves from other minorities in Greece, such as Slavophone populations in Macedonia whose cause had been coopted by EAM. To demonstrate their loyalty to Greece: they would exclaim, "It would be a blatant crime for us to be considered suspect and targeted for assimilation."[24]

The General Inspectorate of Foreign and Minority Schools (ΓΕΞΜΣ) responded to minority objections by outlining the Greek administration's management framework: educational affairs were to revert to their pre-war status, and the designation "Turkish minority" was no longer acceptable. Nor was the existence of Muslim organizations acceptable, as the main goal of minority leadership was the recognition of the Management Committees for Minority Property (ΔΕΜΠ) – its self-governing body, a Legal Entity of Public Law, which undertook to perform certain state functions and services. Additionally, the ΔΕΜΠ's work was limited to the management of the minority's finances. Those elected to the Committees during the short-lived EAM administration were to be replaced by prior members who served "before April 1941." Thus, the State was asserting its sovereign right to control minority education, per its treaty obligations, without conceding to any changes that undermined national interest. According to the General Inspectorate (ΓΕΞΜΣ), the minority's reactions reflected the influence of the modernizers that grew over time. At that point dominant in urban centers, they sought to extend their influence in rural areas, where traditional Islamic views prevailed. Ultimately, the modernizers' expansionist strategy was attributed to Turkey's incitement, although the involvement of the revolutionary Left was also considered.[25]

In the first post-War Greek national elections (31 March 1946), the majority of minority leaders elected to Parliament expressed modernist views. This pattern was replicated in subsequent elections until 1951.[26]

The Civil War (1946–1949) did not alter the balance of power inside the minority. The overwhelming majority of minority members avoided becoming involved in the war and attempted to maintain their distance from both sides. They did not take advantage of the turmoil to turn against the national majority, and when they organized, they did so with the intent of defending their villages. Most minority members tried to avoid conscription to either military forces: the revolutionary Left's or the Hellenic armed forces. Quite a few chose to immigrate – legally or illegally – to Turkey, an action largely tolerated by the official Greek government. Despite its military and political activity in the region, the Left failed to win over the minority. Turkish leftist Mihri Belli ("Captain Kemal") joining the Democratic Army of Greece's (ΔΣΕ) and the creation of an "Ottoman regiment" consisting exclusively of minority members in the Democratic Army did not bring the anticipated results, since most participants were serving under duress.[27] There were several reasons for the Left's failure to mobilize the minority, including the minority's isolation from the rest of Greek society, the Turkish State's anti-Communism, and the Greek Left's inability to articulate a compelling narrative about how minorities would be integrated in a future, inclusive Greek society. Additionally, in the mountainous regions of Xanthi and Rhodopi controlled by the Left, the latter's narrative failed to win over the minority population since much of what it professed and tried to implement (adoption of the Latin alphabet, publication of school textbooks and newspapers in Latin script, dispensing justice through "people's courts" and bypassing Sharia) was secular in nature and found little appeal among the conservative and deeply religious Pomak populations, who found them more off-putting than inspiring.[28]

The late 1940s found the modernizers the winners in their struggle with the traditionalist "Old Muslims." To a large degree, the Civil War redefined Greek national identity by integrating the majority elites' disapproval of the Left to its traditional ethnolinguistic and religious redlines. Leftists and ethnic Others were deemed suspect and dangerous, and even more so when they declared a non-Greek national affiliation. In this political climate, anti-minority rhetoric intensified by constructing minorities as a threat to national security. Minority claims and aspirations were branded communist and attributed to minorities' alleged collusion with the revolutionary Left. Wary of the mounting influence of the Turkish-affiliated modernizers, Greek authorities continued to support Thrace's "Old Muslims."

Despite its wariness of Turkish influence in Western Thrace, the Greek State was still unwilling to integrate the Muslim minority into Greek society. It was understood as an Orthodox Helleno-Christian structure, by definition incompatible with Otherness in general and Islam in particular. For the Greek State, either inclusion in, or exclusion from, the "ethnos" was foremost determined by religious criteria. In fact, by drawing a divide between the Orthodox Christian majority and the Muslim minority, the state was unwittingly reinforcing the modernizers' argument that the minority's social development aspirations could only be fulfilled by adopting its secularist, Turkish-culture agenda. The resulting prevalence of modernist ideas among Western Thrace's Muslims signified the transformation of the minority's identity from "pre-national" to "national" and strengthened its Turkish cultural and political dimensions.

The end of the Civil War appeared to be a point of inflection: elections were held to choose the leadership of the Management Committee for Minority Property (ΔΕΜΠ), a contest won by the modernizers (Nikolakopoulos, 2002: 120–121). The Greek administration proceeded to satisfy many of the minority's demands: it withdrew its unequivocal support of the Old Muslim faction and accepted the use of "Turkish" as a descriptor of the minority. At the same time, it also approved the circulation and use in minority schools of textbooks published in Turkey. Further, it sanctioned the exchange of educators (with revocable status) teaching in minority schools of the other country, allowed minority teachers to be educated in Turkish teaching institutions, and formally considered allowing minority teachers to receive pedagogic training by Turkish counterparts and fully adopting the Latin alphabet. Finally, in 1952, in Komotini, the Greek administration chartered the first minority secondary school (Petridis 2013: 71–129).

These shifts in Greek State policies influenced the positions of the Old Muslim faction. Islam continued to serve as its primary signifier; however, religion was no longer perceived as antithetical to Turkish identity, as Kemalism had defined the latter. Instead Islam served as the expression of a broader sense of conservatism. At the same time, it functioned as the main qualifier of the traditionalists' relationship – and also that of the modernizers' – with the Greek authorities. For its part, the Greek administration's support for the Old Muslim faction waxed and waned over time, as it sought to mobilize it variously as a counterweight to the modernizers (Mavrommatis, 2007: 43, 246–248, Troubeta, 2001: 43–50). The repositioning of the Old Muslim faction reflected changes in Turkey's politics and the brief détente in Greek-Turkish international relations. Concerning Greek politics, extreme anti-Turkish viewpoints had been disavowed, and new moderate voices called for a rapprochement with Turkey (Nikolakopoulos, 2002: 121–122, Aarbakke, 2000: 112–114).

The minority considered the years 1950-55 the Golden Age of relations with the Hellenic State. Greece's security concerns, articulated as "the threat from the North," focused on adversaries along its northern frontier, especially Bulgaria, which had emerged in the Cold War as a Warsaw Pact frontline state. Further, the joint accession of Greece and Turkey to NATO in 1952 appeared to signal a new era of Greek-Turkish amity. Simultaneously, the surfacing of the Cyprus issue – its decolonization struggle against the British tied to nationalist Greek and Greek-Cypriot aspirations for the island's union with Greece – signified a new terrain of possible conflict with Turkey. In light of the growing inter-communal tensions between Greek and Turkish Cypriots, the Greek Government sought to project an image of tolerance toward and respect for its official minorities, in contrast to Turkey's actions on these matters. In this manner, Greece wished to communicate to Turkish Cypriots that they had nothing to fear from Cyprus's potential incorporation to Greece.

The Istanbul anti-Greek riots of 1955 greatly influenced Greece's approach to Western Thrace's minority affairs. However, the Greek administration did not engage in direct reprisals against the Muslim minority. Two main factors determined its course of action: on the one hand, with the unfolding Cyprus crisis in mind, it still sought to brand itself as a polity that respected its minority communities and, on the other, it weighed the negative impact such actions would have precipitated in Turkey against the still considerable Greek Orthodox community of Istanbul. The only reasonable course of action was to keep open lines of communication to the Turkish government.

Faced with this new crisis, those in the Greek administration charged with minority affairs questioned whether the permissive standing policies should be changed. Specifically, they suggested that Greece should cease recognizing the minority as Turkish and revert to its long-standing characterization as Muslim; withhold any new concessions to the minority (unless a concession was dictated by reciprocal extension by the Turkish administration to the Greek-Orthodox community of Istanbul; limit Turkey's involvement in minority affairs in Western Thrace; and develop a comprehensive minorities policy. Further, it resuscitated the strategy of leveraging any and all intra-minority divisions between Old Muslims and modernizers to its advantage. Most importantly, it sought to coopt the Bulgarophone Pomaks.

In September 1959, Greece's minority policy was in the hands of the Ministry of Foreign Affairs and the Coordinating Council of Thrace (CCMPT) (Συντονιστικό Συμβούλιο Μειονοτικής Πολιτικής Θράκης (ΣΣΘ)). Shifts in the region's geopolitics appeared to justify the hardening of Greek positions in Western Thrace anew. The creation of a Cypriot state in 1960 by the Treaty of London, the sovereignty of which was to be guaranteed by Britain, Greece, and Turkey, signaled Greece's failure to incorporate the island to Greece.

With the prize of Cyprus lost and intercommunal strife on the island rampant, so was removed the diplomatic imperative that Greece behave irreproachably towards its minority. In 1962, however, concerns over the preservation of the Greek-Orthodox minority and the Ecumenical Patriarchate in Istanbul compelled Greece to maintain a reciprocity regime on minority affairs. However, as the size of the Istanbul-based minority dwindled, so did the motive for moderation.

In that same year, the mandate of the Coordinating Council of Thrace as policy vehicle for minority policy in Western Thrace was extended to include "...the coordination, compliance, and proper application of competences in the implementation of each of the policies defined by us": the Council monitored minority affairs at both regional and local scales and had final say on whether subsidiary authorities were fully and faithfully carrying out the dicta of the government: if the Council determined "that the orders of Ministries or Offices are contrary to the minority policy followed, it may suspend, by its order, their local application in Thrace."[29] The Council operations were confidential and most members of the administration, apart from the few who were directly implicated in its operations, were unaware of its existence. It was abolished in February 1969, and its competences transferred to the Ministry of North Greece. In October 1971, the military Junta (21 April 1967-23 July 1974) charged the Regional Administration of Eastern Macedonia and Thrace with the oversight of minority policy, and in May 1973, this responsibility was transferred to the Cultural Relations Offices of the Prefectures of Thrace.

The creation of the Council made a reality the long-standing demand of hardliners for a unified and coordinated minority policy. Critically, the

State's systematized oversight and surveillance of the minority, and the subordination of its communities to a regime of control and discipline, signaled its official decoupling from the rest of Greek society. Greece's treatment of the Thrace minority became a permanent feature of Greek-Turkish international relations, was (and continued to be) hinged on Turkey's policy on the Greek-Orthodox minority of Istanbul, and complicated Greece's and Turkey's entanglement in the Cyprus issue. In periods of tension and détente, the Western Thracian and Istanbul minorities emerged, respectively, as laboratories of Greek and Turkish cultural politics.

By 1965, the primary concern of Greek policy was to halt and contain the perceived threat that the adoption by many members of the minority of Turkish identity represented. The Greek state perceived the shift to an identity affiliated with its historic adversary as an act of disloyalty and as a potential danger. The existence of the Greek-Orthodox minority and the Ecumenical Patriarchate in Istanbul compelled Greece to adopt reciprocity (1962) as its official policy paradigm, considering that targeted measures would have their own impact on the minority of Thrace, would ensure the preservation of Greek-Rum populations in Turkey. However, because their numbers in Istanbul began to diminish more and more since the mid-1960s, Greek authorities considered that they could abandon any self-imposed restrictions and limitations on the purview of the state on minority affairs and adopt more stringent policies towards the minority. The presence of the minority in Thrace was now considered a "national security" issue, which could create significant problems for Greece. As such, Greece saw its containment as necessary. Meanwhile, minorities and their attempts to defend their rights were perceived as evidence of Turkish expansionism and a threat to Greece's constitutional and territorial integrity. Therefore, defending the national interest was deemed an issue that superseded existing laws on (and by extension the rights) of minorities.

In such an environment, Greek policy aimed, among other things, to control minority education to the maximum extent possible and to sever its links to Turkey. It sought to suspend the right of minorities to supervise minority schools, silently supported the emigration of members of the minority (mainly to Turkey), and revoked the Greek citizenship of many, creating a large number of stateless persons. Many of them continued to live in Greece without being officially citizens (Sitaropoulos, 2004: 205–223). It continued to support the traditionalist Old Muslims, although it acknowledged the ineffectiveness of this policy. In turn, the state tried to engage the minority Roma, and, above all, the Pomaks, who were considered to possess distinct ethnic consciousness. It was asserted that, in the best-case scenario, Pomak identity could be subsumed by Hellenism, or, alternatively, be recognized as an ethnicity distinct from the Turkish-defined one espoused by a growing number of minority members. At the same time, the Greek administration made efforts to support the ethnic Greek majority population of the region structurally.

During its seven-year run (1967–1974), the Junta's minority policies did not markedly differ from what was in place before it, rather the opposite. In the early years of the dictatorship, the Junta governments adopted a conciliatory policy toward Turkey, which led to the signing of the Greek-Turkish protocol of 1968. This improvement in Greek-Turkish relations temporarily reduced tensions between the Greek State and the minority, although this did not amount to a notable departure from previous positions. One achievement of the minority was the establishment in 1968 of the Special Pedagogical Academy in Thessaloniki (ΕΠΑΘ), which was to train minority teachers.

The Turkish invasion of Cyprus in 1974 caused the downfall of the Junta, the return of democracy to Greece, and the election of Konstantinos Karamanlis as Prime Minister of a pro-European center-right government. Turkey's occupation of Cypriot territory, which endures to this day, again contributed to raising the majority's fear of the Muslim minority. Importantly, it helped consolidate in wider Greek society that the compromise of rights and the marginalization of the minority were acceptable.

This public disapproval of the minority contributed to its exclusion from the social goods of democratization and liberalization in post-Junta Greece. The minority now faced continuous impediments to, and bans of, almost all of its activities. Importantly, these repressive measures, continuous restrictions, and the curtailment of minority rights ("administrative harassment," as those in the Greek administration involved with the minority called them), did not represent official Greek policy towards the minority. In the manner redlining, blockbusting, and urban market steering worked in American cities before civil rights gains in the 1960s, the harassment occurred off the record. Impediments to the purchase of land and other real assets and the construction and maintenance of buildings, dealt a blow on quality of life. Bank deposits and lending were unduly hindered, and few business grants and loans were ever approved, hurting minority entrepreneurship. The purchase of cars and agricultural machinery was banned (an especially critical penalty on agrarian communities). All public references to a Turkish minority were banned and were now prosecuted. Also, minority graduates of Turkish universities would face obstacles and delays in the recognition, by the competent Greek authorities, of their degrees and qualifications (Yagcioglu, 2004: 169–192, Heraclides, 2001: 310–311, Anagnostou, Triantafyllidou, 2007: 5–8, Kostopoulos, 2009: 108, Hristopoulos, 2006: 1–7). The minority found itself under continuous surveillance. The Greek administration treated with much suspicion any contact with Turkey or any activity that could substantiate a relationship with the Turkish government.

In the 1980s, especially, this harsh treatment of the minority cultivated a sense of indignation and precipitated its radicalization. It was clear to the minority that it was persecuted not because of what its members did but because of who they were or aspired to be. It amplified the most extreme voices, which in most cases self-identified as Turkish and identified the minority as a whole with Turkey. Several members of the minority chose to immigrate to Turkey, despite the Turkish state's reluctance to support emigration. Turkey remained the only option for those members of the minority who wanted to pursue university studies. The majority of the minority population, however, remained in Greece and lived through these troubles.

The polarization of the minority and the Greek State prevented the emergence of more moderate minority voices (and also majority ones). Concerning minority moderates, those who wished to engage the Greek administration in negotiations feared they had little room to do so. Specifically, they were afraid that their actions would signal their break with the Turkish side and precipitate their isolation.

The minority began to systematically react to the situation mainly after the mid-1980s (Yagcioglu, 2004: 193–197, Aarbakke, 2000: 206–263, 299–356, 430–442, 504–556). The minority's mounting reaction and legal challenges, and the great changes that occurred internationally following the end of the Cold War, led the Greek government to revise its minority policy in 1990 (Frangoudaki, 2008: 495–497). The decision to cease the harassment of the minority as of 1990 reveals the state's contradictions and geopolitical dilemmas. Greece's desire to cater to its national mythology, often in contradiction to building a full liberal democracy and a civil state in the Europeanist mold, has shaped its long-fraught relationship with its Muslim communities. The linchpin is still "Lausanne." Minority policy is where the need to protect the territorial status quo in Thrace and the Aegean Sea and shield the remaining Greek-Orthodox population of Istanbul makes it a priority to defend the Treaty of Lausanne. Should Turkey depart from it, the region will be gravely destabilized. Lastly, the minority is both clay and sculptor in Hellenic statecraft. It is shaped by and shapes state agency with geopolitical consequences that extend far beyond Greece's contestatory domestic politics of difference.

Conclusions

Since the partition of Thrace among Greece, Turkey, and Bulgaria (1919–1923), state strategies have variously sought to predetermine minority identities and weaponize minority issues in the service of statecraft, nationcraft, and regional security. The Greek State's long-standing reluctance to grant rights to its minorities first appears as a dilemma between democracy and national security. However, the dilemma was also profoundly entangled with racialized views of non-Hellenes and, reflexively, a sense of exclusive Hellenic exceptionalism. For that to be the case in Western Thrace, two assumptions would have to be valid: first, minority identities and aspirations are axiomatically contrary to the imperative of protecting the Hellenic-ontological, constitutional, and territorial integrity of a nation-state. Second, a minority's extraterritorial, trans-boundary, and international linkages are, prima facie, suspect and subject to containment. The first assumption closes the door to the full incorporation of minorities into the polity and charter society.

Interactions between the Greek State and Western Thrace's Muslim minority strongly suggest that pre-1990 Greek administrations were disinterested, if not openly hostile, to integrating minorities into national society as full and equal members. Apart from Eleftherios Venizelos' liberal minority policies and the brief "golden age" of Greek-Turkish relations in 1952–1955, the state vacillated between exclusion and détente, without any discernable movement toward inclusion. By 1990, legal challenges, pressure from European institutions, a greater sense of regional security post-Cold War, and the gradual emergence of civil society, opened the door to the minority's incorporation in Greek society and polity. This process is not complete.

In the case of Western Thrace, the second assumption that minorities with trans-boundary affiliations to an adversarial state are, essentially, Trojan horses, became a self-fulfilling prophecy of sorts. There are no official démarches that claim the minority is either disloyal or colludes with either Turkey or Bulgaria to the detriment of Greece's territorial integrity. However, the Greek State's sustained marginalization and, at times, outright suppression of the minority, provided cause to members of the minority to seek patronage and support abroad, and opened the region to Turkish influence. Under Turkish President Recep Tayyip Erdoğan's "Neo-Ottoman" security dogma, long-held apprehension about Turkish interference in the domestic politics of Balkan states is now in full display. The question remains whether the State will fully embrace the civil society turn to develop a geopolitics of difference that would inoculate it from external interference.

The Venizelist aspiration of incorporating Greece's Jewish and Muslim minorities into a more expansive definition of Hellenism was annihilated post-1923 in the fires of nationalist zeal. Ultimately, in its harsh treatment of the Muslim minority, the Greek State undermined national interest in four ways: it compromised socio-spatial and political cohesion in the vulnerable northeastern border region and provided openings to Turkish and Bulgarian interference. By frustrating the non-territorial aspirations of the minority for representation, access to the economy, and political participation, it provided radicalized minority elements a platform for articulating territorial aspirations, very much to the benefit of Turkish diplomacy.[30] By limiting the minority's socio-spatial mobility and its full access to the national economy, it undercut the economic development of the region and, by extension, the state as a whole. And by using the Treaty of Lausanne as justification for denying the ethnization of some of the minority, it opened the door to international criticism and diplomatic démarches that undermined the State's international prestige and the effectiveness of its diplomacy.

The ground shifted in the 1990s and opportunities have since abounded for the state to repair the relationship between the charter society and the minority. Athens has been projecting soft power and deepening its governmentality in Western Thrace by expanding the social state, new infrastructure investment, even in remote areas of the Prefectures of Xanthi and Rhodopi, and mobilizing EU Structural Fund resources to re-animate a region that has long lagged behind. For its part, the minority is divided between suspicion and hopefulness. In the next chapter, we bring this Muslim minority case study to the present time by focusing on Western Thrace's Pomaks.

Notes

1. The inhabitants of the islands of Imvros and Tenedos were not included in the Convention for the exchange of populations, but were later exempted *post factum* by Article 14 of the Treaty of Lausanne. The Greek Orthodox of Constantinople belonged to two categories: former Ottoman subjects and Greek subjects. Notably, among those who were, ultimately, not exchanged were the Muslim Albanian speakers of Epirus (Τσάμηδες (Tsamides)) and the Orthodox speakers of Arabic in Cilicia ("Antiochians") (Kouidis, 2009: 38-39, Alexandris, 1988: 39).
2. European Union, National Policies Platform, Euridice file://eurydice__ppopulation_demographic_situation_languages_and_religionsp_-_2019-06-05.pdf (accessed on 7 August 2019).
3. Minority groups characterized period 1950–1955 as the "golden age" in their relations with the Greek State. During this period, among other things the Greek-Turkish cultural agreement of 20 April 1951 was signed, the first minority secondary school was founded (the "Celâl Bayar" gymnasium in the Prefecture capital of Komotini), and Greek authorities stopped subsidizing the "Old Muslims" who were hostile to Turkey and closed down their newspaper. On 28 January 1954 the Governor-General (General Administrator) of Thrace G Fessopoulos ordered that in all inscriptions the terms "Muslim" (noun, adjective) be replaced by the terms "Turk, Turkish" (Andreadis, 1956: 9–10, Akgönül, 1999: 52–53 Aarbakke, 2000: 134–135, Nikolakopoulos, 2002: 121, 123).
4. Perhaps the most characteristic expression of the ban of the term "Turk" and its derivatives in public use was the names of the various minority associations, since their statutes had to be approved by the Greek courts. In 1928, the "Union of Young Turks"; in 1936, the "Union of Turkish Teachers of Rhodopi – Evrou"; in 1938 the "Union of Turkish Youth of Komotini" was founded in Komotini (having operated unofficially since 1928); and in 1948, the "Turkish Union of Xanthi" was founded. See Andreadis 1956, p. 25, Panagiotidis 1995, pp. 153–155, Tsitselikis 2002α, pp. 471–475, and Frangoudaki 2008, pp. 492–494, where there is a presentation of the "war of names," i.e., the entire effort by the Greek administration to forbid through court action the use of the terms "Turk" and "Turkish" for minority organizations and clubs/associations.
5. One of the more indicative examples of this policy is the Greek translation of the villages in the Treaty of Lausanne, which refer to the minority in Thrace. In the original French and English texts of the Treaty, references to Turkey normally speaks of non-Muslim minorities ("minorités non-musulmanes," "non-Moslem minorities") and when it refers to Greece, to the Muslim minority ("minorité musulmane," "Moslem minority") (Baltsiotis, Tsitselikis, 2001: 33, 61-62). This differentiation is suppressed in the Greek text of the Treaty, which references "Muslim minorities" both in Greece and Turkey. In contrast, in the Turkish translations this differentiation exists ("Müslüman olmayan azınlıklar," "Müslüman azınlık").
6. In the public's imagination, Western Thrace's plains Turkish-origin minority stands in for all Muslims in Greece, thereby often determining attitudes

towards Muslims of other sectarian identities and geographical origins. The building of the Athens Mosque in the Votanikos quarter – the first built after the Greek State's creation – met resistance from nationalists and the Orthodox Church. It was perceived as a threat to the nation-state, although the Athens Mosque mainly serves Pakistani, Indian, Afghani, and Senegalese Muslims.

7. Analytically here on the General Inspectorate of Foreign and Minority Schools (ΓΕΞΜΣ); for the Coordinating Council of Thrace (ΣΣΘ) see below.
8. Coordinating Council for Thrace/President in "Αναφορά του προέδρου του Συντονιστικού Συμβουλίου Θράκης προς τον Υπουργό Παιδείας" [Report by the President of the Coordinating Council of Thrace to the Minister of Education], protocol number ΜΘ 386, ΑΣΦΜΣ φ. 10).
9. See below.
10. Askouni, 2006, pp. 39-42. Cf. also Hersant 2008.
11. See section 2.7.β, this volume.
12. After the military coup of 21 April 1967, which unseated Greece's democratically elected government, and until the reestablishment of democracy on 23 July 1967, only one relevant reference is found in the related archival material. In a reassertion of the official position of the Treaty of Lausanne on minorities, the Junta systematically defined the minority as "Muslim": "Although they [the minority teachers who studied in Turkey] are Greek subjects, not only do they not serve National Interests, [emphasis in the original source] they have overtly sided with the policy of the Turkish Consulate. ., and contributed to the corrosion of the Minority Element in favor of Kemalist Principles and Turkism, thus working consciously in favor of their state of origin" [Prefecture of Rhodopi/Office of Cultural Relations, Minutes of the Local Meeting of February 1974 (27.2.1974), ΑΣΓΜΣ, φ. 12].
13. According to the provisions in Paragraph 3 of Ministerial Decision Φ. 152.11/Β3/790 of 28.2.1996 (issued in application of the provisions of Law 2341/1995, article 2, para. 1): "*[T]he candidates who come from the Muslim community of Thrace and wish to participate in the general examinations, are required to submit above and beyond the supporting documents required by the pertinent regulation, a certificate from their municipality or community in (one of) the prefectures of Xanthi, Rhodopi, or Evros which certifies their enrollment in the municipal or community* population registry." According to said provision, if a member of the minority community had settled permanently outside Thrace, and had transferred his family registration to his new place of residence, he would be deprived of this possibility. Likewise, if a member of the minority wished to settle outside Thrace, he was obliged to maintain his rights in his former place of residence in Thrace so that he (or his children) could make use of this provision in the future. In the end, this limitation was lifted though only in part; according to Ministerial Decision Φ.151/17104/Β6 of 17-2-2006, article 1, paragraph 7, a member of the minority who was transferred to another municipality in Greece could benefit from the quota measure provided he demonstrates that he was registered in a Thracian municipality, i.e., that he belonged to the "Muslim Community" of the region.
14. A Mufti is an Islamic jurist and scholar responsible for assessing and reporting the compliance of lay laws with Islamic law.
15. Ν 1242Β of 24.6.1919 "Concerning the codification of the provisions pertaining to secondary education," *ΦΕΚ* 190α/28.8.1919, Ν 3179 of 5.8.1924; "Concerning the personnel of schools in Thrace," *ΦΕΚ* 186α/7.8.1924.
16. "Μουσουλμανική Κοινότητα Κομοτηνής Ξάνθης και Διδυμοτείχου (Muslim Community of Komotini, Xanthi and Didymoteicho), 'Υπόμνημα προς την Γενικήν Διοίκησιν Θράκης," (Memorandum to the General Administration of Thrace), 20.9.1945, ΑΣΓΜΣ, φ. 95, σ. 11–12.

17. Featherstone et al., op. cit. pp. 101–119, 131–157 and Huseyinoglu, op. cit. pp. 145–154.
18. Γενική Διεύθυνση Αλλοδαπών (General Directorate of Aliens), "Έκθεσις περί των εν Ελλάδι βιούντων Μουσουλμάνων" (Report on Muslims living in Greece), 31.7.1952, ΑΣΓΜΣ, φ. 95, p. 8, Συντονιστικό Συμβούλιο Θράκης/Πρόεδρος (Coordinating Council of Thrace/President) "Διαχειριστικαί Επιτροπαί Περιουσίας των τέως Μουσουλμανικών Κοινοτήτων" (Management Committees of the former Muslim Communities) to the Minister of Education, Research and Religions (ΥΠΕΠΘ) and the Minister for Coordination, 4.9.1967, ΑΣΓΜΣ, φ. 11, p. 2; Featherstone et al., op. cit. pp. 174–191.
19. Report of 31.7.1952, op. cit., pp. 14–15.
20. Memorandum of 20.9.1945, op. cit., p. 1.
21. Memorandum of 20.9.1945, op. cit., pp. 7-12; see also Mavrommatis, op. cit., pp. 121–123.
22. Memorandum of 20.9.1945, op. cit., p. 4.
23. Memorandum of 20.9.1945, op. cit. pp. 10–11.
24. Memorandum of 20.9.1945, op. cit., p. 3.
25. "Note by the General Inspector of Foreign and Minority Schools no. 122/20.9.1945, memorandum on the Muslim communities of Komotini, Xanthi, and Didymoteicho," 28.9.1945, ΑΣΓΜΣ, φ. 95, pp. 1-6.
26. Ilias Nikolakopoulos, "The path towards the autonomous political formation of the Muslim minority in Western Thrace," *Μειονότητες στην Ελλάδα (Επιστημονικό Συμπόσιο, 7-9 Νοεμβρίου 2002)*, Athens, Εταιρεία Σπουδών Νεοελληνικού Πολιτισμού και Γενικής Παιδείας (Modern Greek Culture and General Education Studies Society), 2002, pp. 119–120, Aarbakke, op.cit., pp. 109–112, Featherstone et al., op. cit., pp. 186–187.
27. Featherstone et al., op.cit., pp. 193–229, 259–277, Mihri Belli, *"Καπετάν Κεμάλ." Αναμνήσεις από τον ελληνικό εμφύλιο ("Captain Kemal." Recollections from the Greek Civil War)*, İstanbul, Telos International, 2009, pp. 57–61, 78–81, 85–86, 93–94, 99–103, 127–128, 140–141, 173–174, 222–223, 226–227, 241.
28. [xxviii] Report of 31.7.1952, op.cit. pp. 15-16; Featherstone et al., op.cit., pp. 242–244, 250–258, 284–288, 297–301, also Belli, op.cit., pp. 49–50, 112–116.
29. «Απόφαση Υπουργού Εξωτερικών Ευάγγελου Αβέρωφ-Τοσίτσα για ανασύσταση Συντονιστικού Συμβουλίου Μειονοτικής Πολιτικής Θράκης», αρ. απόφασης 1946/16, 16.5.1962, σ. 2, ΑΣΓΜΣ φ.17n [Decision of the Minister of Foreign Affairs, Evangelos Averof-Tositsas for the resuscitation of the Coordinating Council of Thrace].
30. By claiming collusion between minority elements and the revolutionary Left, right at the onset of the Cold War, the Greek State appeared to cast its minority policy in East-Western terms. However, that construction did not apply to the Muslim minority. Instead, it was those who self-identified as Slavo-Macedonians who were considered directly entangled in the broader global conflict.

Sources

Aarbakke, Vemund. *The Muslim Minority of Greek Thrace.* Unpublished doctoral dissertation Vols. 1-2. University of Bergen, Bergen 2000.

Akgönül, Samim. *Une Communauté, Deux États: La Minorité Turco-Musulmane de Thrace Occidentale*, Istanbul : Εκδ. Isis, 1999.

Anagnostou, Dia; Triantafyllidou, Anna. *Regions, Minorities and European Integration: A Case Study on Muslims in Western Thrace, Greece.* Αθήνα: ΕΛΙΑΜΕΠ, 2007 http://www.eliamep.gr/wp-content/uploads/en/2006/05/Case_study_report_Thrace.pdf

Arar, Ismail. "Atatürk' ün Günümüz Olaylarina Işik Tutan Bazi Konuşmalari" (Some Homilies by Ataturk that Shed Light on our Era). *Belleten* Vol. 45-6, No. 177 (1981): 23–4.

Belli, Mihri. *«Καπετάν Κεμάλ». Αναμνήσεις από τον Ελληνικό Εμφύλιο ("Captain Kemal." Reminiscences from the Greek Civil War*). Athens: Editions Telos International, 2009.

Birtek, Faruk. «Ο Ελληνικός Ταύρος στο Υαλοπωλείο της Οθωμανικής "Μεγάλης Χίμαιρας". Ο Πόλος της Ελλάδας στην Κατασκευή της Σύγχρονης Τουρκίας» (The Greek bull in the Ottoman 'chimera''s china shop. Greece's role in the construction of modern Turkey). In *Ελλάδα και Τουρκία. Πολίτης και Έθνος-Κράτος*, (Θάλεια Δραγώνα (Dragona), Φαρούκ Μπιρτέκ (Birtek), eds. Αθήνα: Εκδ. Αλεξάνδρεια 2006: 84–103.

Dalègre, Joëlle. « Entre Grèce et Turquie : Une Situation Délicate, les Minorités." *Cahiers Balkaniques* 33 (2004): https://journals.openedition.org/ceb/4474#text

Dalègre, Joëlle. *Populations et Territoire en Thrace Grecque Depuis 1878, Tome 2 : Cartes, Documents, Fiches Statistiques, Bibliographie* (published doctoral dissertation), Dèpartment de Geographie, Paris: Universite de Paris X-Nanterre, 1995.

Featherstone, Kevin; Papadimitriou, Dimitris; Mamarelis; Argyris; Niarchos, Georgios. *The Last Ottomans. The Muslim Minority of Greece, 1940-1949.* Hampshire: Palgrave Macmillan, 2011.

Hersant, Jeanne. "Surveillances Croisées et Rivalité Gréco-Turque en Thrace Occidentale: Entre Coercition et Contrôle Social." *European Journal of Turkish Studies* Vol. 8 (2008) http://ejts.revues.org/2693

Hobsbawm, Eric. «Επινοώντας Παραδόσεις», *Η Επινόηση της Παράδοσης* (The Invention of Tradition). Αθήνα: Εκδ. Θεμέλιο, 2004: 9–24.

Huseyinoglu, Ali. *The Development of Minority Education at South-Easternmost Corner of the EU: The Case of Muslim Turks in Western Thrace, Greece.* Unpublished doctoral dissertation, University of Sussex, 2012.

Jenkins, Richard. *Κοινωνική Ταυτότητα* (Social Identity). Αθήνα: Εκδ. Σαββάλας, 2007

Keyder, Caglar 2006, «Ιστορία και Γεωγραφία του Τουρκικού Εθνικισμού» (History and Geography of Turkish Nationalism). In Θάλεια Δραγώνα (Dragona), Φαρούκ Μπιρτέκ (Birtek), *eds. Ελλάδα και Τουρκία. Πολίτης και Έθνος-Κράτος*. Αθήνα, εκδ. Αλεξάνδρεια 2006, σ. 25–49

Kızılyürek, Niyazi 2006, *Κεμαλισμός. Η Γένεση και η Εξέλιξη της Επίσημης Ιδεολογίας της Σύγχρονη Τουρκίας*. Αθήνα: Εκδ. Μεσόγειος, 2006

Mazower, Mark. "Minorities and the League of Nations in Interwar Europe." *Daedalus*. Vol. 126, No. 2, Human Diversity (Spring, 1997): 47–63.

Memişoglu, Hüseyin. *Pages of the History of Pomaks Turks*. Ankara: 1991.

Poulton, Hugh. *Ημίψηλο, Γκρίζος Λύκος και Ημισέληνος* (Top hat, Grey Wolf and Sycle Moon). Αθήνα: Εκδ. Οδυσσέας, 2000.

Riga, Liliana; Kennedy, James. "Tolerant majorities, loyal minorities and 'ethnic reversals': constructing minority rights at Versailles 1919." *Nations and Nationalism* Vol. 15, No. 3 (2009): 461–482.

Sitaropoulos, Nikolaos. "Freedom of Movement and the Right to a Nationality v. Ethnic Minorities: The Case of ex article 19 of the Greek National Code." *European Journal of Migration and Law* Vol. 6 (2004): 205–223.

Smith, Anthony D. *Εθνική Ταυτότητα* (National Identity). Αθήνα: Εκδ. Οδυσσέας, 2000.

Soner, Cagaptay. "Race, Assimilation and Kemalism: Turkish Nationalism and the Minorities in the 1930s." *Middle Eastern Studies* Vol. 40, no. 3 (2004): 86–101.

Tsibiridou, Fotini. *Les Pomak dans la Thrace Grecque. Discours Ethnique et Pratiques Socioculturelles.* Paris: Εκδ. L' Harmattan, 2000.

Tsitselikis, Konstantinos. "The Pending Modernization of Islam in Greece: From *Millet* to Minority Status." *Südesteuropa* Vol. 55, No. 4 (2007): 354–372

Tsitselikis, Konstantinos. "Aspects of Legal Communitarianism in Greece: Between *Millet* and Citizenship." *Oñati Socio-legal Series* Vol. 2, No. 7 (2012): 106–118 http://ssrn.com/abstract=2050345

Yagcioglu, Dimosthenis. *From Deterioration to Improvement in Western Thrace, Greece: A Political Systems Analysis of a Triadic Ethnic Conflict.* Unpublished doctoral dissertation, George Mason University, 2004.

Zürcher, Erik Jan. *Σύγχρονη Ιστορία της Τουρκίας* (The Modern History of Turkey). Αθήνα: Εκδ. Αλεξάνδρεια, 2004.

Αλεξανδρής (Alexandris), Αλέξης. *Οι Ελληνοτουρκικές Σχέσεις 1923-1987 (Greek-Turkish relations 1923-1987)*, Αθήνα: Γνώση, 1988.

Ανδρούσου (Androusou), Αλεξάνδρα. *«Πώς σε λένε;». Διεργασίες μιας Επιμορφωτικής Παρέμβασης στη Μειονοτική Εκπαίδευση* (What's your name? Processes of a training intervention in minority education). Αθήνα: Gutenberg, 2005.

Ασημακοπούλου (Asimakopoulou) Φωτεινή, Λιοναράκη (Lionaraki) Σεβαστή. *Η Μουσουλμανική Μειονότητα της Θράκης και οι Ελληνοτουρκικές Σχέσεις* (The Muslim minority of Thrace and Greek-Turkish relations). Αθήνα: Α. Λιβάνη, 2002.

Ασκούνη (Askouni), Νέλλη. *Η Εκπαίδευση της Μειονότητας στη Θράκη. Από το Περιθώριο στην Προοπτική της Κοινωνικής Ενταξης* (The education of the minority in Thrace. From the margins to the perspective of social inclusion). Αθήνα: Αλεξάνδρεια, 2006.

Βεντούρα (Ventura), Λίνα. «Κράτος, Έθνος και Ομογένεια: 1974 – 2001» (State, nation, and Greek diaspora: 1974-2001). In Μίλτος Παύλου, Αθηνά Σκουλαρίκη (eds). *Μετανάστες και Μειονότητες. Λόγος και Πολιτικές.* Αθήνα: Βιβλιόραμα, 2009: 105–140.

Γιώλτζογλου (Yoltzoglou), Σταύρος Χ. *Οι Ελληνοτουρκικές Σχέσεις (1922-1930)* (Greek-Turkish relations (1922-1930)). Διδακτορική διατριβή, Αριστοτέλειο Πανεπιστήμιο Θεσσαλονίκης, Τμήμα Ιστορίας – Αρχαιολογίας, Τομέας Νεότερης Ιστορίας και Σύγχρονης Ιστορίας και Λαογραφίας, 2006 (http://ikee.lib.auth.gr/record/100698/files/Gioltsoglou.pdf, Accessed on 8 August 2020).

Γκιουβέν (Güven), Ντιλέκ. *Εθνικισμός, Κοινωνικές Μεταβολές και Μειονότητες. Τα Επεισόδια Εναντίον των μη Μουσουλμάνων της Τουρκίας (6/7 Σεπτεμβρίου 1955)* (Nationalism, social change and minorities. Actions against the non-Muslims of Turkey (6/7 Σεπτεμβρίου 1955)). Αθήνα: Εστία, 2006.

Δραγώνα (Dragona), Θάλεια. «Εκπαιδεύοντας τον "Ανοίκειο Αλλο": Το Παράδειγμα της Μειονοτικής Εκπαίδευσης» (Educating the "unfamiliar other:" The example of minority education), *Ψυχολογία* Vol. 11, No. 1 (2004): 20–33.

Δραγώνα (Dragona), Θάλεια. «Εκπαιδεύοντας τον Ανοίκειο Αλλο: Ταυτότητες, Ψυχικοί Μηχανισμοί και Ιδεολογία» (Educating the "unfamiliar other:" Identities, psychological mechanisms and ideology). In Θάλεια Δραγώνα (Dragona), Άννα Φραγκουδάκη (Frangoudaki), (eds). *Πρόσθεση, Οχι Αφαίρεση – Πολλαπλασιασμός, Οχι Διαίρεση. Η Μεταρρυθμιστική Παρέμβαση στην Εκπαίδευση της Μειονότητας στη Θράκη.* Αθήνα: Μεταίχμιο, 2008: 423–435.

Δραγώνα (Dragona), Θάλεια. «Όταν η Εθνική Ταυτότητα Απειλείται: Ψυχολογικές Στρατηγικές Αντιμετώπισης.» In Άννα Φραγκουδάκη (Frangoudaki), Θάλεια Δραγώνα (Dragona)(eds). *«Τι είναι η πατρίδα μας;»* (When national identity is threatened: Psychological coping strategies in "What is our homeland?"). *Εθνοκεντρισμός στην Εκπαίδευση* (1997): 72–105, Αθήνα: Αλεξάνδρεια.

Ηλιάδης (Iliadis), Χρήστος. *Η Εθνική Ταυτότητα της Μουσουλμανικής Μειονότητας και η Εκπαιδευτική Πολιτική. Μελέτη Αρχειακών Πηγών. 1945-1967* (The national identity of the Muslim minority and educational policy. Studies of archival sources 1945-1967), Μεταπτυχιακή διπλωματική εργασία, Εθνικό και Καποδιστριακό Πανεπιστήμιο Αθηνών, Πρόγραμμα Μεταπτυχιακών Σπουδών «Πολιτική Επιστήμη και Κοινωνιολογία.» Αθήνα, 2004.

Ηρακλείδης (Heraclides), Αλέξης. «Τα Αίτια της Αντιμειονοτικής Στάσης της Ελλάδας. Ορισμένες Υποθέσεις για Ερευνα» (The causes of anti-minority attitudes in Greece. Some case studies). In *Μειονότητες στην Ελλάδα* (Πρακτικά επιστημονικού συμποσίου, Αθήνα 7-9 Νοεμβρίου 2002). Αθήνα: Εταιρεία Σπουδών Νεοελληνικού Πολιτισμού και Γενικής Παιδείας Σχολής Μωραΐτη, 2002: 43–57.

Ηρακλείδης (Heraclides), Αλέξης. *Η Ελλάδα και ο «Εξ ανατολών κίνδυνος»* (Greece and "the peril from the East"). Αθήνα: Πόλις, 2004.

Καλαντζής (Kalantzis), Γιώργος, «Το Μουσουλμανικό Εκπαιδευτικό Δίκτυο της Δυτικής Θράκης κατά την Περίοδο 1923-1928: Ορισμένα Στατιστικά Στοιχεία» (The Muslim educational network in Western Thrace during 1923-1928. Some statistics). *Παιδαγωγική Επιθεώρηση* Vol. 38 (2004): 9–28.

Κιτρομηλίδης (Kitromilidis), Πασχάλης. «"Νοερές Κοινότητες" και οι Απαρχές του Εθνικού Ζητήματος στα Βαλκάνια» ("Imagined communities" and the origins of the national question in the Balkans). In Θάνος Βερέμης (ed.), *Εθνική Ταυτότητα και Εθνικισμός στη Νεότερη Ελλάδα.* Αθήνα: Μορφωτικό Ίδρυμα Εθνικής Τραπέζης, 1997: 53–131.

Κονόρτας (Konortas), Παρασκευάς. «Από τα Μιλλέτ στα Εθνη: Διαμόρφωση Συλλογικών Ταυτοτήτων στη Θράκη (19ος–αρχές 20ού αιώνα)» (From *Millet* to nations: Formation of collective identities in Thrace (19th – early 20th Century). In *Θράκη. Ιστορικές και Γεωγραφικές Προσεγγίσεις.* Αθήνα: Εθνικό Ίδρυμα Ερευνών, 2000: 169–190.

Κουίδης (Kouides), Βασίλης. *Οι Ελληνοτουρκικές Σχέσεις και η Εκπαιδευτική Πολιτική του Ελληνικού Κράτους Απέναντι στη Μουσουλμανική Μειονότητα της Δυτικής Θράκης κατά την Περίοδο 1974-1996* (Greek-Turkish relations and the Greek State's education policy on Western Thrace's Muslim minority during 1974-1996), Μεταπτυχιακή διπλωματικής εργασία, Αριστοτέλειο Πανεπιστήμιο Θεσσαλονίκης, Παιδαγωγικό Τμήμα Δημοτικής Εκπαίδευσης, 2009 (http://ikee.lib.auth.gr/record/123843/files/koud.pdf, Accessed on 8 August 2020).

Κωστόπουλος (Kostopoulos). Πέτρος, Το *«Μακεδονικό» της Θράκης. Κρατικοί Σχεδιασμοί για τους Πομάκους (1956-2008)* (The "Macedonian" [question] of Thrace. Government plans for the Pomaks). Αθήνα: Βιβλιόραμα, 2009.

Λέκκας (Lekkas), Παντελής. *Η Εθνικιστική Ιδεολογία. Πέντε Υποθέσεις Εργασίας στην Ιστορική Κοινωνιολογία* (Nationalist ideology. Five working hypotheses in historical sociology). Αθήνα: Κατάρτι, 1996.

Λεπίδα (Lepida), Στυλιανή. *Η Μουσουλμανική Μειονότητα της Δυτικής Θράκης μετά τη Συνθήκη της Λωζάνης. Η Διένεξη Νεωτεριστών – Συντηρητικών κατά το Μεσοπόλεμο* (The Muslim minority of Western Thrace after the Treaty of Lausanne. The conflict between modernizers and conservatives in the Interwar Period), Μεταπτυχιακή Διπλωματική Εργασία, Αριστοτέλειο Πανεπιστήμιο Θεσσαλονίκης, Τμήμα Ιστορίας – Αρχαιολογίας, Τομέας Νεότερης και Σύγχρονης Ιστορίας, Λαογραφίας και Κοινωνικής Ανθρωπολογίας, 2011 (http://ikee.lib.auth.gr/record/129021/files/GRI-2012-8494.pdf, Accessed on 8 August 2020).

Λιάζος (Liazos), Νικόλαος. *Τα Τουρκικά Σχολικά Εγχειρίδια της Μειονοτικής Εκπαίδευσης στην Ανάπτυξη Διαπολιτισμικών Σχέσεων μεταξύ Χριστιανών και Μουσουλμάνων στη Δυτική Θράκη* (Turkish textbooks for minority education in the development of intercultural relations between Western Thrace's Christians and Muslims), Διδακτορική διατριβή, Αριστοτέλειο Πανεπιστήμιο, Θεολογική Σχολή, Τμήμα Ποιμαντικής και Κοινωνικής Θεολογίας, 2007 (http://thesis.ekt.gr/thesisBookReader/id/14508#page/1/mode/2up, Accessed on 8 August 2020).

Μαλκίδης (Malkidis), Θεοφάνης. «Η Διαμάχη "Παλαιομουσουλμάνων" – Κεμαλικών στην Ελλαδική Θράκη την Περίοδο του Μεσοπολέμου» (The conflict between "Old Muslims" and Kemalists in the Interwar Period), *Νέα Κοινωνιολογία* Vol. 39 (Autumn 2004): 73–87.

Ματάλας (Matalas), Παρασκευάς. *Έθνος και Ορθοδοξία. Οι Περιπέτειες μιας Σχέσης. Από το «Ελλαδικό» στο Βουλγαρικό Σχίσμα* (Nation and Orthodoxy. Adventures in a relationship. From the "Greek" to the Bulgarian schism), Ηράκλειο: Πανεπιστημιακές Εκδόσεις Κρήτης, 2003.

Μαυρομμάτης (Mavrommatis), Γιώργος. *Εθνικισμός και Ιστορία Εκπαιδευτικής Πολιτικής: Η Εκπαίδευση των Θρακιωτών Μουσουλμάνων Μειονοτικών 1945-1975* (Nationalism and the history of educational policy: The education of Thracian Muslim communities 1945-1975), Διδακτορική διατριβή, Πάντειο Πανεπιστήμιο Κοινωνικών και Πολιτικών Επιστημών, Τμήμα Πολιτικής Επιστήμης και Ιστορίας, 2007 (file:///C:/Users/User/AppData/Local/Temp/11DID_MavrommatisGe1.pdf, Accessed on 8 August 2020).

Μήλλας (Millas), Ηρακλής. *Εικόνες Ελλήνων και Τούρκων. Σχολικά Βιβλία – Ιστοριογραφία – Λογοτεχνία και Εθνικά Στερεότυπα* (Images of Greeks and Turks. Textbooks – Historiography – Literature and national stereotypees). Αθήνα: Αλεξάνδρεια, 2001.

Μούδουρος (Moudouros), Νίκος. *Ο Μετασχηματισμός της Τουρκίας. Από την Κεμαλική Κυριαρχία στον «Ισλαμικό» Νεοφιλελευθερισμό* (The transformation of Turkey. From Kemalist domination to "Islamic" neoliberalism). Αθήνα: Αλεξάνδρεια, 2012.

Μουζέλης (Mouzelis), Νίκος; Παγουλάτος (Pagoulatos), Γιώργος. «Κοινωνία Πολιτών και Ιδιότητα του Πολίτη στη Μεταπολεμική Ελλάδα» (Civil society and citizenship in post-war Greece). In Θάλεια Δραγώνα (Dragona) and Φαρούκ Μπιρτέκ (Birtek) (eds.), *Ελλάδα και Τουρκία. Πολίτης και Εθνος-Κράτος.* Αθήνα: Αλεξάνδρεια, 2006: 169–202.

Μπαλτσιώτης, (Baltsiotis) Λάμπρος; Τσιτσελίκης (Tsitselikis), Κωνσταντίνος. «Η Μειονοτική Εκπαίδευση της Θράκης: Νομικό Καθεστώς, Προβλήματα και Προοπτικές» (The minority education of Thrace: Legal status, problems and perspectives). In Θάλεια Δραγώνα (Dragona) and Άννα Φραγκουδάκη (Frangoudaki) (eds.), *Πρόσθεση, Οχι Αφαίρεση – Πολλαπλασιασμός, Οχι Διαίρεση. Η Μεταρρυθμιστική Παρέμβαση στην Εκπαίδευση της Μειονότητας στη Θράκη.* Αθήνα: Μεταίχμιο, 2009: 59–90.

Μπαλτσιώτης (Baltsiotis), Λάμπρος; Τσιτσελίκης (Tsitselikis), Κωνσταντίνος. *Η Μειονοτική Εκπαίδευση της Θράκης. Συλλογή Νομοθεσίας – Σχόλια* (Minority education in Thrace. Legislative corpus – Commentaries). Αθήνα–Κομοτηνή: Εκδ. Αντ. Ν. Σάκκουλα, 2001.

Μπαλτσιώτης, (Baltsiotis) Λάμπρος. «Εθνικοί Μύθοι και Ιστορικές Κατασκευές στις μη Ελληνόφωνες Ομάδες. Το Παράδειγμα των Αρβανιτών και των Βλάχων» (National myths and historical constructions in non-Greek speaking groups. The cases of the Arvanites and the Vlachs). In *Μειονότητες στην Ελλάδα* (Πρακτικά επιστημονικού συμποσίου, Αθήνα 7-9 Νοεμβρίου 2002). Αθήνα: Εταιρεία Σπουδών Νεοελληνικού Πολιτισμού και Γενικής Παιδείας Σχολής Μωραΐτη, 2002: 337–366.

Μπαλτσιώτης, (Baltsiotis) Λάμπρος. «Ελληνική Διοίκηση και Μειονοτική Εκπαίδευση στη Δυτική Θράκη. Αιτιότητες και Αντιφάσεις μιας Εκπαιδευτικής Πολιτικής» (Greek administration and minority education in Western Thrace. Causes and contradictions of education policy). In Κωνσταντίνος Τσιτσελίκης (Tsitselikis) and Δημήτρης Χριστόπουλος (Hristopoulos) (eds.), *Το Μειονοτικό Φαινόμενο στην Ελλάδα. Μια Συμβολή των Κοινωνικών Επιστημών*. Αθήνα: Κριτική, 1997: 315–348.

Νικολακόπουλος (Nikolakopoulos) Ηλίας. «Η Πορεία προς την Αυτόνομη Πολιτική Συγκρότηση της Μουσουλμανικής Μειονότητας στη Δυτική Θράκη» (The path to the autonomous political formation of Western Thrace's Muslim minority). In *Μειονότητες στην Ελλάδα* (Πρακτικά Επιστημονικού Συμποσίου, Αθήνα 7-9 Νοεμβρίου 2002). Αθήνα: Εταιρεία Σπουδών Νεοελληνικού Πολιτισμού και Γενικής Παιδείας Σχολής Μωραΐτη, 2002: 119–162.

Παπαργυρίου (Papargyriou), Όλγα. Ο Θρησκευτικός Χαρακτήρας της Μουσουλμανικής Μειονότητας της Δ. Θράκης και η Εκπαίδευσή της στα Πλαίσια του Ελληνικού Κράτους (The religious character of Western Thrace's Muslim minority and its state-established education), Μεταπτυχιακή διπλωματική εργασία, Αριστοτέλειο Πανεπιστήμιο Θεσσαλονίκης, Θεολογική Σχολή, Τμήμα Ποιμαντικής και Κοινωνικής Θεολογίας, 2007 (http://ikee.lib.auth.gr/record/100968/files/gri-2008-1129.pdf, Accessed on 8 August 2020).

Παύλου, Μίλτος; Σκουλαρίκη, Αθηνά (eds). *Μετανάστες και Μειονότητες. Λόγος και Πολιτικές* (Immigrants and minorities. Discourse and policies). Αθήνα: Εκδ. Βιβλιόραμα, 2010.

Πετρίδης (Petridis), Τριαντάφυλλος. *Η Δευτεροβάθμια Εκπαίδευση των Μειονοτικών Μουσουλμάνων της Δυτικής Θράκης (Δεκαετία 1950-Δεκαετία 1970)* (Secondary education of Western Thrace's minority Muslims), Μεταπτυχιακή διπλωματική εργασία, Πάντειο Πανεπιστήμιο Πολιτικών και Κοινωνικών Επιστημών, Τμήμα Πολιτικής Επιστήμης και Ιστορίας, 2013 (http://pandemos.panteion.gr/index.php?op=record&lang=el&pid=iid:6507).

Πλεξουσάκη (Plexousaki), Έφη. «Το Ονομα του "Άλλου": Θέσεις και Αντιπαραθέσεις κατά τη Σύνθεση της Κατηγορίας "Μειονότητα" στη Θράκη» (The name of the "Other:" Positions and controversies in the composition of the category "minority" in Thrace). In Αλεξάνδρα Ανδρούσου (Androusou) and Νέλλη Ασκούνη (Askouni) (eds.), *Πολιτισμική Ετερότητα και Ανθρώπινα Δικαιώματα. Προκλήσεις για την Εκπαίδευση*. Αθήνα: Μεταίχμιο, 2009: 223–243.

Σκουλαρίκη (Skoulariki), Αθηνά. «Δηλώσεις Εθνικών Φρονημάτων: Ο Πολιτικός Λόγος και η Δημόσια Αντιπαράθεση στην Ελλάδα σχετικά με τον Αυτοπροσδιορισμό των Μειονοτήτων και την Υποψηφιότητα της Γκιουλμπεγιάζ Καρά Χασάν» (Declarations of national allegiance: Political discourse and public controversy in Greece regarding the self-determination of minorities and the candidacy of Gülbeyaz). In Μίλτος Παύλου and Αθηνά Σκουλαρίκη (eds.), *Μετανάστες και Μειονότητες. Λόγος και Πολιτικές*. Αθήνα: Εκδ. Βιβλιόραμα, 2010.

Τρουμπέτα (Troubeta), Σεβαστή, *Κατασκευάζοντας Ταυτότητες για τους Μουσουλμάνους της Θράκης. Το Παράδειγμα των Πομάκων και των Τσιγγάνων*. Αθήνα: Κριτική, 2001.

Τσιούμης (Tsioumis), Κωστής. «Η Διαχείριση της Μειονοτικής Ταυτότητας στον Ελληνικό Χώρο και η Διαμόρφωση της Εκπαιδευτικής Πολιτικής: Η Περίπτωση των Μουσουλμάνων της Θράκης 1923–1974» (The management of minority identities in Greece and the shape of educational policy: The case of Thrace's

Muslims 1923-1974). In Κωνσταντίνος Δημάδης (Dimadis) (ed.), *Ταυτότητες στον ελληνικό κόσμο (από το 1204 έως σήμερα)* (Πρακτικά Δ′ Ευρωπαϊκού Συνεδρίου Νεοελληνικών Σπουδών, Γρανάδα, 9-12 Σεπτεμβρίου 2010). Αθήνα: Ευρωπαϊκή Εταιρεία Νεοελληνικών Σπουδών, 2011: 577-590 (https://www.eens.org/wordpress/wp-content/uploads/2012/05/Identities-in-the-Greek-world-Granada-2010-Congress-Vol_1-2011-isbn_978-960-99699-3-21.pdf Accessed on 8 August 2020).

Φραγκουδάκη (Frangoudaki), Άννα. «Η Θράκη Αλλάζει: Επίμετρο Σχόλιο για τις Προοπτικές και τα Εμπόδια» (Thrace is changing: An afterword on prospects and obstacles). In Θάλεια Δραγώνα (Dragona) and Άννα Φραγκουδάκη (Frangoudaki) (eds.), *Πρόσθεση, Οχι Αφαίρεση – Πολλαπλασιασμός, Οχι Διαίρεση. Η Μεταρρυθμιστική Παρέμβαση στην Εκπαίδευση της Μειονότητας στη Θράκη*. Αθήνα: Μεταίχμιο, 2009: 483–499.

Χατζίκος (Hatzikos), Γεώργιος. *Η Επίδραση του Ισλάμ στον Κοινωνικό Ιστό και το Εκπαιδευτικό Σύστημα της Μειονότητας της Δυτικής Θράκης* (The impact of Islam on the social fabric and educational system of Western Thrace's minority), Μεταπτυχιακή διπλωματική εργασία, Αριστοτέλειο Πανεπιστήμιο Θεσσαλονίκης, Θεολογική Σχολή, Τμήμα Ποιμαντικής και Κοινωνικής Θεολογίας, 2009 (http://ikee.lib.auth.gr/record/113462/files/Binder1.pdf, Accessed on 8 August 2020).

Χριστόπουλος (Hristopoulos), Δημήτρης. «Θράκη: Όψεις της Μεταπολίτευσης στη Μειονοτική Πολιτική» (Thrace: Government positions in the post-Junta democratic era). *Θέσεις* Vol. 96 (July-September 2006): 1–7 (http://www.theseis.com/index2.php?option=com_content&do_pdf=1&id=952, Accessed on 8 August 2020).

Χριστόπουλος (Hristopoulos), Δημήτρης. *Η Ετερότητα ως Σχέση Εξουσίας. Όψεις της Ελληνικής, Βαλκανικής και Ευρωπαϊκής Εμπειρίας* (Otherness as a power relationship. Aspects of the Greek, Balkan and European experience). Αθήνα: Κριτική, 2002.

6 Pomaks, "our forgotten siblings"

From suspect aliens to fulcrum of statecraft

Building on our study of Western Thrace's Muslim minority, we focus on one of its constituent ethnic groups, the Bulgarian-speaking Muslim Pomaks. We posit how Greek minority policy, built on the «Lausanne» legal-administrative framework and Hellenic nationalist geopolitics, has shaped the spatial logics of Pomak life, identity, and territoriality: the group's forced spatial sequestration at different times, its efforts to integrate village economies to regional, national, and international markets controlled by others, its inclusion in competing Greek and Turkish spaces of communication flows, and the community's politics of mobility, settlement, and built environment. Further, we consider how in the last one hundred years, the Pomaks have been variously constructed by the Hellenic State as either assets or hostiles in its efforts to counter perceived threats, primarily from Turkey. In contrast to Turkey, Bulgaria has been reluctant to assume the role of kin-state for Thrace's Pomaks, although it has readily done so for the region's minority Turks (Tsitselikis, 2004: 429). Lastly, we discuss the state's inclusionary policy turn after 1990, the degree to which it has met the Pomak community's expectations, and its material impacts in select settlements covered in our fieldwork.

Pomak population and settlements

Out of Western Thrace's population of 361,208 (3.35 percent of the national population), approximately 35 percent belong to minority groups. Great efforts were made between 1919 and 1995 to determine the precise sizes of the minority population and those of its constituent ethnic groups (Kostopoulos, 2003: 70). The 2017 Pomak population estimate of the Hellenic Statistical Authority was 36,000. The settlement-level count of Kara Hussein Ikmet was 27,307 (approximately 21 percent of the total minority population) (Ikmet, 2017: 28).[1] The discrepancy may reflect the impact of internal and external migrations in other regions of Greece and European Union Members States.

Sparsely populated Western Thrace had an urbanization rate (or degree of urbanization) of 55.3 percent, featuring only intermediate density areas and thinly populated ones.[2] Its settlement geography includes 498 settlements

(Hellenic Statistical Authority in Ikmet). Wars and forced migrations have led to the abandonment of many villages, mainly in montane regions, as late as the Civil War (De Jong, 1980, xvi). Only six urban places have a population greater than 5,000, four of which are located in the Prefecture of Evros and fairly evenly spaced along the Greek-Turkish border and the Evros River. We address the phenomenon of Evros' borderland urbanization in a subsequent chapter. Out of the six urban places, three are prefecture capitals with modest populations of about 55,000. Rhodopi, with Komotini as capital, is a majority Sunni Muslim Prefecture (54.77 percent minority population). Xanthi also has a significant minority population (42.19 percent), while Evros is a strong majority regional state (6.65 percent). 213 villages and hamlets are settled by minority populations, 110 of them Pomak (Figure 6.1).[3] Their settlements are commonly very small (mean population: 288; minimum: ten; maximum: 2,580 (Kentauros)), although their close spatial clustering (with often fewer than three kilometers of distance between communities) allows cooperative economic activities and cultural reproduction (Figure 6.2). For the most part, these are socially conservative places, as evidenced by the presence of traditional, patrilineal/patrilocal, multigenerational households, and strictly regulated gender relations

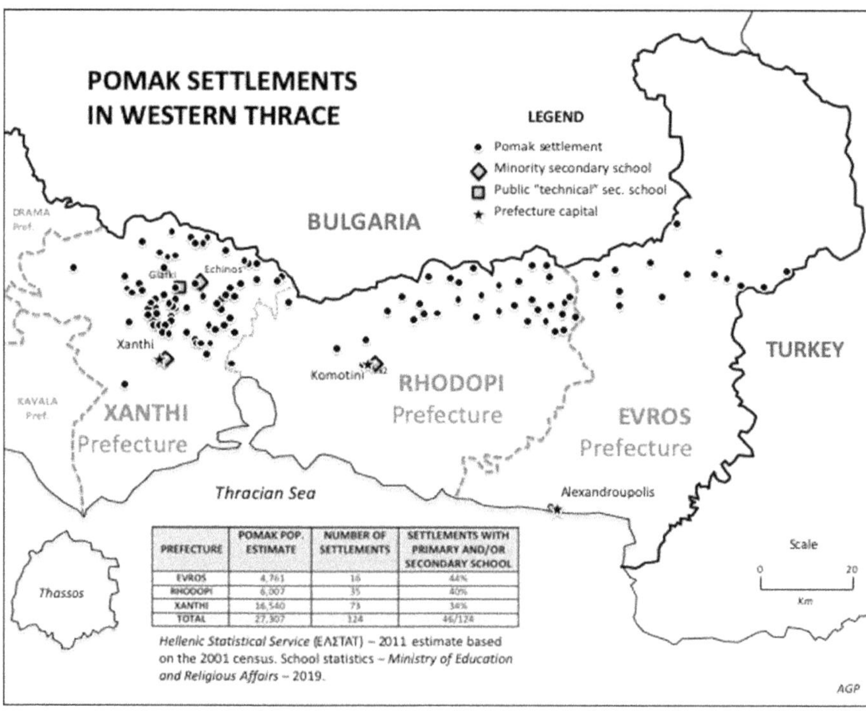

PREFECTURE	POMAK POP. ESTIMATE	NUMBER OF SETTLEMENTS	SETTLEMENTS WITH PRIMARY AND/OR SECONDARY SCHOOL
EVROS	4,761	16	44%
RHODOPI	6,007	35	40%
XANTHI	16,540	73	34%
TOTAL	27,307	124	46/124

Hellenic Statistical Service (ΕΛΣΤΑΤ) – 2011 estimate based on the 2001 census. School statistics – *Ministry of Education and Religious Affairs* – 2019.

Figure 6.1 Geographic distribution of inhabited Pomak villages, including census statistics.

Figure 6.2 Echinos is the largest settlement in Xanthi's Pomak-inhabited region. A historic epicenter of the region's tobacco industry, it now seeks to diversify its economy by capitalizing on European Union regional and social development funding (Source: authors).

and mobility in public space (Brunnbauer, 2012: 327; Varvounis, 2003: 276; Adamou, 2012: 15). Since the 1980s, significant numbers of minority Thracians have out-migrated to Thrace's plains, Greece's bigger cities, and abroad, including to Turkey. They have thus provided new sources of income to their village economies via flows of remittances from Germany, Sweden, and other developed European economies (Dimitriou, 2013: 17).

Pomak communities have long been economically disadvantaged, as also evidenced by the historical morphology of their settlements. Kostas Lalenis et al. note that the political-economic peculiarities of the three minority groups "has forced them to produce different spatial patterns," also how public and private (residential) spaces are constructed and used (Lalenis et al., 2012: 117). For example, some built environment variation can be attributed to the physical geographic context (Ibid, 134). Their relative isolation in the Rhodopi highlands has curated a vernacular morphology adapted to the mountain relief and distinguished by organic town plans featuring narrow and sinuous streets. Overall, they are more compact, dense, and neatly bounded, relative to their agricultural hinterland, than settlements of other minority groups in the plains of Thrace. Pomak settlements commonly have a single public space, the site of the village mosque, the school (if one exists there), a few cafes, and, possibly, a general store. Two- or three-story stone buildings with roofs made of wood distinguish their local vernacular architecture style. A class gradient was evident as homes range from very modest to extensive and elaborate (Ibid: 125–127). However, more influential on community welfare and settlement development have been the restrictive and punitive regulations imposed upon Pomak communities as part of the state's exclusionary minority policies (1936–1996). Thus, the built environment and the deficits and structural pathologies that distinguish it, reflected the political territorializing of Pomak identity and community as it has been unequally negotiated between Pomaks and the state since the 1930s. These conditions were greatly ameliorated after the lifting of the last restrictive regulations in 1996 (Figure 6.3).

Figure 6.3 Urban development in Echinos reflects the economic openness the Pomak region has enjoyed since the 1990s and the benefits of a strong remittance-based economy. Traditional housing is replaced by denser, often up-market, housing (Source: authors).

Pomak communities are predominantly located in the montane regions of the Prefectures of Xanthi and Rhodopi. There are also a few communities in the uplands of the Prefecture of Evros. Pomaks are predominantly speakers of a Bulgarian vernacular (now, with the unofficial blessings of the state, a formal script, Pomak language), and also speakers of Greek and increasingly of Turkish. Evangelia Adamou notes that language shift, in the case of Western Thrace's minorities, works counter to the dominant hypothesis, which anticipates that minority languages would experience [and be handicapped by] "the reduction of the functional domains in which they were traditionally used [i.e., the marketplace] and the addition of new domains [i.e., engagement with a distant regulatory state] from which [they] are excluded." Therefore, instead of a decisive shift to Greek, the official language of state administration, Romani and Pomak-speaking communities shifted to a local minority language, Turkish (Adamou, 2012: 10). The relative remoteness of their montane settlements and intermittent interaction with plains towns had, historically, limited their "institutional contact with Turkish, resulting in more traditional codes switching and emblematic lexical borrowings" (Adamou, 2010: 147).

Since the Treaties of Neuilly and Lausanne, they have been nationals and citizens of Greece and subject to these treaties, but also aggressively courted by Turkey, which was granted by the Treaty of Lausanne, a «droit de regard» over the affairs of the Muslim minority (as was Greece with regard to the Greek-orthodox (Rum) community of Istanbul). Geopolitical circumstance and Greece's ethnic identity politics have forced the Pomaks to choose among and negotiate three self-essentializing identity and historical imaginaries: "as Pomaks, as Turks, or as descendants of the ancient Greeks,"[…] "three basic processes of uncertainty" (Tsibiridou, 2004: 227; Herzfeld, 2009: 324).

Pomaks are Sunni Muslim. During the Interwar period, the overwhelming majority of Pomaks sided with the "paleo-" or "Old Muslims" (Lepida 2011: 90), but they have gradually transitioned from traditionalism to a general acceptance of Turkey's mixing of Kemalism and Islam (Akturk, 2009: 897). Their peripheral status in relative to the Greek polity and economy and the European Union, and their geographic situation and cultural liminality (with their Bulgarian idiom and Muslim religion pointing to sites of signification outside the Hellenic State) made them vulnerable to political pressure by the littoral states (Tsibiridou et al.). For the state, the intersection of the signifiers that define Pomak identity – Slavism and Islam – represented a perfect geopolitical storm: during the Civil War "the peril from the North" (ο εκ Βορρά κίνδυνος) described the threat of Communism territorialized as a menacing (and Slavic) Bulgaria. Perennially, but especially since 1923, Turkey represented "the peril from the East" (ο εξ ανατολών κίνδυνος). If Thrace's minority Turk community triggers thoughts of the latter, the Pomak community has been cast at different times as the point of convergence of both threat sources.

Pomaks as the Hellenes' "lost siblings"

At this inflection point, every effort was made to highlight the ethnic characteristics of Pomak identity. The objective was to differentiate Pomaks from the rest of the minority and especially from Muslims who chose to self-identify as minority Turks. The assertion that the Pomaks "have greater racial affinity to us than to the Turks" had already been expressed in the mid-1940s. In 1946, General Inspector of Foreign and Minority Schools, Athanasios Papaevgeniou, in his book *Northern Greece – Minorities,* articulated the relevant view: the Pomaks "are probably of the same race [as] … their physical characteristics, morals and customs, and the fact that they call themselves "Achrian" ("Αχριάν"), suggests that they are descendants of the ancient Thracians, the "Agriani"(p.23).[4] Gradually, this origins theory informed Greek authorities' Pomak policy and was promoted as the official view. With folk culture as further evidence, the Achrian-Agrian theory contributed one more story to Greek national mythology by making place and time for, and integrating the Pomaks into the Greek nation's imaginary.[5]

The state narrative below lays out the history of Pomak origins, not in the Pomak voice or with documented Pomak input. Greek and Bulgarian scholarship and United Nations fact-finding supplants them.

> The Pomaks live in the montane region of the Prefectures of Xanthi and Rhodopi. It is a race of undetermined origin. According to some, they are Bulgarians who were Islamized by the Turks following their occupation of Thrace. Others claim that they are descended from the tribes of the KUMANS and PECHENEGS [emphasis in the original document] who, during the eleventh century, crossed the Danube and settled in Bulgaria and Thrace, to be eventually Islamized. However, it is more likely that the Pomaks are the descendants of the ancient Thracians, the warrior tribe of Agrians. This is what they believe themselves today by calling themselves "AGRIAN." This view is further supported by a settlement in Xanthi by the same name («Agrian – Achrian»). These Agrians (Pomaks) accepted Islam between the 15th and 16th centuries to preserve life and keep property away from the Turks' reach. Their designation as "Pomaks" is attributed to Bulgarians, who called them «Pomagom» (пОМОщ) from the Bulgarian word for "help": in 1876, Agrians helped the Turks suppress a Bulgarian insurrection, during which many Bulgarians were slaughtered. However, others claim that the name Pomak is [derived from] the Turkish words "Pamu-ak" (white cotton) on the account that during their conquest of Thrace they observed Pomaks wearing white cotton leggings. According to the United Nations Special Committee on the Balkans (1948) Report, the Pomaks are considered, with respect to their ethnicity and language, as Bulgarians forcibly converted to Islam during the seventeenth century.[6]

Gradually, a supporting historical and ethnological literature emerged, which reproduced the narrative concerning the origins of the Pomaks and linked them to the Hellenic national narrative. A notable contribution to this tradition is Petros Theocharidis' book *Pomaks. The Muslims of Rhodopi (History, origin, language, religion, folklore).* Theocharidis was a majority Greek teacher in Echinos' Madrassa in the period 1964–1968. His book, an anthology of essays on Pomak folk culture, was completed in 1968, the year he applied to the Coordinating Council of Thrace for public financing of the publication. Despite an initial positive response, he did not receive funding. According to the minutes of the Council, the book's publication, scheduled for 1968, was at the initiative of the writer and not of a state body. Its publication was eventually canceled following the intervention of the Greek Foreign Ministry (Kostopoulos 2009: 90–91).

The reticence of the Council is revealing. Events in and about Thrace's minority can be best understood as systemic phenomena of Greek-Turkish geopolitics. Although Theocharidis' theory about the origins of the Pomaks, was consistent with the state's general strategy of driving a wedge between them and the broader minority, it also had the potential of further alienating Turkey, which considered all Muslims (Bulgarophone Pomaks included) a kin-group and a lever in its diplomatic relations with Greece. At the time, geopolitical tensions over Cyprus were at their highest. Unprepared, misinformed, and miscalculating Turkey's intentions about Cyprus' sovereignty, in September of 1967, the Junta called for a bilateral meeting at the symbolic Evros River crossing to discuss Cyprus' future. The meeting was disastrous. The Junta's proposal that Greece should annex Cyprus was rejected outright. Its misreading of Turkish foreign and security policies signaled to the Turkish delegation that the Colonels were entirely out of their depth (Sirigos, 2014: 316–318). Following further inter-communal trouble in Cyprus in October and November, Turkey was emboldened, and the Junta beat a retreat.

Theocharidis' manuscript was submitted to the Coordinating Council of Thrace during a crisis that could have escalated to another Greek-Turkish war – a prospect that the Junta greatly feared (Ibid). The Junta's opposition to Theocharidis' book was not limited to denying it funding, but extended to canceling its publication altogether. That decision was written into an agreement between Greece and Turkey regarding the status of Western Thrace's minority, and was formalized in the Protocol of the Greek-Turkish Educational Committee on 20 December 1968 (Kostopoulos 2009: 90–91). Ultimately, Theocharidis' work was published in 1995 by the Cultural Development Center of Thrace with the Thrace Fund's sponsorship.[7] It was on that year that the last "μπάρα" ("bára": road barrier), controlling road access to Xanthi's Pomak towns and villages, was removed. A year later, the book received the Academy of Athens' First Prize. The enthusiastic – if not over the top – review of Theocharidis' book by Georgios Pavlos, professor at Democritus University of Thrace, signals the mainstreaming of the thesis:

> The grand scientific and intellectual scope of this endeavor is made ever more manifest when one considers that teams of researchers backed by grants of many millions could not replicate Petros Theocharidis' findings. In contrast, working without fanfare and without financial support for a long time [and] drawing on his salary, he succeeds in completing a research program demonstrating the sort of scientific integrity, competence and originality rarely encountered in similar research undertakings [trans] (Pavlos for biblionet online, 1995).[8]

Theocharidis' fieldwork and research on Pomak culture, its scientific rigor notwithstanding, was undertaken in good faith and was, clearly, not commissioned by the state. His thesis attempts to address the two pillars of signification, necessary for the inclusion of Pomaks in Hellenic national ideology: language and religion. In his extension of the Hellenic national myth, he considers the Pomaks of Greek Rhodopi an aboriginal population of Thrace. They are of Aryan or Indo-European descent, as evidenced, he claims, by their physical characteristics, which differentiate them from the Bulgarians and the Turks. The latter belong to the Ural-Altaic Group and, per Theocharidis, are of Mongolian provenance (that racist claim has gained traction in Greek nationalist websites).[9] The self-identification of the Pomaks, or at least some of them, as "Achrian," is considered a vestige and evidence of their descent from the Achrians/Agrians of antiquity, one of the local Thracian tribes. He further alleges Achrians were Hellenized in the fourth century BCE and joined the military campaigns of Alexander the Great. The peoples of Rhodopi were Christianized at the same time as the rest of the Balkan region. Their fortunes were subsequently connected with the Byzantine Empire.

At that time the region of Rhodopi was called Achrido (Αχριδώ) and its inhabitants Achrians/Agrians or Rhodopians. The region's eventual conquest in 1344 and its incorporation into the Second Bulgarian Kingdom (1186–1393) precipitated the linguistic Bulgarization of its inhabitants (Theocharidis, 1995: 49–51, 74–75). The conversion of the Pomaks to Islam occurred during the reign of the Ottoman Sultan Mehmet IV (1648–1687) under orders of Grand Vizier Mehmed Köprülü (1656–1661). In response to their persecution by the Turks and the Bulgarians, the Agrians converted to Islam en masse to share the privileged status of that reigning community. In the period marking the emergence of Balkan nationalisms, from the second half of the nineteenth century onwards, the Pomaks generally demonstrated anti-Bulgarian sentiments. They greatly appreciated the religious freedom they enjoy as Muslims in Greece since Western Thrace became incorporated Greek territory in 1920 (Theocharidis 1995: 53–76, 373–378, 431).

A more in-depth study of the term "Achrian" reveals that it goes out of use at the beginning of the millennium. At the beginning of the fifteenth century, the term "achrian" appears in Ottoman sources in reference to all inhabitants of the Balkan region and describes uninitiated Muslims (those who have not had a proper Islamic education) as well as Islamized Christians. Its etymology might be related to either the medieval toponym of Rhodopi, Achrido, or to the Slavic word «hrid», meaning rock (Chatzidimitriou, 2005: 51, 74–78, Papadimitriou, 2008: 301–302, Kostopoulos, 2009: 136–137). The construct emphasizes origins and territoriality. Consequently, the Pomaks are incorporated into the Greek national imaginary as a native, ancient, Hellenized population of Thrace, Christianized alongside all other Greek populations in the first centuries of the Common Era. This view highlights Hellenicity and Christianity as the oldest and most foundational elements of Pomak identity. It renders this interpretation of Pomak identity more *historically* resilient against rival national imaginaries (specifically, the Bulgarian and Turkish), which are also claiming paternity over the Pomaks.

The Bulgarian national narrative, which privileges language as the keystone of Bulgarian ethnonationalism (and certainly so in including Slavic populations of Macedonia in the Bulgarian imaginary), considers Pomaks to be Islamized Bulgarians. The Turkish nationalist imaginary of Pomaks, much like its Greek counterpart, posits that the answer to who the Pomaks are rests on the discovery of their origins story. According to it, the Pomaks are an old Turkish people (some referencing the Pechenegs, others the Cumans but also other Turkish peoples) who arrived in the region ahead of the Seljuk and Ottoman Turks and chose to stay put. They gradually lose their early-Turkish mother tongue, while at the same time, they are Christianized – an inevitable consequence of their incorporation in the Byzantine Empire. Notably, this extension of the Turkish national myth concerning Pomaks, does not focus on the subsequent Ottoman legacy of the region and the fact that relationships among Muslim populations (regardless of language and origin story) are built during Ottoman dominion and social organization. Nor is it based on the Ottoman-era links between these peripheral populations and Turkey, the successor state of the Ottoman Empire (Chatzidimitriou, 2005: 78–80, Papadimitriou, 2003: 198–209, Tsibiridou, 2000: 56–59).

In sum, in Greek assimilationist discourses, like Theocharidis', the only door open to Pomaks desiring membership in charter society required tailoring Pomak culture and history to the measure of Hellenism. Pomaks are described as an Indo-European people, already Hellenized in the fourth century BCE, their fortunes linked to those of "Hellenism" during most of their "historical" life: Greek-speaking and Christian for centuries, their linguistic Bulgarization occurs later, while their Islamization is even more recent – over the past 350 years. Although the thesis suffers from significant contradictions and requires leaps of faith, and even more so, is scientifically questionable, Theocharidis' construct confers to Hellenism and

Christianity the requisite temporal and spatial primacy over Slavic culture and Islam. By contesting Pomak culture's Bulgarian and Islamic historical connections and, instead, claiming to reveal vital and long-standing connections to Hellenicity, Theocharidis' Pomak identity can be subsumed into Hellenic national identity as this was conceived and constructed in the nineteenth century by Konstantinos Paparrigopoulos and other nationalist scholars.

Exclusion of Pomaks as "suspect others"

During the Nazi Occupation of Greece, and because of the Pomak idiom's affinities to the Bulgarian language, the Bulgarian collaborationist state attempted to co-opt them with the ultimate objective of incorporating them and their region to the Bulgarian State. During the Greek Civil War, many mountain communities were occupied and used as strongholds by the revolutionary Left. In the years that followed its end in 1949, Pomaks started out-migrating from their mountain communities to the plains to improve their standard of living and avoid their involvement in the conflict. Out-migration was especially prevalent in the Prefecture of Xanthi. By migrating to towns and villages in the Xanthiot plains and to the Prefecture's capital, both of which were (and continue to be) settled primarily by the majority and by significant numbers of minority Turks, Pomaks encountered and selectively adopted aspects of Turkish culture.[10] Speakers of the Slavic Pomak vernacular, as well as Greek, many also adopted Turkish in the custom of Western Thrace's minority Turks. Pomak culture in villages and Pomak culture in the plains and urban centers became increasingly differentiated through contact with other minority communities – especially those who were self-identifying as Turkish (Trubeta, 2001: 87–98). Since the 1980s, Pomaks have been out-migrating to Turkey, Germany, the Netherlands, and Sweden, among other international destinations, and large cities in Greece, notably Athens (Antoniou, 2005: 86; Michail, 2003: 141–142).

In the post-Civil War era, Pomaks were gradually abandoning their pre-national Muslim identity and, to the consternation of the Greek State, were increasingly adopting Turkish national identity, as did other minority groups in Western Thrace. Until the mid-1950s, the measures suggested by Greek administrators in minority affairs focused on improving the Pomaks' living conditions and enhancing their access to education (Kostopoulos, 2009: 41–50). Accordingly, the state sought to staff schools in the Pomak region with teachers from the majority community and reopened in 1956 the County capital Echinos' Medresse. The aim of these interventions was twofold: to communicate the Greek state's interest in the Pomaks and affirm the centrality of the traditional religion, thereby limiting the spread of modernist ideas.

After 1955, following the anti-Greek riots in Istanbul and the deterioration of Greek-Turkish relations over Cypriot statehood, relations between the Greek State and the Muslim minority became more antagonistic. The Greek administration redoubled its efforts and expanded the range of measures in support of Pomak communities. Policy related to the Pomaks became a distinct part of any Greek minority strategy, the main aim of which was to treat Slavic minority communities differently from Muslim ones and leverage their linguistic and cultural diversity to geopolitical ends. The state appeared substantially less interested in assimilating them into Hellenism than denying their assimilation into Turkish national culture.

> Turkey's fundamental objective was – and is – the unification of the Minority [emphasis in the source], namely its Turkification. In contrast, our own effort, despite the lack of a clear policy, is to maintain the distinctiveness of the constituent communities of the Minority according to race (as Turks-Pomaks-Roma) and support the separation between, or even disrupting the merging of, Modernizers and Paleo-Muslims, that is, [securing] the non-unification of the Minority.[11]

In support of these objectives, the Greek administration carried out small-scale infrastructure projects in the Pomak region, distributed food and animal feed regularly to families experiencing shortages, and Pomaks were usually excluded from "repressive measures" (κατασταλτικά μέτρα) against the broader minority (Kostopoulos, 2009: 73–82). At the same time, however, the state sought to spatially segregate Pomak communities from the rest of the Prefecture and, especially, from settlements with Turkish populations, less to penalize them by limiting their mobility and more to limit, if not eliminate, their inter-communal interactions.

Concerning education, existing schools were supported and new ones built with housing provisions for majority community teachers who, as the case may be, relocated to the Pomak region from other parts of the country. A leap forward in funding the social state in the region would only take place in the 1990s. Grants and aids were made available to Pomak schools. The state provided support to minority teachers who self-identified as Pomak and were willing to teach in Pomak villages, thus barring the placement of minority Turks, perceived as Turkophiles (Iliadis, 2004: 56–59). Moreover, the state decided to financially support high performing pupils studying in public secondary schools and streamlined their entrance to Greek universities. The hope was that as graduates of Greek universities, they would form a pro-status quo elite, which would serve as a counterweight to the already formed minority Turkish elite (Kostopoulos, 2009: 85–86). Furthermore, the majority of the Pomaks studying at the Special Pedagogical Academy of Thessaloniki (ΕΠΑΘ) were considered by the Greek administration to be "loyal," that is to say they were willing to cooperate with Greek authorities and were not affiliated with contrarian and Turkophile elements in the minority community.

Importantly, the state attempted to dilute the influence of modernizer education elites by piloting a parallel school system targeting the Pomak population. The Coordinating Council of Thrace considered two plans for creating public high schools in the Pomak region; both operationalized outside the existing minority education framework. The first plan (1965) funded four gymnasiums (Middle School is the closest equivalent), including boarding houses for students, in the towns of Oraio and Echinos in the Prefecture of Xanthi, and in Organe and Kechros in the Prefecture of Rhodopi. Education in these schools would have been delivered in four grades, the first of which would have been preparatory with the main objective, students' mastery of the Greek language. Courses in the remaining three grades were to be taught in Greek, except for courses on religion for which the language of instruction remained unspecified. While the curriculum was tailored to the specifications of the public school system, as applied in the rest of the country, provisions were made to include courses on farming. The graduates of these schools would have been able to continue their studies in high school and university at state expense, as long as they demonstrated philhellenic sentiments. The expectation was that at least initially enrollments would be modest. The administration, however, was optimistic that numbers would grow over time. The government would have covered all costs related to this plan. The primary aim of the project was to "maintain the racial (φυλετική) distinctiveness of the Pomak race," since "otherwise there [was] a risk of wholesale erosion of the Pomak element at a significant cost to the nation". Ultimately, these gymnasiums were not established, but this plan was not abandoned altogether (Baltsiotis, 1997: 333–334).

A year later (1966), the Coordinating Council of Thrace considered a second plan, which provided for the establishment of two rural Gymnasiums, one in Xanthi and one in Rhodopi, along the same lines as in the previous plan, with one important distinction: while enrollment would, again, be for 4 years, the schools would include boarding facilities for students, the lessons would be taught in Greek, and the curriculum would include courses in farming. The Koran and other religious subjects would be taught in the Pomak language. The main objective in this case was to nurture "through such measures ... elite families with a purely Pomak conscience."[12]

A third plan (1972), this one conceived during the Junta years, visualized the establishment of an entry-level technical school in the town of Echinos in Xanthi. This plan was abandoned as well on the grounds that existing technical schools in the cities of Xanthi and Komotini could enroll any interested student. However, it was scuttled primarily because the state preferred such students to study at equivalent technical schools in other parts of Greece. Clearly, the political rationalization was that distance from Thrace and immersion in majority Greek culture would lessen the influence of the minority Turk community. The question arose, however, as to whether such transfers were feasible mainly for economic reasons. A move to southern or western Greece to attend technical school imposed a considerable, if not unbearable, a burden upon the families of minority students, who were not affluent. Further, a potential exodus of perspective minority artisans from the local labor market could be seen as an attempt to restrict competition for the benefit of majority artisans.

The state's containment of Turkish influence included the sequestration and segregation of Pomak communities from the rest of the prefectural territory and, especially, from settlements with Turkish populations. This strategy intended to limit, if not eliminate, Pomaks' interactions with the Turkish population in the plains. "Surveillance zones" extended across Greece's northern frontiers at different times since the Metaxas Dictatorship established them with Emergency Law 736/1936 "Concerning the security of fortified positions" («περί μέτρων ασφαλείας οχυρών θέσεων») (Figure 6.4). Pomaks and other members of the minority who inhabited the zone were subjected to Orwellian torments:

- Entry into surveillance zones was allowed only by police authorization. An "internal" passport, also granted by the police (the so-called "white papers"), was required
- Inhabitants of such zones needed to carry a special identification card in addition to the police identification card carried by

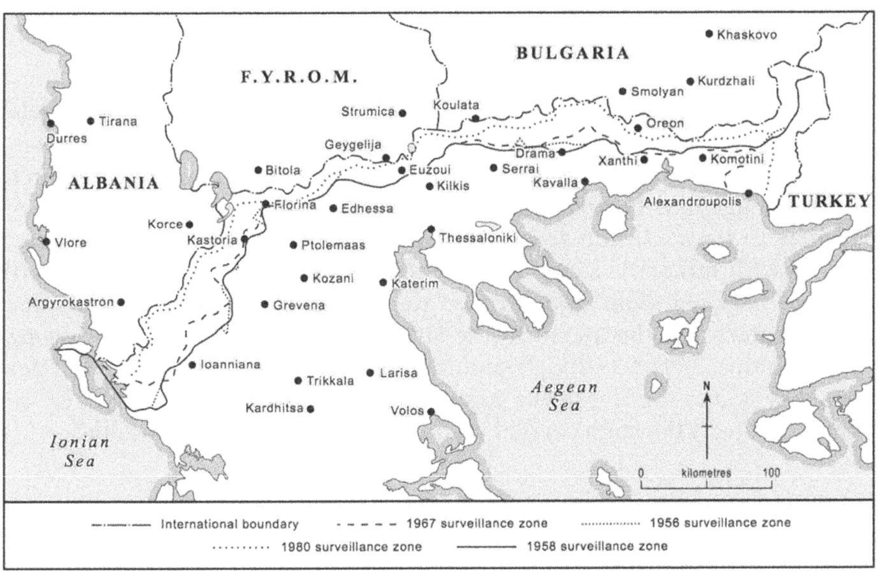

Figure 6.4 Exclusion zones extended across the entire northern borderlands of Greece. Since 1974, the exclusion zone has been gradually reduced in area. In 1996 it was finally abolished as a result of a significant policy turn with respect to the minority question and the state's perception of border security in the post-Cold War era (Source: Labrianidis, 1999).

all nationals. Presentation of a "loyalty certificate" (πιστοποιητικό κοινωνικών φρονημάτων), issued by the police, was required for every legal transaction in the area (1948–1974). Lois Labrianidis notes that such certificates were widely used in Greece at that period

- In-migration to surveillance zones was not allowed, and, further, inhabitants were not permitted to change their residence from one village to another (a restriction that directly undermined the Pomaks' patrilocal tradition and sabotaged economic development). Exception could be made by permission of the Prefecture, which was required to vet the request with the Committee for Military Security (ΕΣΑ)
- Inhabitants were not allowed to change professions unless they secured ΕΣΑ's permission
- ΕΣΑ was vested with the authority to ban from the zone, for five years, any person deemed suspect of compromising the security of the region
- All public servants were required to collaborate and assist ΕΣΑ in its mission under threat of dismissal
- A zone-wide curfew was to be observed every night from midnight to five in the morning
- Natural resources and property located in a surveillance zone were at the disposal and under the jurisdiction of ΕΣΑ and could be exploited (i.e., leased out) without complying with national regulations on competition (meaning that ΕΣΑ was rendered financially autonomous by exploiting state resources for the fulfillment of its policing and surveillance mission)[13]

Such zones were gradually abolished after 1974, following the Colonels' Junta's collapse and the reestablishment of democracy, except in Thrace. Access was controlled at different points across the regional road network. Signage and barriers (the infamous «μπαρα») marked the points of entrance to the exclusion zone. This apartheid regime restricted Pomaks' contact with minority Turks and, theoretically, averted the cultivation of relations between them. It controlled the former's travel to the plains and urban centers and the ascent of the latter to the montane Pomak villages (Kostopoulos, 2009: 99–105). It simultaneously segregated the Pomaks from the majority and the rest of the country, thus undermining whatever development projects the state carried out in support of their communities.

Nurturing and shepherding Pomak identity as policy pivot

Following the fall of the Junta in 1974 and throughout the 1980s, Greek authorities sustained that the Pomaks should constitute a community apart from the broader Muslim minority. Importantly, it continued their preferential treatment. Again, education policy was at the heart of the state's efforts to build a more accommodating relationship with the entire minority.

In the mid-1980s, four public high schools were established in different Pomak towns and villages in the Prefecture of Xanthi, and a fifth one in Rhodopi Prefecture. The official justification was that these schools provided alternative means by which the Greek administration would satisfy the Minority's demand for expanded minority secondary education. A sixth school planned for the town of Kechros, Rhodopi was not established. Given the general animus with which the state approached inter-community relations up to that point, the whole enterprise was perceived by local communities as suspect. However, after the change of Greek policy towards the minority in the 1990s, these schools were, indeed, established and flourished. The education infrastructure was further expanded with the establishment in 1995 of a Lyceum (high school would be the closest US equivalent) and a Technical or training High School in 2001, both in the Village of Glafki (Baltsiotis, 1997: 333–334).

The enduring policies of segregation and sequestration, apart from being undemocratic, were also not entirely effective. Case in point the voting score in eight Pomak villages in the Prefecture of Rhodopi during the Greek national elections of 20 November 1977. In five of these villages, Pomak voter support of majority community candidates was considered by the Greek administration to range from "satisfactory" to "very good." In the remaining three villages, however, support for majority candidates was nearly non-existent.

Most alarming was the trouncing of majority candidates in the Village of Ano Vyrsini, a significant settlement located very close to the Greek-Bulgarian border and considered «a bastion of the nation» (προπύργιο της Ρωμιοσύνης) in Greece's geopolitical imaginary. The cause for this electoral setback was geographic: state law ties voting to the voter's registered address. Unless a voter actively moves their registration, they need to return to their original locality to exercise their voting rights. Pomak voters who had out-migrated in past years to the Rhodopi plain and to the capital of the Prefecture, Komotini, and who were substantially Turkified, returned to their home village of Ano Vyrsini to exercise their voting rights, thereby upsetting electoral projections. Furthermore, minority candidates supported by the local minority teacher and the teacher's family surged in the poll. The returning Ano Vyrsini voters, therefore, their mind «polluted» by minority Turk beliefs, voted counter to custom. In contrast, secure in their belief of the majority's political dominance, local majority teachers did not extend themselves in support of majority candidates. In reaction to the election results, local authorities responsible for monitoring minority affairs proposed a range of measures, sure to alienate further and entrench the Pomak community:

> In order to win back the political ground we lost in the villages..., we must (a) rescind the [white] passcards[14] of Ano Vyrsini Pomaks, who reside in Komotini but vote in their [uplands] hometown. This way,

> we keep them from communicating with their peaceful and good-natured compatriots and eroding their national [convictions, spirit, or identity, implied]. (b) To impose sanctions on the representatives of the movement contrary to the nation, and especially [...] on the imam of the village and the minority teacher. (c) Replace the lazy, indifferent, and damaging to the national interest State [employed] teachers of the Pomak villages of Megali Ada and Sarakini. d) Cease the distribution of animal feed and other [provisions] to the Pomak families who have been protagonists in the anti-national campaign, both during the pre-election period and during the summer[.] [During this period] the Turkish Consulate General of Komotini and the Mufti attempted to subvert by all means of misinformation and intimidation the enrollment of new students at the Komotini Medresse and the [enrollment of the] graduates of [that institution] to the University of Thessaloniki[.] Their efforts failed because they encountered our sustained opposition, and new, effective legislation in support of minority members who embrace the Greek government with love and trust.[15]

The revanchist policies proposed by local majority community operatives in 1977 – three years after the return of democracy in Greece – revealed what will become a growing ideological divide between Athens and the conservative Thracian periphery in the 1990s. Hypervigilance against suspect behavior (often over something as trivial as enrolling in foreign language lessons) and the suppression of civil rights were a constant feature of regional governance under the Junta. Greece, however, experienced a seismic political paradigm shift after its fall. The establishment of free elections in a multiparty democratic system, the development of a liberal economy, and the desire to accede to the European Communities (a goal achieved by 1981) were increasingly incompatible with sustained sanctions and surveillance of the Pomak community. It took some time for change to occur.

Between 1979 and 1995, mobility restrictions in the Pomak region were gradually abolished. There were three sentry points and barriers («μπάρες») on the 28-kilometer road linking the capital of the Prefecture, Xanthi, and the large Pomak settlement of Echinos to the north. The barrier closest to Xanthi was taken down in 1979, the second in 1993, and the last one, 21 kilometers north of Xanthi, on 17 November 1995 during the visit to the Pomak region by the Minister of Defense, Gerasimos Arsenis.

Minister Arsenis ordered the Commander of the Fourth Army Corps, Lieutenant General Paragioudakis, to have the remaining barrier immediately removed. He said, «This barrier, which separated this region from the rest of Xanthi and isolated the villages, is, from this point on, history. Until today a special permit was required to cross it. Therefore, those who hold the «White Pass permits» should keep them as heirlooms of the past era. We live in a new era, and our country's security rests on social cohesion and society's economic development. Communication is free [unrestricted].»[16]

In the 1980s, Law N.289/76 "Concerning incentives and policies related to development of border regions" provided the legal basis for support of industrialization, commercial agriculture, and services, including the building of Democritus University of Thrace. Notably, it also mandated the Fourth Army Corps' creation, armed forces playing a significant logistical role in Western Thrace since 1919 (Academy of Athens, 1995: various). At the implementation stage, perennial deficits in national and regional governmentality resulted in few tangible benefits to the region, which, as we profiled earlier, remains to date a European and national economic periphery. Law N.1892 *(1990) "Incentives for the achievement of productive investment,"* articulated a strategic direction for the liberalization of the national economy mapped on the Single European Act. Significant Community and Greek government resources were earmarked for social and regional development without any territorial qualification. Peripheries "lagging behind," such as Western Thrace, were to benefit from extraordinary investment (Hellenic Parliament ΦΕΚ 1892). It was under the auspices of that law and other legislation supporting regional development that harassing and restrictive regulations targeting the minority were suspended (not necessarily repealed) in the 1990s (Labrianidis, 1999: 84).[17]

The 1990s saw positive policy shifts in other areas, notably in the removal of Article 19 of the Hellenic Nationality Code.[18] Promulgated in 1955, used after 1967 by the Junta to target Western Thrace's Muslims specifically, and abolished in 1998, Article 19 was used as a tool for bloodless ethnoracial cleansing. By racializing citizenship and the right to it, it differentiated citizens into two groups: those of Hellenic ethnic origin (ομογενής) and those of «other» origin (αλλογενής). It stated, "persons of other [non-Hellenic] origin leaving Hellenic territory without intent to return can be declared as having lost their Hellenic citizenship." As intent to return could not a priori be established, the arbitrary regulation rendered stateless approximately 60,000 Greek citizens, 50,000 among them were Thracian Muslims, some of whom were guest workers in Germany and other developed European economies (Sitaropoulos, 2004: 205–206; Kostopoulos, 2003: 53–54). It structurally excluded members of the minority from full access to the economy, civil association, and cultural expression at a more granular scale.

In the mid-1990s, the state realized that lifting the last restrictions on Pomak communities was necessary. While its strategy promotes social development as an argument for national cohesion, it has also encouraged, rather than officially sponsored, an assimilationist approach: defining and promoting a Pomak ethnolinguistic identity and grafting it to Hellenism as a means of incorporating the Pomaks into *national* society. Embarrassingly for the promoters of the strategy, the idea of Pomaks-as-Hellenes coheres to the neo-Nazi "Golden Dawn" party's ideology and modus. On 27 April 2016, Ioannis Lagos, Member of Parliament and vice-commander of "Golden Dawn," toured the Pomak town of Glafki "to support Hellene-Pomaks," and organized the event "Hellene-Pomaks say 'No' to Turkish propaganda."[19] Sparsely attended and much protested, but widely publicized, the visit accomplished "Golden Dawn"'s goal of promoting its racialized vision of Hellenism. The Pomak community did not rise to the bait. "Golden Dawn" no longer sits in the Hellenic Parliament, although populist, nationalist speech can be easily found in Greece's blogosphere and fringe press.

State and community interests intersect in Pomak heritage preservation: language, symbols, customs, and norms - Pomak social facts. There is support for formalizing the Pomak language, which, up to that point, existed as an oral tradition. There is an effort to collect and curate a broad range of folk culture, such as songs, tales, legends, and material culture. Press in the Pomak language covers topics that demonstrate culture distinctiveness. Moreover, assistance is available to set up associations, which would be active in all things Pomak. Nevertheless, there is still ambivalence among Pomaks about the sincerity of the state's commitment. On its part, the state continues to raise the bar of its commitment by supporting the right to education in the Pomak language.[20]

The heritage enterprise of revealing and then assembling cultural expressions of Pomakness into a concrete Pomak culture complex has been ontologically problematic. Its political intentionality notwithstanding, the invention of Pomak heritage was based on the presumed existence of a Pomak cultural type with discernable contours. For the observer, however, even the most remote montane Pomak settlements were continuously and synchronously exposed to Greek, Turkish, and Bulgarian culture signs, among many others. The wish of the strategy's promoters was to reveal 'authentic' Pomak culture as fundamentally distinct from the cultures of Romani Muslims and minority Turks. Notably, private Greek television networks did not serve Western Thrace until the launch of Channel R in Komotini in 1991, granting de facto monopolies to Greece's national ERT network and public and private programming in Turkish from across the border (Figure 6.5).

The default position of Greek governments has historically been that the cultural drift of the Muslim minority toward Turkey and Turkishness represents an existential threat to state and nation. Across the border, especially since 1955, Turkey has been providing structural support for that shift with diplomatic demarches, support of Turkophile associations, the marshaling of its press, and the expansion of Turkish satellite programming across the region. Pomaks are, at best, the minor partner in these efforts. Greek *statism* and Turkey's interventionism in the minority's affairs, undermine the civil right of a people to self-determination, a right both Greek and Turkish governments claim to support from their national perspective.

Mary Neuburger describes Pomakness as "inbetweenness," as it does not conform to the classic profile of national identities in the Balkans. Moreover, the absence of a significant mass movement among Pomaks advocating for the recognition of their Pomakness, has allowed Greece,

Figure 6.5 The endurance of Greek State radio and television monopolies until 1989, and the late launching of Greek satellite television ten years later, created an opening for the early introduction of Turkish satellite television in Western Thrace. Whereas we registered only Turkish satellite television dishes in Pomak villages in our 1996 survey, in 2019 the market has become more competitive with Greek and Bulgarian brands now available. The Pomak home receives satellite television programming from Greek, Bulgarian, and Turkish providers (Source: authors).

Turkey, and Bulgaria to variously contest its definition (Neuburger, 2004: 68). Caught between powerful and, historically, coercive national states, she writes, "Pomaks' fate in Balkan history seems to be primarily as pawn in Bulgarian and other Balkan national rivalries and domestic designs. Pomak history is, more often than not, the story of the centre looking to the margins and imposing its own designs (Neuberger, 2018: 181). Long subjected to shifting boundaries, forced loyalties, and capricious state policies, Pomaks developed "a dual-layered identity: Pomakness as a private identity and Turkishness as a public identity" (Boboc-Cojocaru, 2013: 344). The Greek State is attempting to intervene in this duality (also polarity) by two not entirely cohering strategies: first, by informally supporting efforts to incorporate Pomaks – historical contortions included – into the axiomatically exclusionary definition of Hellenism. Second, by writing them into a newer civil, as opposed to national, definition of the state, in which they are guaranteed rights and equality under the law. The former is patently contradictory to the latter, as inclusion in the ethnonational family of Hellenism should not be a prerequisite for full participation in civil society and the state. Notably, the Pomak heritage project is mainly territorialized in the Prefecture of Xanthi. Pomak settlements «preserve significant traditional social elements» (and therefore, are likely to resist the allure of a secular Turkey), unlike the more modern Pomak communities in the Prefecture of Rhodopi. The latter «can be described as possessing social characteristics of civil society, albeit reflecting local particularities, and is in a position to engage in a public conversation on its national identity and sense of belonging,» meaning they are much more exposed to informational flows from Turkey (Baltsiotis, Tsitselikis, 2008: 75). However, our 2019 fieldwork in the Xanthi's Pomak region suggests that accelerating social development in the region is gradually erasing such differences.

For their part, the majority of Pomaks see the heritage enterprise of Pomak national awakening with suspicion and mistrust. Pomak elites who ascribe to Turkish culture, reject it outright as a state stratagem. They criticize the majority's motives as disingenuous. Further, they claim that it creates, in practice, as many problems to the community's daily existence – by favoring one faction over the majority – as it claims to seek to remedy. For the time being, the absence of trust between the minority and the majority undermines the full enjoyment of minority rights. Consequently, confidence-building measures need to come before the further "top-down" policies for Pomak communities. Simply put, based on its track record, the state is now damned either way.

For the majority of Greeks outside Western Thrace, the "discovery" of the Pomak nation as a "forgotten sibling," the construction and promotion of that national identity as evidence of the state's beneficence, as well as the publicity that surrounded the effort, told a story of national reconciliation. Moreover, it accounted for how a minority identity could be integrated – however implausibly – into the national imaginary of Hellenism. By either silencing or disregarding dissenting voices and casting the Pomaks as a constituent community of the Hellenic nation, the majority could discount its collective memory of the Pomaks' religious Otherness and the record of state prosecution of the community. The fullness of Pomak worldviews, in its majority still critical of the state, does not receive airtime outside the region. Despite its inherent contradictions, the Pomak project on a complementary national narrative has held majority public opinion in thrall.

Hellenic nationalism was now "embracing its ancient, forgotten siblings," who lost the filial language and religion at some point in their long history. The encounter and reconciliation were read by majority society as a presumption of 'forgiveness of sins' for its past repression and exclusion of the Pomaks.

With significant daylight between the majority's perception of minority affairs in Western Thrace and conditions on the ground, it was hard for the state to claim quick success in its courting of the Pomaks. It has taken more than 20 years for a noticeable shift in tone to occur. In fairness, however, progress has been made in minority civil rights. Brutally put, "the transition from the pre-national to the national state of being has been more intimately tied to the concept of [national] interest than democracy" (Ibid). The reading of the Pomak past and its sense of place in terms of Greek history, and the leaps of faith and gaps in logic it represents, has not given pause to the majority public. In fact, the Hellenic reading of the Pomak past now holds sway in the public marketplace of ideas, its validity underwritten by discourse alleging scientific support.[21]

Whatever the state's opinion of the Pomak national awakening approach to reconciliation, Greece's liberal establishment, in and out of government, has strongly supported the civil society "turn" in the Europeanist mold. The Pomak heritage strategy notwithstanding, the state's stated objective has been to establish and support equality before the law (ισονομία) and equal citizenship rights (ισοπολιτεία) as expected in a healthy democracy and Member State of the European Union. However, the unresolved contradictions of minority policies represent a continuing challenge for Pomaks, as for the state and Greek civil society.

Pomak settlements in Xanthi today: Sminthi, Echinos, Medousa

Pomak communities have historically been overwhelmingly rural and economically marginal in their home prefecture, Greece, and the European Union – indeed, they represent a triple-nested periphery. Compared to other European Union regions, Eastern Macedonia and Thrace, as a combined NUTS2 unit,[22] ranked, in 2017, 284th out of 315 regions. The region's Gross Domestic Product (GDP) per inhabitant in Purchasing Power Standard (PPS) was EUR 11,500, equal to 46 percent of EU-28's mean value. The EU ranks the combined regions of Eastern Macedonia and Thrace 13th out of 13 Greek regions by the measure of GDP in PPS.[23] Zooming in to the prefectural level (NUTS3), we observe that Xanthi and Rhodopi are ranked 51st and 49th, respectively, out of 52 Greek prefectures. Evros fairs better, ranked 29th. Importantly, *within* these prefectures, the Pomaks have been, and to an extent, continue to be, especially disadvantaged.

Most Pomaks work in the primary sector – agriculture and husbandry – and they had historically developed a specialist regional economy based on the cultivation of tobacco (Trubeta 2001: 113–133, Aarbakke, 2000: 67–80, Tsibiridou, 2000: 45–56; Akgönul, 1999: 24–36). Scaling Pomak economic territoriality up to 'the global,' the Pomak economy was long-linked to tobacco brokerage houses in New York City, London, and Richmond, VA ("Turkish tobacco" brand often includes Thracian Pomak leaf tobacco) (Figure 6.6). Since the end of EU price supports for tobacco, that culture has been substantially diminished.[24]

Since the elimination of the last punitive regulations in 1996, there have been strong signs of social development, which have an impact on the Pomaks' perceptions of Greece and Turkey. In our field visits to the Pomak villages of Myki, Sminthi, Glafki, Echinos, and Medousa, we observed a

Figure 6.6 The Village of Myki is surrounded by small-scale tobacco farming lots and drying sheds. Individual tobacco growers do have much bargaining power when negotiating with brokers. Historically, growers' cooperatives have been an important remedy, but they have, also, become disempowered as tobacco prices have fallen substantially in the last ten years (Source: authors).

gradient of development. Situated twelve kilometers north of the Prefecture capital of Xanthi on good roads, Sminthi is one of the largest Pomak settlements in Xanthi with a 2011 population of 211, and the seat of the Municipality of Miki. The 2001 census recorded a population of 363. The significant reduction in population is consistent with our findings regarding internal and international out-migration. A large mosque on the left bank of Megalo Rema River dominates the traditional village. The center village has a vibrant retail ribbon with cafes (including The Espressionist Coffee Company store, which belongs to a small regional chain of coffee shops), restaurants, and electronics stores. Off the main street, there are several auto repair shops. There is significant investment in new construction. Therefore, lively commercial activity and new residential built fabric suggest that, first, barriers to entrepreneurship have been lifted and, second, capital has been flowing to the village from out-migrants in the form of remittances and other types of investment.

The Village of Medousa (Memkova in the Pomak language) is geographically remote, and compared to Sminthi and other in-network villages, still substantially traditional. The social state is present in the form of a minority elementary school. Without athletic facilities, soccer is played in the minuscule courtyard where a single set of goalposts is painted on the back wall. It is situated at the end of a tertiary-grade road in the far montane part of Xanthi's Pomak region and approximately two kilometers on foot, over the ridge, from the Greek-Bulgarian border. As most villages in the region, it has experienced significant population losses in the last twenty years (from 311 inhabitants recorded in the 2001 census to 195 inhabitants in the 2011 census). According to a local informant who runs the combination general store and cantina, the village has an enduring kinfolk relationship with the Pomak village of Zlatograd on the Bulgarian side of the border. The village has a considerable amount of historic built fabric in the form of whitewashed stone and stucco single- and two-story buildings. There is some evidence of abandonment; however, there is also evidence of investment in residential real estate, albeit of a much smaller magnitude than Sminthi or Echinos.

The Village of Echinos is situated approximately midway between Sminthi and Medousa on a quality secondary road (Figure 6.1). A monument to the fallen troops who defended the Echinos fort against German forces in April of 1941 sits on a prominent point overlooking the village. In 1936, the fort was one of the strong points of Metaxas' line of fortifications, which automatically situated Echinos in that era's exclusionary surveillance zone. One of the largest Pomak settlements, its population has seen a small increase between censuses (2,232 in 2001 and 2,486 in 2011). An extensive Islamic cemetery sits at the foreground between the municipal road and the town. Three mosques, one of which was expanding its community facilities at the time of the field visit, define the skyline. At the road junction that leads into the town, in a fringe belt, state services are on plain display: the modestly-sized Church of Agios Georgios and directly across the street, the police station, the branch office of the Hellenic Army, and the campus of the public minority high school.

The community president, Kopel Ahmet, profiled the town in a free-form interview as follows: since 1996, the local economy has diversified, although tobacco is still the region's signature crop. Its cultivation is in decline as market returns to the producer are very modest at EUR 2-3 per kilo. It remains a premium product sought after by tobacco manufacturers (notably, Philip Morris). With less acreage dedicated to its cultivation, other kinds of farming are increasing (cattle farming). Transhumant animal farming is still practiced, which means that the Greek-Bulgarian border is not an obstacle (made frictionless by a borderless European Union). Micro- and small-scale plots and the natural contours of the terrain make difficult economies of scale in cultivation. Individual tobacco producers work with intermediaries (brokers) from Xanthi, other parts of Greece, and abroad to sell their crops. Disagreements among tobacco producers undermined the development of cooperatives, thereby undermining the producers' bargaining power (hence the low price/unit).

Although Western Thrace ranks low in the economic and social development rankings of the European Union's statistical arm, EUROSTAT, the EU *as a whole* is a zone of prosperity. The European Union's cohesion, regional and social development, and Common Agricultural policies have had a significant impact on the region. Regional and rural development funding and financing are visible in the town, having underwritten projects, such as the construction of a waste disposal site in the periphery of the town. The EU has also supported the development of hot springs tourism in the region, although there is little tourism investment in the town proper. Most importantly, Common Agricultural Policy (CAP) funding in the form of payments for soil conservation and animal welfare has been the salvation of Pomak farmers, who would otherwise not be viable as producers.

There has been out-migration from Echinos to Xanthi, to the rest of Greece, and abroad (mainly Germany) as in other Pomak settlements. Given capital scarcity, remittances from abroad have been a very significant source of capital for Echinos.

Mr. Ahmet noted that the state's facilitation of students' access to higher education has been critical to the community. The importance of this emerging class of entrepreneurs and professionals is not lost to Echinos. The President remarked that they do come back to visit and the interaction invigorates the community.

In closing, Mr. Ahmet alluded to difficult times in the past and remarked that the situation in which Pomaks found themselves today has much improved, the prolonged recession notwithstanding. He remarked that further diversification of local and regional economies are urgently needed. Tourism is one prospective industry of interest. When asked if the community wanted tourism (as opposed to needing it), Mr. Ahmet answered in the affirmative. There are, however, no hotels planned at the time of the visit.

We discussed the loss of a substantial part of the village's traditional built fabric to redevelopment and the role historic vernacular architecture may have played in attracting tourism. He recommended a visit to Medousa to witness what traditional settlements looked like in the past.

Mr. Ahmet went to great lengths to impress upon us that Echinos is a peaceful and trouble-free community. They cherish the pristine state of their physical environment. He attributed the absence of forest fires to their stewardship of forests. Nonetheless, there is concern that the community (or, rather, none of their communities) has a fire department. He made no direct references to the Greek State, politics, Turkey, or civil rights.[25]

The interview and other field-based information suggest that political and economic liberalization since 1996 and Greek State and European Union structural programs and resources have reduced the appeal of a Turkish cultural and political turn.[26]

Conclusions

The relationship between the Greek State and Western Thrace's Muslim minority traces Greece's twisting road from illiberal and nationalist to democratic and civil statehood. It also examines the State's willingness to rework its historical and territorial imaginaries of Hellenism to accommodate differences. On the one hand, regional security concerns – traditionally, the "perils" from the North and East, and more recently, the "Muslim arc" – have tested that resolve. On the other hand, democratization, since 1974, and political and civil Europeanization, since 1992, have resulted in "stop-and-go" liberalization by the majority political establishment. In part compelled by the European acqui and in part responding to modernizing norms established under prosperity, the state has been establishing in law, and increasingly in practice a more inclusive politics of difference.

For majority and minority communities in Thrace, the starting point of this journey was set during the Balkan Wars and then the First World War at the closing of the age of Europe's Great Power geopolitics. The Greek-Turkish War and the Treaty of Lausanne re-territorialized identity politics in partitioned Thrace, not based on either Wilsonian principles of self-determination or nativist-national ideologies, as one should have expected, but by retreading the Ottoman *Millet* system (a neo-Millet system, perhaps) in which religion was again the operative signifier of inter-communal and politics and geopolitics of difference.

By obscuring and re-scripting ethnic as religious difference, the "Lausanne" regime provided geopolitical stability in the region until the Cyprus crisis and the Istanbul anti-Greek riots of 1955 upended it. The eviction of nearly the entire Greek-Orthodox community of Istanbul undermined the numerical balance and legal equity that religious minorities enjoyed under "Lausanne" in Greek Thrace and Istanbul. Political modernization/secularization with strong nationalist undertones pulsing into Western Thrace from Turkey, divided Western Thrace's Muslim community into traditionalists (paleo- or Old Muslims) and modernizers – the former cleaving to the "Lausanne" reading of their political-cultural positionality in the Greek State and the latter cleaving to Turkey, which was promising opportunities for belonging and social development, seemingly unobtainable in Greece.

In the face of this ontological transformation of the region's geopolitics, Greece continued to steadfastly support the old school traditionalists, as well as follow the letter of "Lausanne." It simultaneously excluded the minority from the full benefits of majority society, economy, and polity. Siding with traditionalists and denying the existence of nationalist forces in the minority was a losing proposition as the tide was decidedly in favor of Turkish modernizing forces. After the fall of the Colonels' Junta (1967-1974), Greece started elaborating a more nuanced minority policy that progressively integrated – albeit, first, reluctantly – an understanding of the difference and Minority diversity.

Importantly, until the 1990s, there was scant evidence that the Greek State had any interest in integrating the Pomaks into Hellenism either in the Pomaks' terms or in terms of a civil society standard. In the 1990s, the Greek State saw opportunity in the potential transformation of the Pomaks into an ethnicized group. The Pomaks' linguistic differentiation from other Muslim minority communities, their reduced social and economic circumstances, and their persistently strong relationship to Islam (at least among older Pomaks), provided levers for influencing their social-political positionality. It also made it possible to draw them away from Turkish State agency and the influence of minority Turkish modernizers. The state encouraged actions that differentiated the Pomaks from other groups in the Muslim minority (and in particular the minority Turks) in official political speech. It highlighted their cultural difference with particular reference to their linguistic distinctiveness and folk culture (Theocharidis, 1995). Its Pomak policy (or strategy) supporting the community's national awakening was introduced in the 1990s and is still in effect. In sum, it has demonstrated the State's acceptance, however problematic in its articulation, of ethnic, as opposed to only religious, diversity in the region.

What has been especially vexing regarding the state's civil rights-based policy for the Pomak community, which draws on the principle of self-determination, is its à la carte application: other communities of difference remain submerged. Beyond Greek Thrace, principally in Macedonia, Greek citizens who self-identify as members of a Slavophone community and want to be recognized have a basis for challenging norms, processes, and procedures that refer to their reduced civil state. However, contestation of their exclusion "from below" is a great challenge as there are neither channels nor forums, except, perhaps, at the European level, where contestation can occur. Voicing support for Pomakness while staying silent about the memory, cultural distinctiveness, and right to self-determination of Others, continues to imagine Greece as a homogeneous Helleno-Christian state. It is a challenge to Greece's commitment to liberal democracy and the promotion of equality in a civil society.

Concerning Greece's regional security, Muslim minority policy can be mapped at the intersection of two geopolitical imperatives: first, the containment of an expansionist "Neo-Ottoman" Turkish foreign policy, which challenges Greece's territorial integrity and social cohesion. Second, Greece's deepening commitment to Europeanization as a process for continuing modernization and – as we discussed in Chapter 1– as a potential bulwark of its territorial security.

Embracing difference axiomatically and inviting all Greek citizens into Hellenism's tent would serve democratic and realist ends as groups secure a stake in the state's stability and development. In contrast, by linking national security to a nationally-imagined state paradigm, Greece, like other national states, sees majority-minority relations as a zero-sum game: retreat from the historically- and geographically-grounded national imaginary presents dangers to the territorial and cultural integrity of state and nation.

Notes

1. Ikmet's settlement-level statistical data challenge earlier estimates of 36,000-39,000 Pomaks (Vakalios), 1997; Kostopoulos, 2009). Out-migration to Turkey and labor markets in the EU is the likely cause of population decline.
2. Source: EUROSTAT, JRC, and European Commission, Directorate-General for Regional Policy.
3. «Απογραφή Πληθυσμού - Κατοικιών 2011». Ελληνική Στατιστική Αρχή. 1 Μαΐου 2011 [Census of Population and Buildings, 2001. Hellenic Statistical Authority, 1 May 2011]. Athanasios Karafyllis, 2006, also, http://www.eriande.elemedu.upatras.gr/eriande/synedria/synedrio4/praktika1/karafyllis.htm
4. [Trans.] Athanasios Papaevgeniou, Memorandum concerning the presentation of Mr. D. Bakalbases, former [government] Minister, regarding the needs of the Turkish minority in Thrace; to the attention of the Ministry of National Education and Religious Affairs, protocol no. E178, σ. 7, ΑΣΓΜΣ φ.95.
5. In the 1960s, the Hellenic Folklore Research Centre of the Academy of Athens (Κέντρο Ερεύνης Ελληνικής Λαογραφίας), της Ακαδημίας Αθηνών authorized ten research collections on Pomak topics. The 1967 budget of the Council of Thrace earmarks 100,000 drachma for "Studies on the language and history of Pomaks" («*Μελέται γλώσσης και Ιστορίας Πομάκων – 100.000* [δρχ.].» Council of Thrace, *Proceedings*, 49th Congress, 2 December 1966, p. 9, (ΑΣΓΜΣ φ.10) in Kostopoulos, 2009: 88–90.
6. [Trans.] Directorate-General for Aliens. Report concerning the Muslims living in Greece, 31 July 1952, pp. 1-2, ΑΣΓΜΣ, φ. 95. (Γενική Διεύθυνση Αλλοδαπών, «Έκθεσης περί των εν Ελλάδι βιούντων Μουσουλμάνων», 31.7.1952, σ. 1–2, ΑΣΓΜΣ, φ. 95).
7. The Cultural Development Centre of Thrace (Πολιτιστικό Αναπτυξιακό Κέντρο Θράκης) and Thrace Fund are two associations founded in 1992 and 1994, respectively, and headquartered in the city of Xanthi. They are both very involved in the promotion of a distinct Pomak culture and identity.

8. Biblionet.gr/book/103888/Θεοχαρίδης,_Πέτρος_Δ./Πομάκοι. *Biblionet*, a public-private partnership, was created in 1998 by the *National Book Centre of Greece* (Εθνικό Κέντρο Βιβλίου), a syndicate of Greece's most important presses and publishers' associations. It is underwritten by the private *Hellenic Foundation for Culture* (Ελληνικό Ίδρυμα Πολιτισμού) the official website of which states that it was founded in 1992, is based in Athens, and its "main aim is to promote Greek culture and language throughout the world." *Biblionet* is co-sponsored by the *National Documentation Center* (Εθνικό Κέντρο Τεκμηρίωσης) the mission of which is "to collect, document, manage, disseminate and preserve quality digital content and data produced by the Greek scientific, research and cultural communities. The national infrastructure, which we develop, ensures access to reusable knowledge, the collection of data and the production of indicators and statistics for Research, Development, Technology, and Innovation in Greece." Between the Athens Academy's First Prize and Georgios Pavlos' enthusiastic review, one could, finally, claim that Theocharidis' work has all but the official imprimatur of the State.
9. A sample of websites that repeat the claim that Pomaks are an aboriginal Indo-European group and racially distinct from Turks and Bulgarians includes http://www.nostos.com/history_pomakoi.htm, https://ireporterworld.com/?p=7036, https://www.slideshare.net/NikitasVougiouklis/ss-64786825, https://clioturbata.com/uncategorized/pomakoi/ among many others.
10. As we noted in the previous chapter, we follow here the term "Muslim Turk" to describe those who the State names as "Turks by origin," as this is the term most widely accepted by the academe.
11. Συντονιστικό Συμβούλιο Θράκης, «Πρακτικό της 58ης Συνεδρίασης της 13.8.1968», ΑΣΓΜΣ φ. 11, σ. 4–5 [Council of Thrace, «Proceedings of the 58th Meeting, 13 August 1968» ΑΣΓΜΣ φ. 11, σ. 4–5.
12. Συντονιστικό Συμβούλιο Θράκης, Πρακτικό της 43ης Συνεδρίασης της 4.3.1966, σ. 10, ΑΣΓΜΣ φ. 10. [Council of Thrace. Proceeding of the 43rd Meeting, 4 March 1966, p. 6, ΑΣΓΜΣ φ. 10].
13. Labrianidis, 1999: 83–84.
14. These are, essentially, internal passports checked by the police at the exclusion zone sentry points.
15. Β΄ Εκπαιδευτική Περιφέρεια Μειονοτικών Σχολείων Ροδόπης-Έβρου, «Παροχή πληροφοριών», προς Υπουργείο Εθνικής Παιδείας και Θρησκευμάτων/Διεύθυνση Ξένων και Μειονοτικών Σχολείων, Τμήμα Α΄, αριθμός πρωτοκόλλου: ΕΠ 2, σ. 2, ΑΣΓΜΣ φ. 33 [Education District B – Minority Schools of Rhodopi and Evros, "Report" to the Ministry of National Education and Religious Affairs/Directorate of Foreign and Minority Schools, Section A, protocol no. ΕΠ 2, σ. 2, ΑΣΓΜΣ φ. 33].
16. https://www.efsyn.gr/stiles/apopseis/133691_synora-entos-synoron [accessed 18 July 2019]. Also, Vasilis Kollaros, "When the barriers fell in Thrace" [Trans.], https://slpress.gr/ethnika/otan-epeftan-oi-mpares-sti-thraki/ [accessed 08 August 2019].
17. As noted by Labrianidis, the following restrictive laws were not repealed: Compulsory Law (CL) 376/1936, CL 1237/1938, CL 1405/1938, CL 1921/1939, CL 1839/1939, Presidential Decree (PD) 397/194, L. 2433/1953, L.1952, PD 604/1982.
18. Current Hellenic Nationality Code: http://www.ministryofjustice.gr/site/kodikes/Ευρετηριο/ΚΩΔΙΚΑΣΕΛΛΗΝΙΚΗΣΙΘΑΓΕΝΕΙΑΣ/tabid/256/language/el-GR/Default.aspx. Article 9 of Law 3838/2010 (Α΄ 49/24.3.2010), in the current Code, regarding conditions under which a citizen may lose their citizenship, has replaced Article 19.

19. News of the event was carried by many national and regional press and television outlets, including Turkish news aggregator site *Tourkika Nea*: https://tourkikanea.gr/thraki/xa-20/ and popular Xanthi2.gr https://xanthi2.gr/2016/04/27η-κεερφα-εναντίον-της-επίσκεψης-χα-στη/
20. A notable body of literature that reflects these positions has gradually emerged since the 1990s (Kostopoulos, 2009: 117-144, 155-199, 224-232).
21. The concept of Pomak Hellenism and its historical basis are treated sporadically in postgraduate studies and doctoral dissertations as fact, as in Chatzidimitriou, 2005: 67-84 Liazos, 2007: 15-39, 133-143, Rovola, 2005: 49-63, Papargyriou, 2007: 9-16, 66-69, Kouidis, 2009: 38-48 and Hatzikos, 2009: 16-38.
22. The European Union's Nomenclature of Territorial Units for Statistics (NUTS) is expressed at three scales: NUTS1: Member State, NUTS2: subnational regions (which draw on but not fully duplicate national administrative divisions or departments), and NUTS3: Prefectures.
23. Rankings are based on EUROSTAT's Nomenclature of Territorial Units for Statistics (NUTS) data. "The PPS (purchasing power standard) is an artificial currency that considers differences in national price levels. This unit allows meaningful volume comparisons of economic indicators over countries. "Aggregates expressed in PPS are derived by dividing aggregates in current prices and national currency by the respective Purchasing Power Parity (PPP)" (EUROSTAT).
24. https://ec.europa.eu/agriculture/tobacco_en
25. Unstructured interview of Community President, Kopel Ahmet, conducted by Alex Papadopoulos in Echinos, Prefecture of Xanthi, on 1 September 2019.
26. One telling landscape queue concerns the brands of Echinos' numerous satellite television dishes. In our field visits in 1996, we observed satellite television infrastructure that was entirely of Turkish provenance. In our latest visit in 2019, the vast majority of such infrastructure was stamped "Nova" and "Cosmote," for Greek providers of satellite television.

Sources

Aarbakke, Vemund. *The Muslim Minority of Greek Thrace. Unpublished doctoral dissertation* Vols. 1-2. University of Bergen, Bergen 2000.

Adamou, Evangelia. "Bilingual Speech and Language Ecology in Greek Thrace: Romani and Pomak in Contact with Turkish." *Language in Society* Vol. 39, No. 2 (2010): 147–71.

Adamou, Evangelia. "Social Networks in Greek Thrace: Language Shift and Language Maintenance." In *Balkan Encounters: Old and New identities in South-Eastern Europe, Slavica Helsingiensia*. Edited by J. Lindstedt and M. Wahlström, 7–32. Helsinki, (2012).

Akgönul, Samin. *Une Communauté, Deux États: La Minorité Turco-Musulmane de Thrace Occidentale*, Istanbul : Εκδ. Isis, 1999.

Akturk, Sener. "Persistence of the Islamic *Millet* as an Ottoman Legacy: Mono-Religious and Anti-Ethnic Definition of Turkish Nationhood." *Middle Eastern Studies* Vol. 45, No. 6 (2009): 893–909.

Antoniou, Dimitris. "Western Thracian Muslims in Athens." *Balkanologie* Vol. IX, No. 1-2 (December 2005): 79–101.

Boboc-Cojocaru, Alexandru. "The 'Other Within' in the Balkans – The Case of Pomaks." In *Studies on Literature, Discourse and Multicultural Dialogue – Section: Histor.* Edited by *Iulian Boldea. pp.* 334–346. Târgu-Mureş, România: Editura Arhipelag XXI, 2013.

Brunnbauer, Ulf. "Families and Mountains in the Balkans: Christian and Muslim Household Structures in the Rhodopes, 19th–20th century." *The History of the Family* Vol. 7, No. 3 (2012): 327–350.

De Jong, F. *Names, Religious Denomination and Ethnicity of Settlements in Western Thrace: A Supplement to the "Ortsnamenkonkordanz Der Balkanhalbinsel"*. Leiden: Brill Archive, 1980.

Demetriou, Olga. "Prioritizing 'Ethnicities': The Uncertainty Of Pomakness In The Urban Greek Rhodope." *Ethnic and Racial Studies* Vol. 27, No. 1 (2004): 95–119.

Dimitriou, Olga. *Capricious Borders: Minority, Population, and Counter-Conduct Between Greece and Turkey*. Oxford, UK: Berghahn Books, 2013.

Frangopoulos, Yiannis. (Doctoral dissertation) Une minorité musulmane en Transition: Approche anthropologique des Pomaques Grecs, Université Catholique de Louvain, Faculté des Sciences Economiques, Sociales et Politiques, Nouvelle série-No 273. Louvain-la-Neuve : 1996.

Herzfeld, Michael. "Polyglot Perspectives." *Anthropological Quarterly* Vol. 82, No. 1 (2009): 311–30.

Lalenis, Kostas; Frangopoulos, Yannis; Kiosses, Ioannis; Kipouros, Stefanos. "Ethno-Cultural Diversity, Social Structures and Urban Organisation in Minority Villages of Thrace." *Γεωγραφίες* Vol. 20 (2012): 117–138.

Labrianidis, Lois. "The Impact of the Military Surveillance Zone on the Greek Side of the Bulgarian - Greek Borderlands."*Boundary and Security Bulletin* Vol. 7, No. 2 (1999): 82–93.

Michail, Domna. "From 'Locality' to 'European Identity': Shifting Identities Among the Pomak Minority in Greece." *Ethnologia Balkanica* Vol. 07 (2003): 140–157.

Neuburger, Mary. "Pomak Borderlands: Muslims on the Edge of Nations." *Nationalities Papers* Vol. 28, No. 1 (2018): 181–198.

Neuburger, Mary. *The Orient Within: Muslim Minorities and The Negotiation of National Identities*. Ithaca: Cornell University Press, 2004.

Sitaropoulos, Nicholas. "Freedom of Movement and the Right to a Nationality v. Ethnic Minorities: The Case of ex Article 19 of the Greek Nationality Code." *European Journal of Migration and Law* Vol. 6 (2004): 205–223.

Tsibiridou, Fotini. *Les Pomak dans la Thrace Grecque: Discours Ethnique et Pratiques Socioculturelles*. Paris: L'Harmattan, 2004.

Tsitselikis, Konstantinos. "The Legal Status of Islam in Greece." *Die Welt Des Islams*, New Series Vol. 44, No. 3 (2004): 402–31.

Varvounis, Manolis. "Historical and Ethnological Influences on the Traditional Civilization of Pomaks of the Greek Thrace." *Balkanica* Vol. 34 (2003): 268–283.

Ακαδημία Αθηνών (Academy of Athens). *Η Ανάπτυξη της Θράκης. Προκλήσεις και Προοπτικές* (The development of Thrace: Challenges and perspectives), Αθήνα: Ακαδημία Αθηνών: 1995.

Βακαλιός (Vakalios), Θανάσης. *Το Πρόβλημα της Διαπολιτισμικής Εκπαίδευσης στη Δυτική Θράκη: Η Περίπτωση της Μουσουλμανικής Μειονότητας με Εμφαση στους Πομάκους: Ερευνα (The problem of intercultural education in Western Thrace: The case of thee Muslim minority with special regard to the Pomaks)*. Αθήνα: Guttenberg, 1997.

Ηλιάδης (Iliadis), Χρήστος. *Η Εθνική Ταυτότητα της Μουσουλμανικής Μειονότητας και η Εκπαιδευτική Πολιτική. Μελέτη Αρχειακών Πηγών. 1945-1967* (The national identity of the Muslim minority and educational policy. Studies of archival sources 1945-1967), Μεταπτυχιακή διπλωματική εργασία, Εθνικό και Καποδιστριακό

Πανεπιστήμιο Αθηνών, Πρόγραμμα Μεταπτυχιακών Σπουδών «Πολιτική Επιστήμη και Κοινωνιολογία.» Αθήνα, 2004.

Θεοχαρίδης (Theocharidis), Πέτρος. Πομάκοι. *Οι Μουσουλμάνοι της Ροδόπης (Ιστορία, Καταγωγή, Γλώσσα, Θρησκεία, Λαογραφικά)* (The Muslims of Rhodopi (history, origins, language, religion, folklore), Ξάνθη: Πολιτιστικό Αναπτυξιακό Κέντρο Θράκης, 1995.

Καρά Χουσεΐν (Kara Hussein), Ικμέτ, *Γεωγραφική Απεικόνιση των Πομακικών Πληθυσμών στον Ελλαδικό Χώρο* (Geographical profile of Greece's Pomak populations), Πτυχιακή εργασία, Πανεπιστήμιο Αιγαίου, Τμήμα Γεωγραφίας, Μυτιλήνη, 2017 (https://docplayer.gr/73114394-Geografiki-apeikonisi-ton-pomakikon-plithysmon-ston-elladiko-horo.html Accessed on 8 August 2020).

Καραφύλλης (Karafyllis), Αθανάσιος, «Το Προφίλ του/της Έλληνα/νίδας Μουσουλμάνου/ας Φοιτητή/τρια του Παιδαγωγικού Τμήματος Δημοτικής Εκπαίδευσης του Δημοκρίτειου Πανεπιστημίου Θράκης» (The profile of a Greek Muslim student in the Department of Primary Education at Democritus University of Thrace), στο *Πρακτικά του 4ου Διεθνούς Συνεδρίου «Ιστορίας της Πανεπιστημιακής Εκπαίδευσης» του Παιδαγωγικού Τμήματος Δημοτικής Εκπαίδευσης του Πανεπιστημίου Πατρών*, 2006 (http://www.eriande.elemedu.upatras.gr/eriande/synedria/synedrio4/praktika1/karafyllis.htm Accessed on 8 August 2020).

Κουτσούκος (Koutsoukos), Βασίλης, *Από την Αυτοκρατορία στο Εθνικό Κράτος: Χωρικές Οψεις της Ενσωμάτωσης της Θράκης* (From empire to nation-state: Spatial aspects of Thrace's integration), Διδακτορική διατριβή, Πανεπιστήμιο Κρήτης, Φιλοσοφική Σχολή, Τμήμα Ιστορίας και Αρχαιολογίας, 2012 (https://elocus.lib.uoc.gr/php/pdf_pager.php?filename=%2Fvar%2Fwww%2Fdlib-portal%2Fdlib%2F2%2F4%2Ff%2Fattached-metadata-dlib-1348563823-156717-2883%2F20120627_Koutsoukos_Vasilis_Phd.pdf&rec=&do=&width=595&height=842&pagestart=1&maxpage=839&lang=el&pageno=1&pagenotop=1&pagenobottom=765 Accessed on 8 August 2020)

Κωστόπουλος (Kostopoulos), Τάσος. «Αφαιρέσεις Ιθαγένειας. Η Σκοτεινή Πλευρά της Νεοελληνικής Ιστορίας (1926-2003)» (Deprivation of citizenship. The dark side of modern Greek history (1926-2003)). *Σύγχρονα Θέματα* Vol. 83 (2003): 53–75.

Κωστόπουλος (Kostopoulos), Τάσος. «Η Κληρονομιά του "Εσωτερικού Εχθρού" – Αφαιρέσεις Ιθαγένειας από το Ελληνικό Κράτος (1927-2003)» (The legacy of the 'internal enemy' – Deprivation of citizenship by the Greek State (1927-2003)) στο *Ιστορία και συγκυρία της ελληνικής Ιθαγένειας*, Αθήνα: Ελληνική Ένωση για τα Δικαιώματα του Ανθρώπου και του Πολίτη, 2005.

Κωστόπουλος (Kostopoulos), Τάσος. *Το «Μακεδονικό» της Θράκης. Κρατικοί Σχεδιασμοί για τους Πομάκους (1956-2008)* (Thrace's 'Macedonian Question.' State plans concerning the Pomaks (1956-2008). Αθήνα: Βιβλιόραμα, 2009.

Μπαλτσιώτης, (Baltsiotis) Λάμπρος; Τσιτσελίκης (Tsitselikis), Κωνσταντίνος. «Η Μειονοτική Εκπαίδευση της Θράκης: Νομικό Καθεστώς, Προβλήματα και Προοπτικές» (The minority education of Thrace: Legal status, problems and perspectives). In Θάλεια Δραγώνα (Dragona) and Άννα Φραγκουδάκη (Frangoudaki) (eds.), *Πρόσθεση, Οχι Αφαίρεση - Πολλαπλασιασμός, Οχι Διαίρεση. Η Μεταρρυθμιστική Παρέμβαση στην Εκπαίδευση της Μειονότητας στη Θράκη.* Αθήνα: Μεταίχμιο, 2009: 59–90.

Μπαλτσιώτης (Baltsiotis), Λάμπρος, «Ελληνική Διοίκηση και Μειονοτική Εκπαίδευση στη Δυτική Θράκη. Αιτιότητες και Αντιφάσεις μιας Εκπαιδευτικής Πολιτικής» (Greek administration and minority education in Western Thrace. Causes and contradictions in educational policy). In Κωνσταντίνος Τσιτσελίκης (Tsitselikis) and

Δημήτρης Χριστόπουλος (Hristopoulos) (eds.), *Το Μειονοτικό Φαινόμενο στην Ελλάδα. Μια Συμβολή των Κοινωνικών Επιστημών*. Αθήνα: Κριτική, 1997: 315–348.

Παπαδημητρίου (Papadimitriou), Παναγιώτης, *Από τις Κλεισούρες στα Δερβένια. Η Σύζευξη της Σλαβοφωνίας και του Ισλάμ στους Πομάκους της Ροδόπης* (From 'kleisoura' to 'derveni'. The coupling of Slavophony and Islam in Rhodopi's Pomaks). Θεσσαλονίκη: Αφοί Κυριακίδη, 2008.

Παπαδημητρίου (Papadimitriou), Παναγιώτης, *Οι Πομάκοι της Ροδόπης. Από τις Εθνοτικές Σχέσεις στους Βαλκανικούς Εθνικισμούς (1870-1990)* (The Pomaks of Rhodopi. From ethnic relations to Balkan nationalisms (1870-1990)). Θεσσαλονίκη: Αφοί Κυριακίδη, 2003.

Παπαευγενίου (Papaevgeniou), Αθανάσιος. *Βόρειος Ελλάς: Μειονότητες από Στατιστικής Απόψεως εν Σχέσει με τον Πληθυσμόν και την Εκπαίδευσιν* (Northern Greece: Statistical perspectives concerning population and education). Θεσσαλονίκη: Σύλλογος προς Διάδοσίν των Ελληνικών γραμμάτων, 1946.

Ροβόλα (Rovola), Γεωργία, *Η Μειονότητα των Πομάκων: Οικονομικές, Κοινωνικές και Πολιτισμικές Απόψεις* (The Pomak minority: Economic social and cultural perspectives), Μεταπτυχιακή διπλωματική εργασία, Πάντειο Πανεπιστήμιο Πολιτικών και Κοινωνικών Επιστημών, Πρόγραμμα Μεταπτυχιακών Σπουδών «Κοινωνικός αποκλεισμός, μειονότητες και φύλο.» Αθήνα, 2005.

Συρίγος (Syrigos), Άγγελος. «Ελληνο-Τουρκικές σχέσεις 1967-1974: Από τη Συνάντηση στον Έβρο στην Εισβολή στην Κύπρο, » (Greek-Turkish relations 1967-1974: From the encounter at Evros to the invasion of Cyprus). In Παύλος Σούρλας (ed.), *Η δικτατορία των συνταγματαρχών και η αποκατάσταση της δημοκρατίας* (Πρακτικά συνεδρίου, Αθήνα, 20-22 Νοεμβρίου 2014). Αθήνα: Ίδρυμα της Βουλής των Ελλήνων για τον Κοινοβουλευτισμό και τη Δημοκρατία, 2016: 313–336.

Τρουμπέτα (Troubeta), Σεβαστή. *Κατασκευάζοντας Ταυτότητες για τους Μουσουλμάνους της Θράκης. Το Παράδειγμα των Πομάκων και των Τσιγγάνων* (Constructing identities for Thrace's Muslims. The cases of Pomaks and Roma). Αθήνα: Κριτική, 2001.

Τσιμπιρίδου (Tsibiridou), Φωτεινή. «Οι Αποχρώσεις μιας "Μειονοτικής Κατάστασης": Η Περίπτωση της Ελληνικής Θράκης Σήμερα» (Shades of a 'minority situation:' The case of Greek Thrace today). In *Μειονότητες στην Ελλάδα* (Πρακτικά επιστημονικού συμποσίου, Αθήνα 7-9 Νοεμβρίου 2002). Αθήνα: Εταιρεία Σπουδών Νεοελληνικού Πολιτισμού και Γενικής Παιδείας Σχολής Μωραΐτη, 2002: 101–118.

Χατζηδημητρίου (Chatzidimitriou), Σεβαστός Κ., *Τοπωνυμιολογική Θεώρηση των Πομάκικων Οικισμών του Νομού Ξάνθης* (Review of the toponymics of Pomak settlements in the prefecture of Xanthi), Διδακτορική διατριβή, Αριστοτέλειο Πανεπιστήμιο Θεσσαλονίκης, Φιλοσοφική Σχολή, Τμήμα Φιλολογίας. Θεσσαλονίκη 2005 (http://thesis.ekt.gr/thesisBookReader/id/19065#page/1/mode/2up, Accessed on 8 August 2020).

7 The Hellenization of borderlands I

Settlement and urbanization as statecraft – 1919–1941

Between 1919 and 1941, settlement, urbanization, borderland-making constructed a security ecosystem that redefined national statecraft. Borderland settlements emerged as proxy terrain for contests between Hellenic and non-Hellenic identities and between neighboring states. Importantly, they operated as actuators of a garrison-state outlook in domestic politics. Tensions at external borders justified creating internal borders. Once in place, internal borders conditioned the politics of the center.

The relationship between the state and Western Thrace's Muslim minority is exemplary of this political territorial culture. Some of these settlements were built, others transformed, or expanded in the aftermath of the Treaty of Lausanne, highlighting the challenges to Hellenic statization. Borderlands figured prominently in the national conversation about Hellenism. Three borderlands case studies in Eastern Macedonia and Western Thrace illustrate how the state and the border were equal partners in making an exclusionary politics of place.

It is unclear whether the increased settlement activity in the borderlands post-"Lausanne" was a stratagem planned by the Hellenic State, or an urbanization phenomenon that emerged organically in the frantic period following the population exchange. If the former were the case, it would brand borderlands settlement as an instrument of geopolitics. Such settlement policies, which would have been inextricable from regional and agricultural development and refugee population relief, would have exemplified the intersection of urban development, the territorialization of state power, and regional geopolitics.

Stratagems and planning decisions in the service of nationalism and borderland security, and the manner in, and the degree to which they were applied, reveal the peculiarities of Hellenic statecraft: First, there was a significant gap between the State's ambition and its capability, which put in doubt the ability of the center to shape facts on the ground in the periphery.

Second, following protracted wars and forced migrations, the fluid state of local societies in the borderlands undercut the role of localities and localism in support of national security and statization. As exemplified by its minority policies in Western Thrace, the state was reluctant to assimilate and incorporate non-Hellene ethnics of the borderlands into the national collective, despite its record of subsuming into Hellenism other groups (Vlachs, Sarakatsani, and, speakers of "Makedhonika," Macedonia's Slavophone population, many if not most of who subscribe to Hellenic identity (Karakasidou, 1997). Anastasia Karakasidou notes the pitfalls of using definitive criteria of ethnicity in Macedonia, while describing the identities and allegiances of its Slavophone population in her study of the inhabitants of the Guvezna area, she notes "[t]hose present-day inhabitants of Assiros who are descendants of Guvezna's Slavic-speakers conceive of and conduct themselves as nothing less than Greek citizens and full-fledged members of the nation of the Hellenes; they maintain strong convictions that Macedonia is Greek, and some believe that they themselves are the Greek descendants of Alexander the Great. Nevertheless, they do not call themselves Macedonians [like Macedonian ethnics of the Republic of North Macedonia]." She continues that while they may have described themselves as Macedonian in the past and their idiom Macedonian (Makedhonika) the name has an entirely new meaning, especially after the break-up of Yugoslavia and the creation of the Republic of Macedonia in 1991 (officially between 1991–2018, Former Yugoslav Republic of Macedonia) (Karakasidou, 1997: xlvi).[1]

Third, settlement and urbanization policies entangled border security and rural and regional development objectives, establishing facts on the ground that promoted Hellenization and statization, however incompletely.

We distinguish three phases of bordering in 1919–1941: the acquisition of Western Thrace and the beginning of its re-Hellenization defines the first phase (1919–1922) – The last forward-thrust. The state sought to coopt Thrace's religious minorities to reassure the Great Powers of its proper stewardship of the region; it contained Bulgarian infiltration of borderlands by expelling Bulgarian "suspects," and initiated urban modernization projects and the expansion of important regional centers, such as the cities of Xanthi, Gumuldjina (Komotini) and Dede'agatch (Alexandroupolis), as bases for Hellenic administration.

Greece's defeat in the Greek-Turkish War, the Treaty of Lausanne, border consolidation at the Evros River, and the reception of exchangees defined the second phase (1922–1936) – Retrenchment. The reconstruction of the State post-"Lausanne" was substantially based on its ability to settle and integrate refugee populations (Hirschon, 2003: 3-22). The population exchange necessitated the resettlement of approximately 700,000 Greek refugees, mainly in the New Lands of Western Thrace and Macedonia and in major urban centers, notably, Athens and Thessaloniki. In this period, State actions reshaped Greece's urban hierarchy, especially in the borderlands. Turkish and Bulgarian villages and urban quarters were evacuated or abandoned to accommodate the refugees and other displaced persons; new village refugee settlements were planned and built, and fringe belt developments were etched in existing cities and villages. Normalcy in international and domestic affairs was restored in 1928 with the return of the Liberal Government of Eleftherios Venizelos. The creation of the Bank of Greece (its first central bank) in 1928 and the Agricultural Bank of Greece the following year demonstrated the thickening of State capacity. The signing of the Balkan Pact in 1934 diffused geopolitical tensions with Turkey and re-oriented Greece's security strategy to the containment of Bulgaria. Refugees started building communities around their collective experience of displacement. Some refugees became mobile, leaving provided accommodations in the New Lands to settle in other parts of Greece.

Ioannis Metaxas' Fascist, nationalist, populist, and agrarianist dictatorship defined the third phase (1936–1941) – Internal bordering and garrisoning. The Greek-Bulgarian and Greek-Albanian borderlands and their settlements emerged as the focus of State security – "the peril from the North" gained a new definition. A secret stratagem operationalized borderland settlement-as-geopolitics by introducing measures that entangled border defense, population settlement, and rural development. The Metaxas regime commanded the construction of an extensive Maginot-type line of fortifications, forced the evacuation of village populations of "suspect national sentiment" (especially in the vicinity of the forts), resettled landless loyalist Greek ethnics from Macedonia and Thessaly on sparsely populated parts of the zone, and sought to promote rural development with objectives population stabilization and border security support. Greece entered the Second World War on 28 October 1940, putting the Metaxas stratagem to the test.

The regional theater of war took a decidedly nationalist turn in 1943 when Greek and Yugoslav communist leaders leveraged the Slavo-Macedonian question to rally that population to secede from Greece. The Greek Civil War broke in the immediate aftermath of the Second World War and reshaped the social and political geographies of Macedonia's and Western Thrace's borderlands. In place of Nazi Germany and collaborationist Bulgaria, the "peril from the North" took the shape of the coalesced efforts of Albania, Yugoslavia, and Bulgaria supporting the Democratic Army of Greece (ΔΣΕ) (the Communist Party of Greece's military wing) to establish a "people's democracy" in Greece. The new Slavo-Macedonian national state would have incorporated Greek Macedonia and the Yugoslav Republic of Macedonia: "The Yugoslavs' immediate purpose in founding the Slavo-Macedonian Popular Liberation front [in 1943] was to inculcate a Slavo-Macedonian national consciousness in the Slavophones of Greek Macedonia and to enlist the Slavophones of Greek Macedonia into the [anti-Nazi] resistance movement in Yugoslav Macedonia; while their indirect aim was to promote Yugoslavia's views on the Macedonian Question" (Sfetas, 1995: 297). In contrast, Bulgaria's view of Slavo-Macedonianism was aligned with its long-standing territorial aspiration of a Greater Bulgaria that reached the Aegean Sea and incorporated Thessaloniki and its hinterland.

A critical obstacle to the broad objective of wresting Greek Macedonia from Greece was Josif Stalin's ambivalence about the gambit. His geopolitical scope of the post-war order he co-constructed with the United States, Britain, and France, was more significant than the fate of Greek Macedonia. On the one hand, while allowing the covert resupply of the insurgency, the Soviet Union recognized that the United States and Great Britain would not tolerate a Communist Greece. On the other hand, it recognized that should circumstances allow it and the Communist Party of Greece (KKE) prevailed, a Communist Greece would inevitably be a satellite of the Soviet Union (Iatrides, 2005).

The Soviet position on the Greek Civil War crystallized between the announcement of the Truman Doctrine on 12 March 1947 and 4 July 1948, when President Harry S. Truman committed the United States to aid Greece and Turkey in their anti-Communist campaigns. The pivotal point in the Civil War came in February-March 1948, when Yugoslavia and the Soviet Union ceased to collaborate in support of the rebellion. Yugoslavia insisted on continuing its expansionist activities in the region, while the USSR wished to deescalate to keep the United States from intervening in Albania and Greece (Perović, 2007: 34–35). On 10 February 1948, Stalin called for an end to the Greek rebellion and demanded that Bulgaria and Yugoslavia cease their support (Djilas, 1962: 181–182). With Stalin prioritizing other geopolitical objectives to the Greek insurgency and splitting from Tito, the KKE vacillated between its desires for logistical support from Yugoslavia and Albania, and its political-ideological commitments to Moscow. The loss of Soviet support tipped the scales in favor of the Athens government and its Western supporters. Tito's gambit that support for the Greek Communist insurgents, half of whom were Slavo-Macedonians, would yield Yugoslavia new territory to the Aegean Sea, was not to pay off (Perović, 2007:53).

Concerning the Slavo-Macedonian nationalist dimension of the conflict, Tito miscalculated their commitment to his vision of a Great Macedonia under Yugoslav leadership (Sfetas, 1995: 317). With Tito's abandonment of the Greek insurgents by 1949, the military phase of the Civil War came to an end. Arguably, Greece continued to fight it by other means until the establishment of a robust democracy in 1974.

Greek Macedonia and Western Thrace evaded the "peril from the North" and, by 1952, entered the geostrategic orbit of the United States and NATO. The intersection of Cold War geopolitics, Left-Right ideological enmity, and regional nationalist territorial contestation created conditions in Athens's political and military establishments that shaped Greece's relations with Bulgaria and Turkey, the settlement of the borderlands, and majority-minority relations. Athens' projection of a national Hellenist political-territorial modus in the borderlands transforms them demographically, ethnographically, and materially. In turn, the perceived state of the borderlands prejudiced the course of domestic public discourse, bending it until 1974, toward an exclusionary politics of place.

Ottoman legacy: security and modernization

Macedonia and Thrace came under Ottoman Rule in the second half of the fourteenth century. By the sixteenth century, expanded Muslim settlement in Thrace and the conversion of some of the Christian population to Islam, resulted in Muslims becoming the majority in the region. Muslim settlement was exceptionally high in the territory anchored by the City of Komotini (Ottoman Gumuldjina) in the West and the River Evros (or Maritsa, in non-Greek sources) in the East. Their dominance was most pronounced in cities, which were seats of military and imperial administrative power: 82 percent of Adrianople's population was Muslim, while its hinterland was 75 percent Muslim (Kalantzis, 2016: 24). Important urban centers, primarily, Constantinople (Istanbul) and Adrianople (Edirne), and secondarily, Philippopolis (Plovdiv), and Kallipolis (Gallipoli or Gelibolu), drew a network of central places that lent great economic and security significance to Thrace (Ibid, 28). The fertile Evros River valley was of strategic importance to the Ottomans. Appropriate for rice cultivation, Thrace provided a strategic alternative source of that principal crop to Egypt, which, under Mameluke rule, was at times tenuously tethered to the Empire (Voyatzis, 1998: 426–427).

By the beginning of the seventeenth century, Thrace's plains and Aegean littoral received migrants from less productive, mountainous regions to the West and South. As late as the beginning of the twentieth century, there were still 12 villages with populations of Peloponnesian extraction, 11 villages of Epirotes, and seven villages inhabited by Macedonians in the vicinity of the City of Didymoteicho alone (Kalantzis, 2016: 26).

Russia as existential threat to the Ottoman Empire

The creation of the Bulgarian Exarchate in 1870 at the cost of Ottoman rule, signaled the expansion of Russian influence in the Balkans. Following Germany's unification in 1871, Russia was able to rally Slav politicians around its Pan-Slavist vision and geopolitics. Russia and successor states-in-waiting, Bulgaria, Romania, and Serbia appeared to have the advantage over the Ottomans. The Church concessions to the Bulgarians were, in part, intended to drive a wedge between the Bulgarian population and the Orthodox Ecumenical Patriarchate in Constantinople. However, it also quickened the Bulgarian national awakening and defined it in geopolitically oppositional terms to the Serbs and the Greeks' nationalisms. The Bulgarian uprising of 1876 and the Russo-Turkish War of 1877–78 that followed it, created the conditions for an autonomous Bulgarian Principality. After the war was settled in the 1878 Treaties of San Stefano and Berlin, Russia, as champion of all Slavs, exerted influence in the region through proxies, Bulgaria and Serbia. Nominally under Ottoman administration, the new state was patently under Russian patronage. The Treaty of Berlin of 13 July 1878 also established an autonomous Eastern Rumelia to the south of the Bulgarian Principality, with Plovdid (Ottoman Filibe, Greek Philippopolis) as capital. That presence of significant Greek, Turkish, Bulgarophone, and Pomakophone Muslim populations, called for different geopolitical treatment by the Treaty. Moreover, the Great Powers, and especially Great Britain, wanted to avert the creation of a Russia-dependent and territorially expansive Bulgaria, that would have taken over much of multicultural Macedonia and Thrace. By 1885, however, Eastern Rumelia was incorporated by the Bulgarian Principality into a de facto union that, in 1908, became the independent State of Bulgaria.

The Congress of Berlin set the stage for new Southern Balkans geopolitics. National identities were fleshed out, new borders were drawn, and ethnic populations started their spatial sorting out according to the geography of new national states. The pre-national *Millet* identities that defined the Ottoman community experience – with religion as the primate signifier and language as a conditional and situational spatio-temporal one – were gradually undermined and replaced by formal national identities crafted by political and intellectual elites in Athens, Sofia, Belgrade, and Bucharest, and enforced by their respective militaries and civil administrations. Although caught in the vice of Great Powers' and national states' strategies, regional and local populations resisted, mediated, or selectively applied regulations, determining for themselves the rate and degree of national statization.

Economic liberalization modernizes and creates dependency

The liberalization of the Ottoman economy started in 1838 with the signing of the Anglo-Turkish Commercial Treaty (Convention of Balta Liman). The Tanzimat Reforms (1839–1878), which soon followed it, framed the Empire's modernization strategy. Modernization had the high political aim to equip the Empire in its economic and geopolitical contests with Great Powers. Expanded port facilities and road and railroad networks, supported increased trade with industrializing Europe. The opening of Ottoman markets to the West created opportunities for the penetration of foreign direct investment in transportation, utilities, and agricultural commerce. Railroad development and trade in primary sector commodities were of great interest to foreign investors, including Americans. In the heavily indebted Empire, such projects were structured as State concessions to international investors and were often carried out by European construction and engineering firms. As of the 1870s, cities as trade and transportation hubs were targeted for expansion and regulatory modernization. Urban modernization in Thessaloniki and cities in Ottoman Thrace, such as Xanthi, Gumuldjina and Dede'agatch, reflected the thickening of the Ottoman state bureaucracy, a measure of industrialization and infrastructure-building, and the expansion of the social state. Enhancements to the built environment of cities included industrial plants, railroad stations and related infrastructure, religious buildings, schools, hospitals, increasingly constructed out of brick and stone (Karadimou-Gerolympou, 2004: 243-262).

British, French, and Austro-Hungarian economic penetration of Turkey-in-Europe, impacted the development trajectories of Ottoman successor states Greece, Bulgaria, Serbia, and Montenegro. Recognizing the state-building potential of railroads, they eagerly embraced them, simultaneously entangling themselves in circuits of finance and debt controlled by Central and Western European states and firms.

On 17 April 1869, the Ottoman State granted to the enterprise of Belgian Baron Maurice de Hirsh – incorporated as an Austrian enterprise with principal objective, the building of the Vienna-Constantinople railroad – the concession to build and exploit the railroad that would connect Constantinople to Sarajevo and the Habsburg border, via Adrianople, Plovdiv, Sofia, Nis, and Belgrade (Jelavich et. Jelavich, 1977: 195–96). The concession also included the development of the junction railroad lines that linked Constantinople to Thessaloniki and Thessaloniki to Monastir – the latter a rail axis that served Austrian commercial interests. The Ottoman State granted such concessions for 99 years. The railroad lines connecting Niç to Thessaloniki and Dede'agatch (Alexandroupolis) to the Constantinople-Adrianople trunk line were in service by 1880. Contracts differentiated the enterprise of infrastructure development from the railroad's operation: accordingly, La Société Impériale des Chemins de Fer de la Turquie d'Europe, received Hirsh's concession and La Compagnie Générale pour l'Exploitation des Chemins de Fer de la Turquie d'Europe was tasked with the execution of projects (Beach et al., 1910: 261–62). The sensitivity of such investments to geopolitics is evident in the shifting regulation of concession contracts. Moody's Magazine reported the restructuring of these enterprises in the last decades of the nineteenth century, following the significant reordering of the region's geopolitics (Figure 7.1). As the result of certain European events about this time, consequent upon the Franco-Prussian War and a revolt in Bosnia, the concession of 1869 was canceled, and the first mentioned company became constructing contractors only for the account of the [Ottoman] government.[2]

The transnational character of these projects illustrates the Ottoman State's dependence on foreign capital and technology. Infrastructure development and urbanization were reliant upon them: the originating investor of the Vienna-Constantinople railroad project was Belgian; the Ottoman Bank provided a significant portion of the financing, and a majority-holding British investor group, the French Banque de Paris et des Pays-Bas, and the Ottoman State managed the project as a joint venture. Importantly, the raison d'être for these projects was as much Ottoman modernization and development as service to mainly Austrian and Western European commercial concerns.

The real property environment also changed when, in 1867, the Empire recognized the right of non-Muslims – Greek-Orthodox, Jews, Armenians,

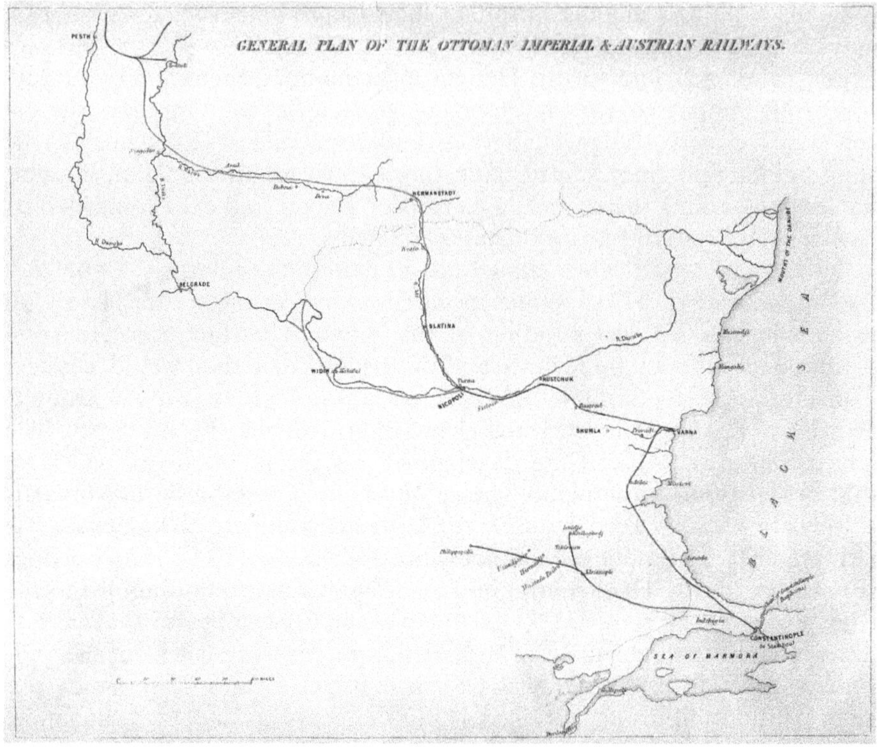

Figure 7.1 By 1880, critical links between Turkey-in-Europe and Central Europe were nearing completion.

Source: General plan of the Ottoman Imperial and Austrian railways, print, map, 1888. The British Museum, with permission.

and foreigners – to own land. A few years later, at the Congress of Berlin (1878), land ownership statistics from Thrace revealed that one-third of the properties in the City of Constantinople were in Greek hands. In the Sanjaks of Radestos, Kallipolis, and Adrianople, Greeks owned approximately one-half, two-thirds, and one-third of land properties, respectively (Kalantzis, 2016: 29; Gondicas and Issawi, 1999: 9–11). In his eyewitness account of Thrace's ethnography, Stylpon Kyriakidis reports that by the time Western Thrace became incorporated into Greece, most urban properties and a significant proportion of rural estates were owned by Greeks (Kyriakidis, 1919: 98–99). However, an often-quoted source, Kyriakidis' exuberant narrative may be suspect of exaggeration. It is difficult to verify the exact extent of the ownership claims, but there is sufficient evidence to support the idea that considerable wealth was transferred from Muslims to non-Muslims after the Congress of Berlin, and many of the marketed properties were urban.

Two crucial takeaways concerning the Ottoman retreat, the Bulgarian surge, the drawing of borders, and modernization in the southeastern Balkans before the First World War: First, the Treaty of Berlin did not satisfy Bulgarian nationalists. They systematically capitalized on opportunities between 1878 and 1919 to expand the territorial extent of the Bulgarian State south of the Rhodopi mountain range, into southern Thrace and, importantly, into Macedonia. Gaining access to the Aegean Sea was a significant objective. A bicoastal Bulgaria, as much as a Greater Bulgaria, was a fixture of its geopolitical imaginary. Even after treaties confirmed the status of Macedonia and Western Thrace as Greek in 1913 and 1919, respectively, and until the outbreak of the Second World War, Bulgarian expansionism and revanchism against Serbia (Yugoslavia after 1919), and Greece was a standard feature of Bulgarian strategy.

Second, Great Power involvement in Balkan geopolitics rises after the Crimean War, as the Ottomans retreated and the Russians surged to fill the power vacuum. Balkan boundaries and identities become currency of Great Power transactional geopolitics.

Third, investor-driven economic modernization and infrastructure building in Turkey-in-Europe and Western Anatolia, would redefine center-periphery relations in the successor states, after the Ottoman retreat is complete. Between 1912 and 1923, the region had become a shatter belt of successor states made up of centers and peripheries that required very different geographies of infrastructure. Macedonia and Thrace were partitioned in a manner that impeded, if not extinguished, long-standing core-to-core and core-to-hinterland functional linkages. Accordingly, in 1923, Adrianople was severed from its economic hinterland to the West and the North. Alexandroupolis, situated at the T-junction of the Thessaloniki-Istanbul trunk railroad and its Adrianople extension, was similarly denied part of its historic hinterland in Eastern Thrace. Soufli's silkworks economy collapsed once access to extensive mulberry orchards across the Evros River was no longer possible.

Censuses and ethnographic mapping of borderlands

Before the breakout of the First World War, territorial and identity-related contests in Eastern and Western Thrace among Turkey, Greece, and Bulgaria, took place under the eye of the Great Powers. In the case of Macedonia, ambiguities in Ottoman census data, and the drawing of new borders across multicultural territories allowed for new territorial claims and animated settlement in contested territories by Greek and Bulgarian populations. Bulgarization through settlement in city and country, and consolidation of existing communities, at times by violent means, intended to establish or crystalize facts on the ground favorable to Bulgarian territorial claims.

After the First World War, Greece and Turkey reported cases of violent Bulgarization to the League of Nations to subvert Bulgarian expansionism to the south and east. The dissemination of such reports had as intended, audiences the diplomatic community and the region's minority populations, which were courted by Greece and Turkey.

Bulgarian campaigns against Greeks and Muslims, which took place in Eastern Rumelia and Western and Eastern Thrace, contributed significantly to anti-Bulgarian sentiments and helped foster a détente, if not closer relations, between Thrace's Greeks and Turks, and Greeks and Muslims more broadly. Bulgarian actions in Eastern Rumelia drove south, into Thrace, much of the Greek-culture population. Many settled in urban centers like Plovdid, Burgas, and towns and villages on the Black Sea coast (modern-day Pomorie, Aheloy, Rousokastro, Nessebar, and Sozopol). Concerning Greece's claims of a robust Hellenic presence in Western Thrace, available information suggested that majority Greek populations settled the Aegean Sea littoral and the broader Evros River valley.

The 1908 map of Bulgarian and Greek schools and churches of the Istituto Geografico de Agostini, Rome (ELIA, Map 003) depicts school enrollment data and church counts at the Vilayet, Sanjak, Kaza, and settlement scales for the Vilayet of Adrianople (Andrianople/Edirne, in the map) (Figure 7.2). The map also includes the European part of the Vilayet of Constantinople. Notably, the Italian map describes the groups according to nationality instead of *Millet* designations. Absent accurate information about the magnitude of Greek and Bulgarian populations, we are left to appreciate the educational enterprises' magnitude in the region of their kin states, Greece and Bulgaria.

Greek school and church presence is strikingly more pronounced in Eastern than in Western Thrace. There are significant clusters of schools and churches in the Sanjaks of Adrianople and Gallipoli, which, at the community level, are coastal, also in the plains (including the Evros River valley), and in urban centers. In the territory that would be incorporated by Greece in 1919 as Western Thrace, there is compelling evidence of Hellenic demography anchored to the north by Adrianople and its significant ethnically Greek suburb of Karagatch, including the significant urban centers of Didymoteicho (Dimotika) and Soufli on the Evros River valley, and extending as far west, across the future Greek-Bulgarian border, as Ortaköy (Ivaylovgrad). There is some Hellenic church-school presence in coastal communities on the Evros River delta and the Western Thracian coast. The Kaza capitals of Xanthi and Gumuldjina also host such institutions, although there is a limited presence of churches and schools in the plains between these cities. There is no church-school presence in the highlands of Xanthi and Rhodopi, which were and continue to be mostly populated by Pomaks. In Bulgarian territory, one school is located in Dari-Dére (Zlatograd), and a cluster is organized around the town of Paşmakli (Smolyan). Overall, the reach of Hellenic education and

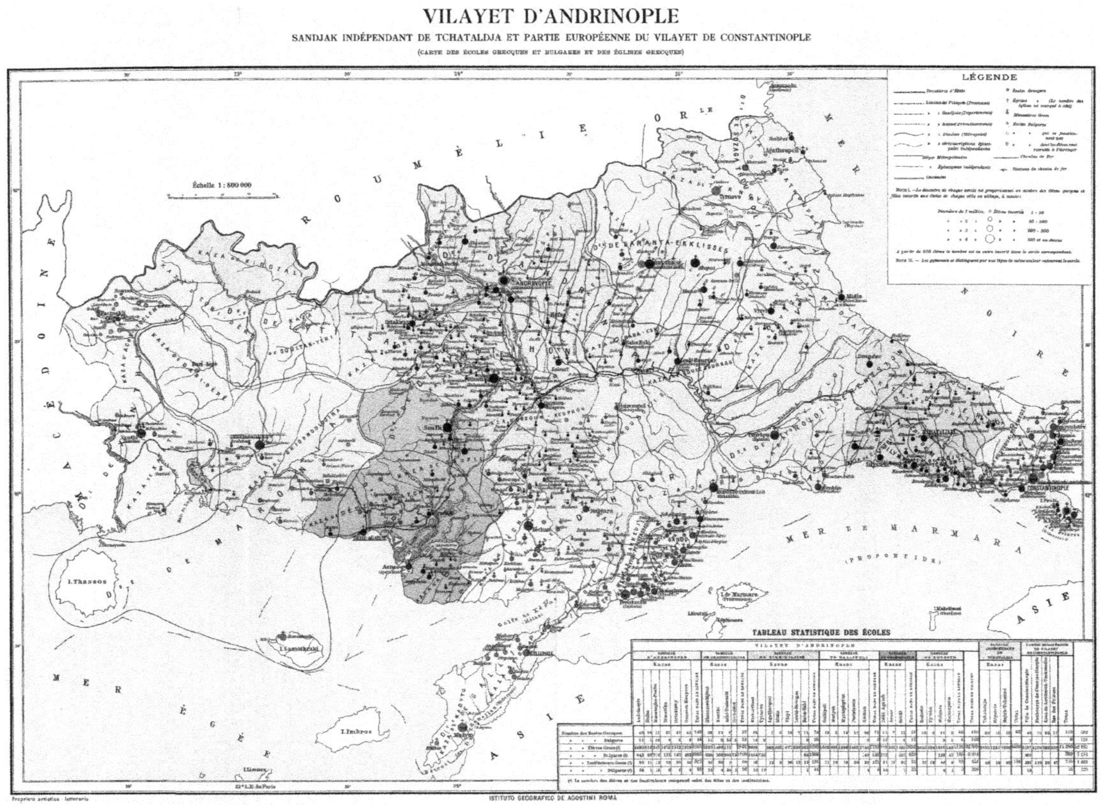

Figure 7.2 Map of Greek and Bulgarian schools and Greek churches in the Vilayet of Adrianople (detail).

Source: Istituto Geografico de Agostini, 1908. ELIA, with permission.

church agency was greater in Eastern than in Western Thrace, and absent in the region of Pomak settlement. As shorthand for the ethnographic geography of that region, the de Augustini map reveals more about Greece and Bulgaria's national and institutional agency in European Turkey than a reliable census of ethnics.

Another 1908 map published by Istituto Geografico de Agostini (ELIA, 1908: MPAGO.001) attempts to provide an ethnographic profile of European Turkey (Figure 7.3). Greek Orthodox and "Musulmans" appear to share Eastern Thrace with small pockets of Bulgarian Schismatics in the northeast and the left bank of Evros River. The legend identifies the groups by religion as "different élements de la Turquie d'Europe" instead of ethnic or ethnonational groups. Further, the map depicts Western Thrace and Bulgaria's southern Rhodopi regions as overwhelmingly settled by Muslims (Roma or "Tsiganes" are identified as a separate element). Pockets of Greek Orthodox are located in Xanthi and Gumuldjina (Komotini) and in dispersed coastal regions, in which the first map locates many Greek schools and churches. The pre-1914 ethnographic map by G. Staggela (likely, pre-1914) is based on Turkish sources and bears out a different geographic distribution of religious groups in the Vilayet of Adrianople. It depicts future Western Thrace as overwhelming Muslim, albeit, at the Kaza level (Xanthi, Gumuldjina, Dédéagatch, and Soufli) minorities of Bulgarian Schismatics are ubiquitous: The Kaza of Demotika (Didymoteicho) is depicted as having an entirely Christian Orthodox population (ELIA, pre-1914: K. No 56–05).[3]

The ethnographic cartography of the Balkans, as we have stressed before, is methodologically problematic, and the de Augustini and Staggela maps are no exception. Nonetheless, the intelligence they provide even in the gross sense, informs our understanding of the demographic impact of the First World War and the population exchange of 1923, the latter underwriting the rapid Hellenization of Thrace and Macedonia by settlement of refugees.

Concerning the Pomaks, when the Bulgarization of their territory commenced, lack of formal infrastructure and institutions afforded them some isolation until the beginning of the Balkan Wars. As of 1919, the Pomak region was partitioned by the international border into Greek and Bulgarian sovereignties, redefining their state of isolation, undercutting (though not eliminating) cross-border cultural communication flows between settlements, and disrupting the Pomak agrarian space-economy, which depended on transhumant livestock farming (Ripoll, 2003: 295). Importantly, it opened their territory up to statization and the political territorialisms of national states. However, it is notable that the Pomak region was excluded from refugee resettlement plans post-"Lausanne," which was, ultimately a double-edged sword: less intrusion and less investment for social development (Koutsoukos, 2012: 490–493). We study the Pomak case in detail in the following chapter.

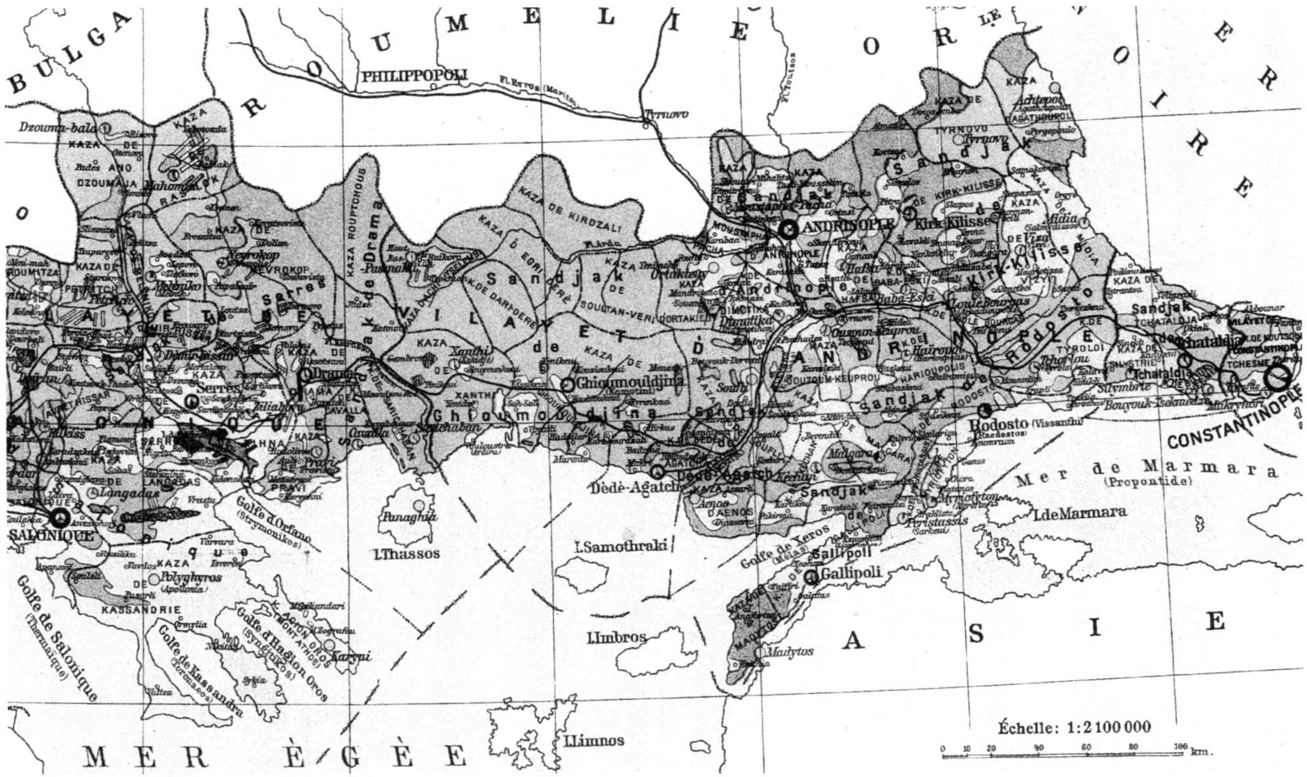

Figure 7.3 "Map indicating the preponderance of different nationalities in various parts of the Vilayets [of Jannina, Monastir, Salonica, and Adrianople]" Ethnological map of Turkey-in-Europe. Greek populations appear in light grey. (detail).

Source: Istituto Geografico di Agostini, 1908. ELIA, with permission.

Bordering in Eastern Macedonia and Western Thrace

The Treaty of Bucharest, which settled the Second Balkan War, awarded much of Macedonia to Greece on 10 August 1913. The Strymon River in Eastern Macedonia defined the new land border between Greece and the Ottoman Empire. By the end of the First World War, the region experienced mass population displacement and the complete or partial destruction of no fewer than 237 villages and the displacement of 92,000 people (Koukoula, 2007). The territory west of the Strymon River became a borderlands zone, and its geography of settlements reflected and catalyzed the new geopolitics of a retreating Ottoman Empire and a surging Greece. Under a Liberal government, the creation of Ministries of Transportation and Agriculture in 1914 and 1917, respectively, signified the thickening of the administrative state and the expansion of State networks and capacity nationally, but critically in the New Lands. A new technocracy, under Transportation Minister Alexandros Papanastasiou, transplanted Western European urban prototypes into Greek planning practice (Sotiropoulos et al., 2008: 138), contributing significantly to the desired de-Ottomanization and de-Bulgarization of the region and settlement in the borderlands. As in the case of the rebuilding of Thessaloniki's urban core after the Great Fire of 1917, foreign engineers and architects, some of whom were attached to Entente forces stationed in the region, received design commissions to design new settlements in Eastern Macedonia. Legislation for the rebuilding of settlements in Eastern Macedonia (N. 1728 and N. 1746) regulated their platting and established eminent domain parameters that allowed the state and its contractor planners to expedite the building of dozens of new villages and towns with a population of up to 5,000 (Kafkoula, 2013: 683). By 1919, and with the prospect of extending the paradigm to Western Thrace, several town plans were ongoing. Exemplary of the promise of the venture was the regular, "garden city" and "city beautiful"-inspired plan for Irakleia in the Prefecture of Serres. Built on the site of war-damaged Tzoumayia, the town plan was drawn by Ernest Hébrard and associate architects John Mawson and Piet de Jong. Overtaken by the events of the Greek-Turkish war and the precipitated refugee crisis, Irakleia's ambitious planning vision was not replicated elsewhere at quite that scale. Nevertheless, by 5 November 1930, Irakleia's first houses are handed over to their owners (Ibid, 695-696).

In 1917, the Allies established an interim government of Western Thrace that relied heavily on the military support and governmentality extended to the region by the Hellenic State. Upon the defeat of the Central Powers, Western and Eastern Thrace became terrains of intense competition among Greece, Bulgaria, and the Ottoman Empire, then on its last days, and a region of contentious border drawing.

Greece incorporated Western Thrace in 1919. The League of Nations ceded Western Thrace to Greece in the Treaty of Neuilly-sur-Seine (1919), with the consent of the Great Powers. Bulgaria was, thus, denied access to the Mediterranean Sea. It was a triumph for the ambitiously expansionist Hellenic State. "Neuilly" transformed Greece's joint occupation of Western Thrace with France into direct territorial control and possession, which allowed statization-as-Hellenization to begin without delay.

An ethnically complex region of great significance to the territorial and economic security of Greece, Bulgaria, and the Ottoman Empire, Thrace became a test case for Wilsonian diplomacy and the effectiveness of the newly founded League of Nations (Geragas, 1925). Thrace's populations of Greeks, Turks, Bulgarians, Bulgarophone Muslim Pomaks, Roma, Sarakatsani, and smaller groups, were variously coerced and coopted by the adversaries under the watchful eye of the Great Powers.

The Powers served their higher-order interests, thereby limiting or promoting the contenders' actions as needed (Kalantzis, 2016). Britain and France regional priorities centered on consolidating the status of the Straits as international waters, establishing their dominion in the Eastern Mediterranean as League of Nations Mandatory powers, and determining the contours of their relationship with the residual Turkish State.

The Greek-Turkish War as Greece's last military forward-thrust

Following the capitulation of the Ottoman Empire to the Entente, convinced by Venizelos of Greece's role as a pole of regional stability, they consented to Greece's annexation of Western Thrace. Thin support in 1920 for further Greek expansion in Eastern Thrace and the littoral zone around Smyrne/ Izmir, quickly vanished after Greece's push into the Anatolian interior

Within four years (1919–1923), Greece experienced the heights of geopolitical success and the lows of a major military defeat in the Greek-Turkish War. The San Remo Conference confirmed Greece's sovereignty over Western Thrace in April 1920. On 10 August of the same year, the Treaty of Sèvres set conditions for further territorial expansion of Greece in Eastern Thrace and western Asia Minor. The realization of the Great Idea appeared to be at hand. By October, however, the fate of the Treaty was sealed when the monarchy in Constantinople and the revolutionary Turkish National Grand Assembly under Mustapha Kemal declared their opposition to its ratification. The Great Powers chose to arrest "Sèvres" implementation.

A string of political and diplomatic reversals were soon to unravel Greece's successes. The surprise defeat of the Liberal Party leader, Eleftherios Venizelos, in the November 1920 elections and the return of King Konstantinos I to the throne on 19 December of that year, following a plebiscite, were contrary to British and French wishes. The memory of the pro-German King's refusal to join the Great War on the Entente side was still fresh in London and Paris. Eleftherios Venizelos had forced the abdication of Konstantinos I in 1917 instigating the "National Schism" (ο Εθνικός Διχασμός) between liberals and monarchists.

With the diplomatic debacle caused by the King's restoration, the two Great Powers found the pretext to open channels of communication to Mustafa Kemal, thus starting to sideline Greece. The diplomatic opening combined with Greece's adventurism in the Greek-Turkish War, increasingly isolated Greece from its Great War allies. The expansion of the Greek military campaign into the Asia Minor interior and great distances from established supply bases and Smyrne (Izmir), proved disastrous. The rout was complete, and the war ended dramatically with the expulsion of the Hellenic Army from Asia Minor and Eastern Thrace and the burning of the Christian quarter of Smyrne on 13 September 1922.

The war alienated Greece's traditional patrons, Britain and France. It allowed the newly created Turkish Republic to leverage its military victory in the field to its statecraft ends: the dissolution of historically cosmopolitan Hellenic-culture communities, such as Smyrne (Izmir) in the Aegean Sea littoral, and Adrianople (but not Constantinople (Istanbul)) in Eastern Thrace, demographically re-profiled these geopolitically important regions as more integrally Turkish. British and French concerns about stability in the region, and growing Italian territorial ambitions in the Eastern Mediterranean, put Greece on the back foot diplomatically. Only Bulgaria found itself in a more disadvantaged geopolitical state than Greece.

The Great Schism between Liberal-Venizelists and conservative monarchists was reaching its final, dramatic conclusion in Western Anatolia. Following its defeat in autumn of 1922, the Hellenic Army rapidly redeployed west of the Evros River.

At the climax of the post-war domestic crisis, Colonels Nicolaos Plastiras and Stylianos Gonatas, aided by Venizelist General Theodoros Pangalos, deposed the royalist government on 27 September 1922 and forced the abdication of King Konstantinos I. In the Trial of the Six, the highest officials of the wartime government were found guilty of treason and four of them, including the wartime Prime Minister, Dimitrios Gounaris, were executed. The monarchy was abolished, and the Second Hellenic Republic proclaimed.

The Treaty of Lausanne, signed on 24 July 1923, profoundly transformed the Hellenic State and society and the State's international relations (Figure 7.4). It established the land border between Greece and Turkey at the Evros River. The dream of extending its territory deeply into Eastern Thrace was irrevocably dashed. The magnitude of the Greek defeat and the emergence of a muscular Republic of Turkey extinguished any prospect of territorial expansion. Greece's foreign and military policies would no longer march to the drum of the Great Idea. From a status quo-disrupting Small Power with outsize territorial ambition, Greece would retrench as a closed, garrisoned state, that counted on regional detent for its security.

The Treaty oversaw a vast population exchange that transformed Greece's demography, while making possible the Hellenization of Greece's multicultural New Lands of Macedonia and Thrace. State control would be sought

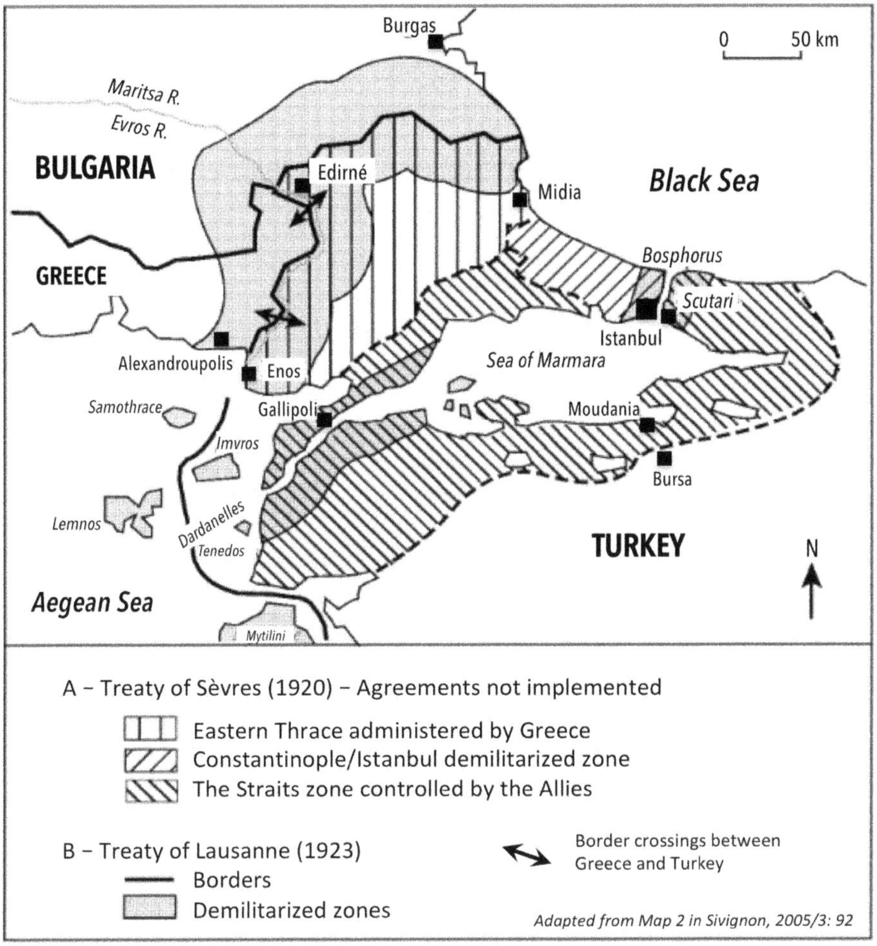

Figure 7.4 "Sevres" and "Lausanne" territorial changes. The Greek-Turkish border is established in 1923.

Source: adapted from Sivignon, 2005.

through the settlement of loyalists and the extension of the administrative state in the borderlands.

Civilian rule returned in January 1924, under the short premiership of the returned Eleftherios Venizelos. There were eight governments between Venizelos' departure from power on 6 February and his decisive return to the prime-ministership on 4 July 1928. Therefore, in the 16 years between the Second Hellenic Republic's proclamation on 25 March 1924 and the outbreak of the Greek-Italian War on 28 October 1940, 17 governments held power, three of which were military, and one dictatorial. The last one was led by Retired Lieutenant General Ioannis Metaxas, the leader of the monarchist, nationalist, agrarianist, and autocratic Freethinkers' Party (Κόμμα των Ελευθεροφρόνων). The new Republic had lasted less than 11 years when, on 3 November 1935, following a rigged plebiscite, King Konstantinos I's heir, ascended to the throne as Georgios II. On 4 August of the following year, Ioannis Metaxas, who served as Prime Minister since 13 April 1936, declared a state of emergency, dissolved the Hellenic Parliament, and established martial law.

New borders, new borderlands, new state

The new border and the associated borderlands became fodder for domestic contests over Hellenic national identity and regional security – a new geopolitics of difference. Drawing the Greek-Turkish border in Thrace by treaty, conventionally delimited the territorial extent of the two states, but did not address other community-based irritants and regional geopolitical anxieties. The specter of lingering "perils" from the East and North justified in the state's eyes, raising internal borders that re-territorialized Greek-Turkish and Greek-Bulgarian enmity on different terms. The external adversary would now be feared to be within the national territory, requiring a firm response by the State. The borderlands were a ready space where that alarmist sentiment could be grounded. Although a long distance from the Capital, the borders with Turkey and Bulgaria, and border security, became an idée fixe for the state, which had to deliver security on the ground. The border as Hellenism's battleground was hard-baked into national ideology and the definition of national interest. Such internal borders, which were deemed critical to the containment of external and internal adversaries, were less visible to, or legible by majority Greek society outside the impacted borderland – life in the surveillance zones along Greece's frontiers becoming an alternative reality for its legacy minority populations. The Pomak case is exemplary in that respect. This harsh, state-driven rejection of difference characterized Greece's cultural politics until the fall of the Colonels' Junta in 1974, and, in the case of the Pomaks, until the 1990s.

The cross-border forced migrations required the comprehensive reordering of Greece's urban geography. The scope of that transformation was statewide but most visible in Western Thrace, Macedonia, and Greece's largest cities, the latter hosting large numbers of refugees. Planned fringe belt developments in these cities became a leitmotif of twentieth century Greek urbanism. However, an important question for the garrisoned state was how settlements would contribute to the borderlands' security ecosystem.

Refugee groups crossing into Greece, already before the signing of the Treaty of Lausanne were confronted with the stark reality of the Hellenic State's reduced circumstances. Although in the early phase of the exchange, the state determined refugees' settlement in ad hoc refugee encampments, refugees had a degree of discretion as to their ultimate settlement. Many

Figure 7.5 Planned refugee settlement in Macedonia.

Source: ELIA, with permission.

found attractive settling in the state's core regions when considering the insecurity of the borderlands. In other cases, refugee groups of common point of origin desired to resettle as close as possible to the border in anticipation of sometime soon returning to their homes (Figure 7.5). In several cases, the engineering and topographic units of the Hellenic Army obliged, by etching out regular town plans upon Evros' fertile left river bank. Accordingly, the settlement pattern reflected the state's managerial and strategic vision, and its capacity to materialize it notwithstanding fully. The settlement geography also reflected the existential anxiety and aspirations of the forcibly displaced, which were qualified by severe limits to credit and capital.

The population exchange required the commitment of extraordinary amounts of resources. This tremendous financial burden was shared by the League of Nations, which counted its refugee assistance and settlement programs in the Balkans as one of its most significant achievements in the 1920s. It supported the settlement of 700,000 refugees in Greece (out of an estimated total of 1.2 million) by extending loans to the state amounting to ten million British pounds. In 1926, Greece was able to return to international capital markets, to secure financing for refugee resettlement and expand its armed forces (Lazaretou, 2005: 225–226). The refugee character of these settlements is evident in their toponyms and their community politics.

The arrival of refugees and their frantic settlement in cities, towns, and villages in the first months after "Lausanne," made plain the physical plant deficiencies of Greek transportation and urbanism. Moreover, in a display of reduced sovereignty, the state's actions were scrutinized by Great Powers and the League of Nations, which provided emergency financing.

The Catastrophe of 1922 defined the discourse between settled national populations and refugees. Although fellow Hellenes by common religion and language, the refugees were, in many ways, strangers. Suddenly, difference was a state of being that was internalized by the nation. Moreover, refugees were a financial liability from the perspective of local communities, especially in rural areas. Animated by conservative parties and the press, anti-refugee sentiments ran high in the first years (Exertzoglou 2012: 195–201; Liakos, 2011: 11). However, in most cases, refugees received care and attention from the communities where they were first installed (Mavrogordatos, 1983, p. 194; 2003: 9-35). By the early 1930s, refugee groups created associations, cultural societies, philanthropic institutions, local museums, and oral history projects that memorialized and preserved the historical and post-1923 legacies of their communities. These sources are important to Hellenism's historiography because they express narratives of community exclusion and inclusion in their own terms.

Census sources show that in 1926 the majority of Greece's population was agrarian, with 3,568,253 persons living in rural settlements and 571,735 in small towns. The urban population was 2,064,634 (Hellenic Statistical Authority). The absence of an adequate network of paved roads and railroads made the mobility of refugee groups difficult. Initially, the geographic configuration of transportation infrastructure provided limited locational options for first settlement and later more permanent resettlement. Western Thrace, Macedonia, and Crete received the most refugees, with their regional urban centers quickly developing, first, large tent cities in the fringes of their urbanized areas, and in the years that followed, new, formally planned communities (Figure 7.6). Many refugees settled in Thessaloniki, the largest city in the New Lands. They joined thousands of persons displaced by the Great Fire of 1917, permanently transforming the city's demographic and cultural geographic character.

In the case of Greece's New lands of Macedonia and Western Thrace, two phenomena were at work: on the one hand, rural cultural practices already established in the pre-national era (and in some cases rooted in ancient practice) would prove to be extraordinarily resilient in the decades that followed the post-Lausanne exchange of populations. Their eventual decline would be as much the result of systematic Hellenization as accelerated modernization in the 1980s. On the other hand, the arrival of waves of refugees, mainly as a result of the Lausanne-brokered exchange, but also

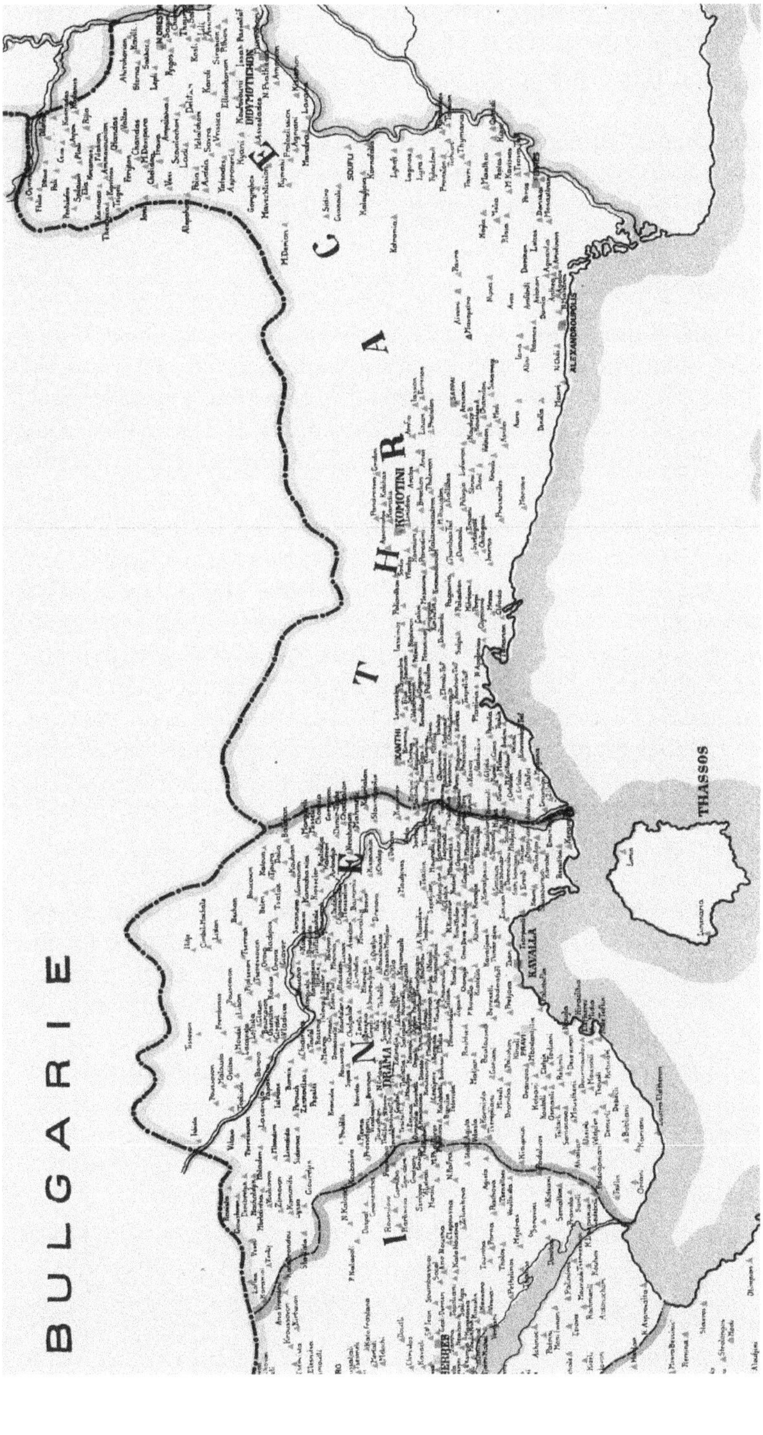

Figure 7.6 Map of rural and urban refugee settlements in Greece (1923–1926) (detail: Eastern Macedonia and Western Thrace). Point symbols mark new agricultural settlements. The montane part of Western Thrace, home to Pomak communities, was not settled by refugees.

Source: Ethniki Hartothiki, with permission.

animated by later geopolitical convulsions in the Black Sea and Eastern Mediterranean regions, would profoundly impact the cultural character of Greece.[4] New definitions of Hellenism, defined in the lost places, in Asia Minor and Eastern Thrace, would be transplanted in new refugee settlements and grafted into existing urban and village communities. The post-Lausanne chapter of nationalist statecraft would be written in major cities like Thessaloniki, and in the legacy and new village and urban settlements in the Macedonian and Thracian borderlands.

Settlement-as-geopolitics.

Expanded urbanization, following the incorporation of the New Lands served the centralization and consolidation of the national state and the "de-Ottomanization" of its urban social structure and built environment. Urbanization in the New Lands had a distinctly political character. Urban plans and regulation, although not attributable to a general national settlement and urbanization program for the consolidation of the state, they were consistent with the logic of political modernization (Gerolympos, 2004).

Two cities on the Greek-Turkish border, Alexandroupolis (Dedeağaç) and Nea Oresteias, the orthogonal town plan of the former laid out by Russian engineers during their brief occupation of the site in 1878, and the latter master-planned soon after 1923 to house refugees from the Karagaç suburb of Adrianople, define a borderlands axis of settlements that included the historic cities of Feres, Didimotihon, and Soufli. The five cities and a network of 72 villages (37 of which were newly platted or significantly expanded). Many of the refugee-settled villages were platted (though not necessarily constructed) by the Hellenic Army's Engineer Corps (Figure 7.7).[5]

A comparison of topographic maps of Thrace at scale 1:400,000 produced by the Geographical Service of the Army in 1936 and 1947 – the critical period during which the surveillance regime of the borderlands was first established – yields three findings: First, by 1936 the campaign for the Hellenization of toponyms, which was underway in 1919, was at an advanced stage of completion. Second, the phase of village and town foundations and extensions to existing settlements, which started under emergency conditions in the autumn of 1922, appeared to have reached a mature state but subsequently stalled because of the Second World War. Last, the state's reach, as illustrated by infrastructure building – specifically a regional transportation road network – did not appear to have grown significantly in this period, despite the region's critical geopolitical significance.

In his memoir, Deputy Governor-General of Thrace, Kostis Geragas, recounts that, as a first step, toponymic changes were made to the region's political territorial subdivisions (prefectures and cantons) and the region's larger urban centers (Geragas, 1925: 123–124). Although, we discussed the history and geopolitics of toponymy in Greece in the third chapter, some discussion of re-toponymy in the New Lands is instructive. When possible,

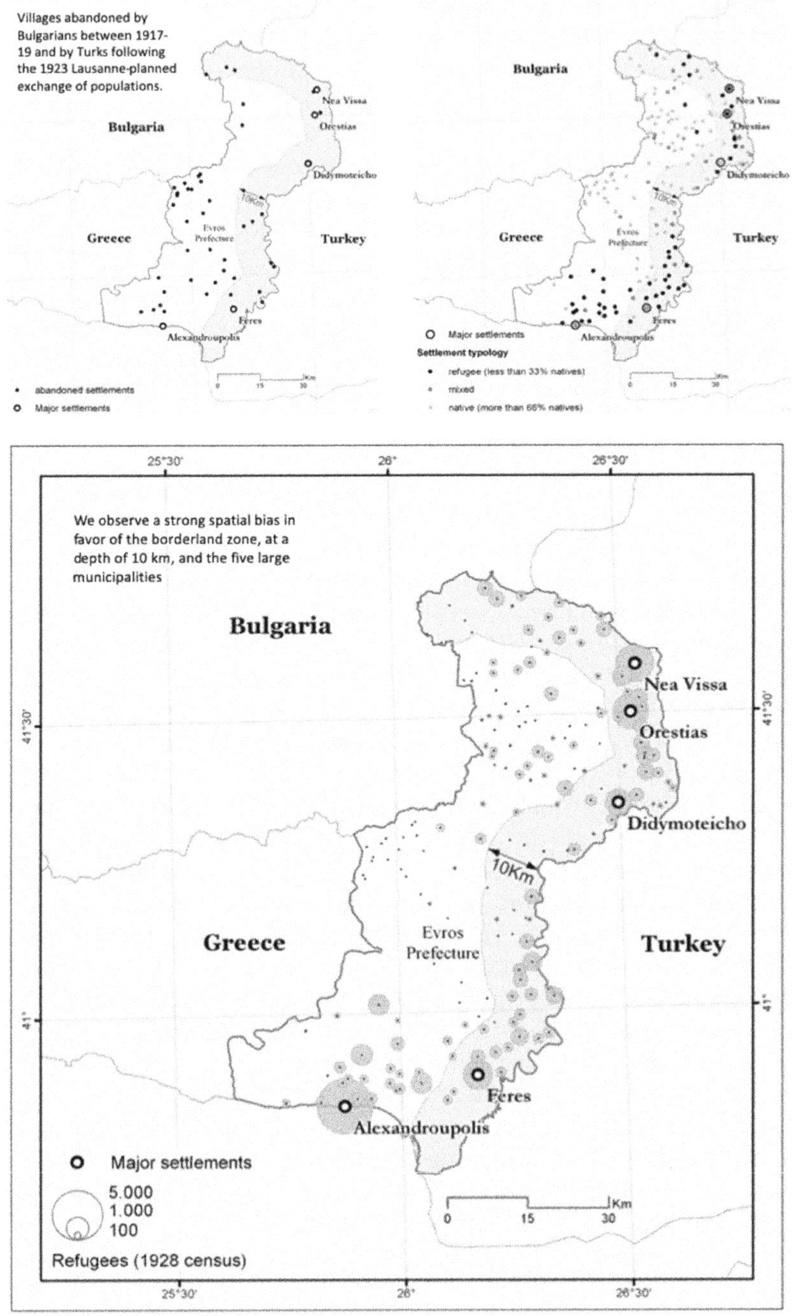

Figure 7.7 Spatial distribution and typology of refugee settlements in the borderlands of the Evros Prefecture.

Source: Refugee census, 1929; Evangelos Papadias, GIS/cartography.

Hellenic toponyms of old were resuscitated, although, in some cases, politicians' and administrators' zeal to erase the Ottoman past drove them to excess and error. A notable example is the unnecessary renaming of the capital city of the Prefecture of Rhodopi, Gumuldjina (an Ottoman loan of the Byzantine place-name Goumoutzina) to the more Hellenic-sounding Komotini (Kyriakidis, 1926: 4). Where the restoration of Hellenic toponyms was not an option, new toponyms were devised by either directly translating the Turkish toponym into Greek (as "Saray" to "Anaktorion" (palace)), referencing Greek mythology (Amphitrite; Kentauros (Centaur)), Greek Orthodoxy (Theotokos (Mother of God); Amphia (vestments)), the site's natural setting (Kallithea (Belview)); its flora (Amigdalia (almond tree)), its geology (Kitrinopetra (yellowstone); Theiafi (sulfur)), fauna (Kyknos (swan); Petinos (rooster)); the built environment (Tris Myli (three mills)), Ohiro (fort)); or an occupation of the inhabitants (Amaxades (carriage builders); Siderades (ironsmiths).

There is an ontological tension between the studied, historicist approach to the restoration of toponyms of long-existing settlements and the lexical ordinariness of invented toponyms. On the one hand, there is a practical need to name new refugee settlements and settlements. On the other hand, there is interplay on display between the state's desire to make Hellenic landscape elements hyper-visible via toponymy, while simultaneously constructing a geography of place-names that express senses of "nowhere" and "everywhere." The proliferation of popular and predictable place-names, such as "Saint Georgios" (there are 66, nationally), "Kampos" (plain) (38), "Saint Dimitrios" (32), "Milies" (apple trees) (27), and "Platanos" (plain tree) (26), further disorient the map reader and the visitor (Kallivretakis, 2003: 19–20). Suppose the resurrected Hellenic toponyms are, in the psychoanalytic sense, expressions of a conscious, performative Hellenic nationalism, the towns and villages that bear names such as "Small Fountain," "Long Bridge," "Oak Tree," and "Elk-village," territorialize the highly contested lands of Thrace and Macedonia in terms of a surface unpolitical, 'unconscious' topography of stone, water, flora, and animalia. In this manner, two territorial qualities are promoted: non-Hellenic toponyms are replaced, and the region is resignified and signed as ordered and domesticated, garden-like and tranquil, Arcadian, and bucolic, and certainly not as a zone of contention and geopolitical engagement. Although the state has advertised since 1920 its openness to local community toponymic selections, contingent upon the Ministry of the Interior's approval, the country's toponymy still reflects the practices of that era.

Second, by 1947 the construction phase of new refugee settlements and the secondary expansion of pre-1923 ones appear to have paused, the settlement hierarchy being mostly unchanged since 1936. Correspondence between the Agricultural Bank of Greece and the Ministry of the Interior and the Army, dated 1936–1938 discusses the implications of the gradual out-migration of refugees from Western Thrace's border zones to Macedonia and central and southern Greece, where opportunities appeared to be more plentiful [Historical Archives of the Army]. The financial crisis of the 1930s, which impacted the state's ability to fund housing and regional development sustainably, was further deepened by Ioannis Metaxas' construction of fortifications along the Eastern Macedonian border with Bulgaria (Kostis). The Second World War and the Civil War that followed delivered the final blow to any formal, long-term settlement and town development program. One of Ioannis Metaxas' propagandist personas was that of the country's "First Farmer," an imaginary that needed the marshaling of precious, residual resources and the mobilization of Agricultural Bank of Greece's network of agents and branches in the service of the borderlands' territorial control. Supports included land, seed capital, and a kit of farming equipment, thereby limiting state resources for funding housing and regional development through grants and loans.

Third, between the state's penury and a decade of war and conflict, minimal new road infrastructure was completed, although the siting of these minimal improvements illustrated the state's perspective on regional security and development: Egnatia Way, the cortical east-west axis originally etched during the Roman period was gradually paved. New paved west-bound roads out Didymoteicho and Nea Orestias linked east-west settlement zones along the Greek-Bulgarian border, to the north-south settlement zone along the border with Turkey. In the Pomak highlands of Xanthi and Rhodopi, the absence of notable improvements suggested isolation and insolation. Except for the completion of the road linking Xanthi to the Bulgarian border at Buyuk-Dere, which passed by Pomak town of Echinos, there was no road construction whatever.

The Statist ecosystem of the borderlands during the Ioannis Metaxas dictatorship entangled refugee resettlement, the extension of technical-financial services for rural development and economic stabilization, and, importantly, the preparation of ground conditions for a robust defense against adversaries to the east and north. These objectives profiled Athens' script for national statecraft in the late phase of the Interwar period. The garrisoning of the borderlands, arrested all efforts to integrate allophone and Muslim communities to Greek society.

In the historical case studies that follow in Chapter 8, we explore the state's strategic calculus and plans in pursuit of these objectives. We distinguished two phases in the transformation of borderlands: in a first, subtractive phase, the state ethnically cleansed them. An institutive phase followed, which saw new infrastructure, services networks, settlements, and state and parastatal capacity extended to the region. In the case study of the City of Didymoteicho, we observe how the co-evolved State-Military-Ecclesiastical complex shapes still today. the borderlands' geopolitics of difference, mobilizing informal, temporary, less regulated, less visible, project-specific coalitions.

Notes

1. Originally constituted as the Republic of Macedonia (a constituent regional state of federated Yugoslavia), Macedonia emerged in 1991 as one of six sovereign successor states of the Federal Republic of Yugoslavia. Greece challenged the choice of "Republic of Macedonia" as country name on, mainly, cultural appropriation and international security grounds and succeeded in forcing the new state to operate in the context of international organizations as "The Former Yugoslav Republic of Macedonia - FYROM." Disallowed to use its chosen constitutional name in formal international settings, the Republic of Macedonia engaged in a long, contentious, and ultimately successful diplomatic process that led to international recognition. Importantly, reconciliation and unity between its Slavic majority and its large Albanian minority populations were prerequisites to any negotiation with Greece and international organizations, such as NATO and the European Union. In January 2015, Greece's newly elected Left government raised the prospect of a solution. In May 2017, Zoran Zaef was elected Premier of Macedonia on a platform advocating for a "responsible, reformist and European government". Zaef formed a coalition government with the Alliance for Albanians party – a consequential act that also led to the recognition of the Albanian language as "co-official" at the state level in 2019. By early 2018, pragmatists in the two governments were converging on the name "North Macedonia." Effective leadership, negotiation, and instrumental logic produced the Prespa Agreement that has been heralded as historic and a point of inflection for the Balkans. Since the ratification of the Treaty, the two countries have signed a broad range of economic and infrastructure agreements, which promise to foster prosperity in the broader region and pave the way to North Macedonia's accession to the European Union. North Macedonia acceded to NATO on 27 March 2020.
2. The line opened to traffic in two stages in 1872 and 1875. The remaining route (except that from Sarambey to Bellova) was completed in 1875, when the construction company ceased to exist. The present company working these lines was reorganized on 31 December 1878, with further changes approved by the Austrian Government on 16 October 1893 and 22 May 1894. This company, called the "Compagnie d'Exploitation des Chemins de Fer Orientaux," has been until very recently an Austrian company, domiciled in Turkey (Moody's 1910, 262).
3. The importance of the "Staggela" ethnographic map, printed in Athens, is its visualization of data from Turkish sources. The cartographer is unknown, as is the date of the map, which is presumed to be pre-1914. "G. Staggela & Compagnie" is identified on the map as "lithographers to the Royal Court," which gives it quasi-official status.
4. Since the end of the Second World War, the Greek diaspora has been defined by two processes: the return of the majority of populations of Greek culture from historical – in some cases, cosmopolitan – centers of Hellenism, such as Istanbul, places on the Black Sea coast of Turkey and the Caucasus, and Egypt (especially, Alexandria and Cairo), and the out-migration of Greeks as economic migrants to North America, Western Europe, and Australia.
5. The authors surveyed town and village settlements on Evros Prefecture's borderlands on 25 to 30 October 2017.

Sources

Beach, Charles F.; Segnitz, Henri. "The Railways of the Ottoman Empire." *Moody's Magazine* 9 (1910): 261–270.

Djilas, Milovan. *Conversations with Stalin*. New York: Harcourt, Brace & World, 1962.

Gondicas, Dimitri; Issawi, Charles Philip. *Ottoman Greeks in the Age of Nationalism: Politics, Economy, and Society in the Nineteenth Century*. Princeton, NJ: Darwin Press, 1999.

Hirschon, Renée (ed.). *Crossing the Aegean: An Appraisal of the 1923 Compulsory Population Exchange Between Greece and Turkey*. Studies in Forced Migration 12. Oxford, New York: Berghahn Books, 2003.

Iatrides, John O. "Revolution or Self-Defense? Communist Goals, Strategy, and Tactics in the Greek Civil War." *Journal of Cold War Studies* Volume 7, Number 3 (2005): 3–33.

Jelavich, Charles; Jelavich, Barbara. *The Establishment of the Balkan National States, 1804-1920*. A History of East Central Europe Volume VIII. Seattle and London: University of Washington Press, 1977.

Kallivretakis, Leonidas. "Historical Study of Settlements in Greece (15th-20th centuries): Research objectives and problems of the sources." In *Πληθυσμοί και οικισμοί του ελληνικού χώρου: ιστορικά μελετήματα*. Athens: Institute for Neohellenic Research/National Hellenic Research Foundation, 2003, pp. 7–24.

Karakasidou, Anastasia N. *Fields of Wheat, Hills of Blood: Passages to Nationhood in Greek Macedonia, 1870-1990*. Chicago: The University of Chicago Press, 1997.

Lazaretou, Sophia. "The Drachma, Foreign Creditors, and the International Monetary System: Tales of a Currency During the 19th and the early 20th centuries." *Explorations in Economic History* Volume 42, Issue 2 (2005): 202–236.

Mavrogordatos, George Th. *Stillborn Republic: Social Coalitions and Party Strategies in Greece, 1922-1936*. Berkeley, CA: University of California Press, 1983.

Perović, Jeronim. "The Tito-Stalin Split." *Journal of Cold War Studies* Volume 9, Number 2 (2007): 32–63.

Ripoll, Manuel Perez. "Preliminary Study of Pastoral Activities among the Pomaks in Greek Thrace." *British School at Athens Studies. Zooarchaeology in Greece: Recent Advances* 9 (2003): 291–295.

Sfetas, Spyridon. "Autonomist Movements of the Slavophones in 1944: the Attitude of the Communist Party of Greece and the Protection of the Greek-Yugoslav Border." *Balkan Studies*, Volume 36, Number 2 (January 1995): 297–317.

Sivignon, Michel. "La Grèce devant l'adhésion de la Turquie." *Hérodote* 118, no. 2 (2005): 82–106.

Βογιατζής (Voyatzis), Γεώργιος. *Η Πρώιμη Οθωμανοκρατία στη Θράκη. Άμεσες Δημογραφικές Συνέπειες* (Early Ottoman rule in Thrace. Direct demographic consequences), Αθήνα: Ηρόδοτος, 1998.

Γεραγάς (Geragas), Κώστας. *Αναμνήσεις εκ Θράκης 1920-1922* (Réminiscences of Thrace 1920-1925). Αθήνα: Τυπογραφείον «Εστία», 1925.

Εξερτζόγλου (Exertzoglou), Χάρης. «Η Ιστορία της Προσφυγικής Μνήμης». In Αντωνης Λιακος (Liakos) (ed.), *Το 1922 και οι Πρόσφυγες. Μια Νέα Ματιά* (The story of the refugee memory in 1922 and the Refugees. A new look). Αθήνα: Νεφέλη, 2011: 191–202.

Καλαντζής (Kalantzis), Γιώργος. *Το Ελληνικό Κράτος και οι Μουσουλμάνοι της Δυτικής Θράκης 1903-1928. Από τη Στρατηγική Συμμαχία στην Εγκατάλειψη* (The Greek State and the Muslims of Western Thrace 1903-1928. From strategic alliance to their abandonment). Αθήνα: Θεμέλιο, 2016.

Καραδήμου- Γερολύμπου (Karadimou-Gerolympou), Αλεξάνδρα. *Μεταξύ Ανατολής και Δύσης. Θεσσαλονίκη και Βορειοελλαδικές Πόλεις στο Τέλος του 19ου Αιώνα* (Between East and West. Thessaloniki and Northern Greek cities at the end of the 19th Century). Θεσσαλονίκη: University Studio Press, 2004.

Καυκούλα (Kafkoula), Κική. «Μια Ευρωπαϊκή Πόλη από Ευρωπαίους Αρχιτέκτονες. Οραματισμός και Τεχνογνωσία στην Ανοικοδόμηση της Τζουμαγιας Σερρών Μετά το 1918» (A European city by European architects. Vision and expertise in the rebuilding of Tzoumagia, Serres). Πρακτικά Β' Διεθνούς Επιστημονικού Συνεδρίου: *Οι Σέρρες και η Περιοχή τους. Από την Οθωμανική Κατάκτηση Μέχρι τη Σύγχρονη Εποχή*, 2013: 681-701.

Κουτσούκος (Koutsoukos), Βασίλης. *Από την Αυτοκρατορία στο Εθνικό Κράτος: Χωρικές Οψεις της Ενσωμάτωσης της Θράκης* (From empire to nation-state: Spatial perspectives on Thrace's incorporation). Διδακτορική διατριβή, Πανεπιστήμιο Κρήτης, Φιλοσοφική Σχολή, Τμήμα Ιστορίας και Αρχαιολογίας, 2012 (https://elocus.lib.uoc.gr/php/pdf_pager.php?filename=%2Fvar%2Fwww%2Fdlib-portal%2Fdlib%2F2%2F4%2Ff%2Fattached-metadata-dlib-1348563823-156717-2883%2F20120627_Koutsoukos_Vasilis_Phd.pdf&rec=&do=&width=595&height=842&pagestart=1&maxpage=839&lang=el&pageno=1&pagenotop=1&pagenobottom=765 Accessed on 8 August 2020).

Λιακος (Liakos), Αντωνης (ed.). *Το 1922 και οι Πρόσφυγες. Μια Νέα Ματιά* (1922 and the refugees. A new look). Αθήνα: Νεφέλη, 2011.

Μαυροκορδάτος (Mavrogordatos), Γεώργιος Θ. «Οι Εθνικές Μειονότητες» (The national minorities). In Χ. Χατζηιωσήφ (ed.), *Ιστορία της Ελλάδας του 20ού αιώνα* B2 (Twentieth century history of Greece, volume B2. Αθήνα: Βιβλιόραμα, 2003.

Κυριακίδης (Kyriakidis), Στίλπων. *Η Δυτική Θράκη και οι Βούλγαροι* (Western Thrace and the Bulgarians). Εν Αθήναις: Βιβλιοπωλείον 'Ιωάννου Ν. Σιδέρη, 1919.

Κυριακίδης (Kyriakidis), Στίλπων. *Οδηγίαι δια την Μετονομασίαν Κοινοτήτων και Συνοικισμών Εχόντων Τουρκικόν ή Sλαβικόν Ονομα* (Instructions for renaming communities and settlements with Slavic and Turkish names). Εν Αθήναις: Εκ του Εθνικού Τυπογραφείου, 1926.

Σωτηρόπουλος (Sotiropoulos), Δ. Π., Παναγιωτόπουλος (Panagiotopoulos), Δ. «"Ειδικοί" Διανοούμενοι και Θύλακες Χειραφέτησης στο Μεσοπόλεμο. Μεταρρυθμιστές Γεωπόνοι και Μηχανικοί στην Υπαιθρο και το Αστυ» ('Specialist' intellectuals and pockets of emancipation in the Interwar Period. Reform agronomists and engineers in the countryside and the city). *Μνήμων* 29 (2011): 121–150 (file:///C:/Users/User/AppData/Local/Temp/7716-13692-1-PB.pdf (Accessed on 18 August 2020).

8 The Hellenization of borderlands II

The Greek-Bulgarian border, Mount Sapka, and the City of Didymoteicho

In the two historical case studies of Eastern Macedonia's and Western Thrace's borderlands, the protagonist was the post-"Lausanne," post-Metaxas-coup authoritarian state. Set in the inter-war period and at a time of significant demographic, economic, ideological, and political regime realignments, these case studies illustrate how the national center and liminal peripheries are equally implicated in making external and internal borders and borderlands. In the third case study of the city of Didymoteicho, we observe Athens, the co-evolved complex of state (now democratic), church (still constitutionally linked to the state), and military (professionalized and under civilian control) engaging in politics of place in the periphery supported by formal, regulated, and visible networks and routines. Athens interlocks with special-purpose, informal, less regulated, and less visible networks of local agents to manage urban change, which explains the geopolitics of difference in the borderlands. In his exploration of 'the territorial trap' in international relations, Agnew posits that "even when rule is territorial and fixed, territory does not necessarily entail the practices of total mutual exclusion which the dominant understanding of the territorial state attributes to it" (Agnew, 1994: 54). Indeed, the borderlands politics of place give substance to his contention. The case studies reveal differences in how borderland territory, identity, and geopolitics are perceived and acted on in the pursuit of Hellenic statecraft and nationcraft.

Bordering and borderlands in the inter-war period

Post-"Lausanne," the state governed with limited public finances and international credit. The onset of the World Economic Crisis of 1929 limited further Greece's public finances. The state was aware that borderland security was entirely contingent upon its ability to draw up a population settlement schema that was sustainable by the measure of its modest resources. It pursued that objective through subsidiary state institutions (the Army and Ministries with relevant portfolios), local elites, and financial institutions (the National Bank of Greece and the Agricultural Bank of Greece). These were dependent on the state, per their charter, and on international credit markets by capital markets circumstance. The banks, then, operated as parastatal entities, at least for the borderlands strategy.

Describing the hierarchical structure and agency of financial networks in inter-war Greece, Catherine Brégianni notes that banking and other technical services and professional networks and how they were organized reflected society's and the state's dominant ideology, the range and positionality of social partners, and the symbols (progress, equity) chosen for their broad legibility (Brégianni, 2019: 14–15). The Agricultural Bank of Greece (ABG) was founded in 1929 to support expansion of credit markets, managed until that time by the National Bank (NB).[1] The two banks established a condominial hold over rural credit through a rapidly expanding network of branches. The cadence of their establishment and their geographic distribution strongly suggest a shift in State policy. While still committed to settlement and rural development as previous governments, the Ioannis Metaxas regime instrumentalized banks and their networks in its pursuit of security objectives.

The Agricultural Bank of Greece experiences explosive growth after the regime's installation in 1935, in part because the worst effects of the World Economic Crisis had abated by then and in part because of the regime's conservative, agrarianist orientation (Ibid: 32). By opening 26 branches in 1935 across all regions, ABG became a national provider of credit (Table 8.1). 14 of these branches were opened Macedonia and Western Thrace, which were also the geographic focus of refugee resettlement (Ibid: 31) (Table 8.2).

The first case study refers to state intervention in the 300 km Greek-Bulgarian borderlands, which extended across the north of Eastern Macedonia and Western Thrace. The fortification of the northern frontier was a major concern of the Ioannis Metaxas authoritarian, monarchist, and nationalist regime. Metaxas came to power by coup d'état on 4 August 1936 with the consent of King Georgios II. The new regime's declared intent was to safeguard Greece from Communism (Cliadakis, 1979: 117) but it was more intent on dealing a blow on Greece's liberals.

The study of the Greek-Bulgarian borderlands in the late 1930s is a showcase of the first, subtractive phase of the state's settlement strategy. Documents in the Hellenic Army Historical Archive dated to 1937–1938 reveal the rationale for the eviction of "suspect" populations from the borderlands, the desired outcomes, and the steps to take in advance of resettling loyalist farmers.

Table 8.1 New Agricultural Bank of Greece branches, 1929-1939 (ABG, 1940: 201; Brégianni, 2002: 292)

Year	*1929*	*1930*	*1931*	*1932*	*1933*	*1934*	*1935*	*1936*	*1937*	*1938*	*1939*
Branches	1	1	7	1	0	5	26	16	28	20	5

Table 8.2 Geographic distribution of Agricultural Bank of Greece branches (ABG, 1940; Brégianni, 2019: 36)

Region	*Macedonia and Thrace*	*Thessaly*	*Peloponnese*	*Epirus*	*Continental Greece and Euboea*	*Crete*	*Islands*
1929	0	0	0	0	1	0	0
1930	1	0	0	1	0	4	1
1931	0	0	0	0	0	1	
1932	0	0	0	0	0	0	0
1933	0	1	0	0	0	0	0
1934	3	1	0	0	1	0	0
1935	14	1	6	1	2	1	1
1936	7	1	3	0	3	0	2
1937	4	4	10	3	3	2	2
1938	6	1	2	2	4	0	5
1939	2	0	2	0	0	0	1
Total	37	9	23	7	14	8	12

Grand total: 110 branches

The second case study, which is paradigmatic of the second, institutive phase of state strategy, studies the Mount Sapka region, which straddles the Greek-Bulgarian border at Rhodopi and Evros in Thrace. Army, Government, and Agricultural Bank of Greece correspondence dated 1937–1939, describes their coordinated strategy for the sustainable resettlement of that region. Reporting the region as ethnically cleansed, the Bank prospects the target region's soil and economic geographies and proposes credit schemas and technical support to prospective settlers. The Army is to facilitate discovery and filter the Bank's recommendations to the Government.

The borderlands strategy was set in motion soon after the installation of the Metaxas regime. However, its regional development dimension faced significant obstacles: bureaucratic inertia at national and regional government levels (the Army had to resubmit the same plan in successive years), difficulty identifying appropriate settler populations willing to relocate into a remote and inhospitable part of the country, and, most importantly, the outbreak of the Second World War on 28 October 1940. When Nazi forces reached Greece's northern borders on 6 April 1941, Metaxas Line's 21 border forts were in an advanced state of completion. The full resettlement of the target regions, intended to support border security, was not. "The enemy from the North" was not Bulgaria, after all, although Bulgaria, as a Nazi collaborationist state, participated in the occupation of parts of Greece during the War. The borderlands strategy, built on the pillars of fortification, resettlement, and rural development, was designed at a measure appropriate to Balkan geopolitics – not a World War. German army superiority overwhelmed Greece's defenses and brought the war south by 9 April.

Critically, state strategy established durable facts on the ground that informed regional policy in the post-Second World War era. Demographic and toponymic Hellenization of settlements, the isolation of "suspect" populations in surveillance zones, and the extension of technical-professional services networks to the periphery served as bases of statization in Eastern Macedonia and Thrace for several decades in the future.

Case study I: the Greek-Bulgarian borderlands of Eastern Macedonia and Thrace

The borderlands security strategy of the authoritarian, agrarianist Government of Ioannis Metaxas was built around Maginot Line-inspired fortifications and bunkers arrayed along the Greek-Bulgarian border (Featherstone et al, 2011: 67). The desire to shield to the extent possible the construction of these installations and then secure them from domestic threats required the political and social stabilization of the zone. The removal of populations deemed potentially disloyal to the state was to be followed by the resettlement of a loyalist population from Greece's historic core regions. The settlers had to be planted on suitable agricultural lands (in short supply in stretches of the Greek-Bulgarian border zone) to make their communities durable and sustainable. Properties in the hands of "suspects" were expropriated. Public land was parceled and made available to the plan. Furthermore, the new settler farmsteads had to be financially and logistically supported by the state, which had rural development as an additional objective to border security.

Where precisely in the target zones and how these actions were to be taken, had to be determined by establishing the physical geographic, demographic, and economic ground truth. The state had assets and resources in the region, most notably an elevated military presence in both urban and rural borderlands settings. The state's geographic reach and capacity were extended by the development of Greece's banking sector post the Greek-Turkish War. Loans provided by international banks and guaranteed by the League of Nations for refugee resettlement and support also underwrote the creation of the national Bank of Greece and the Agricultural Bank of Greece, both of which were instrumental in developing the agricultural credit market and credit markets generally. Expansion of credit was a prerequisite to the state's "security-resettlement-regional development" stratagem.

Army and Department of Agriculture sources do not discuss the out-migration of non-Greek ethnics from Eastern Macedonia and Thrace, to wit whether it expressed chain migration, forced migration, or outright expulsions. However, it is clear that the removal of "populations of suspect national sentiment" was considered a strategic objective

Table 8.3 Estimates of population density in Greek-Bulgarian borderlands areas targeted for State intervention. Low population density was considered a strategic liability

Target Areas	*Approx. area (sq.km)*	*Population total in 1928*	*Inhabitants per sq.km*
Paranestios	770	5250	7
Papadon	530	3100	6
Kato Nevrokopio	800	14500	18
Territory north of Xanthi [Pomak region]	600	12700	21
Territory north of Komotini	500	5700	11.5
Mean national population density			52

of great importance. The rapid and profound depopulation of the region by its non-Greek communities also resulted in reductions in economic activity, itself a corollary of territorial security. Therefore, a fundamental issue for building security capacity in the borderlands was their repopulation.

A secret report by the Army's Section III (Intelligence) dated to 8 December 1937 and signed by the head of the General Staff of the Army (ΓΕΣ), Lieutenant General Alexandros Papagos, laid out the challenges to borderlands military security and made recommendations for meeting them: the challenges were the sparseness of population coupled with the continuing presence of non-Greeks in the region, the nature of the terrain, which limited choices for economic development, and the cost of state interventions (Table 8.3).

> "It has come to our attention that the borderlands, especially those to the North and East of the Nestos River... are either sparsely or very sparsely inhabited, to the degree that a visitor may consider the area abandoned. From a military standpoint, this is very serious.... The thinness of population in these areas has an immediate impact on the security of national territory... The small number of settlements and the numerous border crossings facilitate the entry of intruders from neighboring countries. In a crisis involving Bulgaria, infiltrators will be able to destroy vital installations..." (Ibid: 1–2).

The report identified the need to complete the removal of non-Greek ethnics from villages in the three zones and then resettle the villages with loyalist farmers (also build new ones as needed) (Table 8.4). The objectives of these actions were the built-up of population density commensurate with appropriate logistical support of military activity, and the political stabilization of the region by population reshuffling.

Table 8.4 "Counts of persons of suspect national beliefs in the defense areas," General Staff of the Army, Office III, Section A, in Athens, 15 November 1938

A. Zone anchored by Poroia and the Strymon River

Villages	*Total population*	*Hellenes*	*[Pop.] Suspects*
Neo Petritsi	1594	1512	82
Ano Poroia	1096	1070	26
Platanakia	640	625	15
Leivadia	415	400	15
Makrynitsa	320	310	10
Kato Poroia	542	532	10
Vyroneia	460	450	10
Agriolefka (Oumpagia)	16	6	10
Filyra	117	110	7

B. Zone anchored by Sidirokastro-Achladohorio and Aggistro

Ano Oreini	450	10	440
Aggistron	658	249	412
Ahladochori	940	625	315
Krasochori	281	0	281
Faia Petra	174	21	153
Kleidi	97	0	97
Sidirokastron	4018	3923	95
Ano Vrontou	250	175	75
Karydochori	285	240	45
Vathytopos	430	400	30
Kapnofyton	285	155	30
Damaskinon	29	0	29
Kapnotopos	52	44	8
Katafyton	235	230	5

C. Zone anchored by Kato Nevrokopio and its hinterland and the Greek-Bulgarian border

Kato Nevrokopi	2268	716	1552
Volax	911	300	611
Perithorio	1377	895	482
Pagoneri	393	93	300
TOTAL	*18333*	*13188*	*5145*

Once the subtractive phase was completed, the state envisioned a program of financial supports for settler families in the form of loans and grants, which would generate sustainable rural development, secure the repayment of loans, thereby making the state's treasure whole, and foster growth in commercial agriculture (ΓΕΣ, 1937a: protocol number 36095). That second, institutive phase of the borderlands settlement strategy became fully expressed between 1937 and 1939 in state's plan for the Mount Sapka region in Western Thrace's Rhodopi.

Case study II: the Mount Sapka region, Thrace

The Metaxas regime identified the region as critical to its borderlands resettlement, security, and rural development plans. In 1938, 15 years after the signing of the Treaty of Lausanne and the population exchange, the spatial sorting out of Western Thrace's ethnic Greeks, Turks, Bulgarians, Pomaks, and Roma was ongoing. Reports by agents of the Ministry of Agriculture and the Agricultural Bank of Greece leave little doubt that between 1919 and 1939, out-migration from the borderland region of Mount Sapka of non-Greek ethnics was a mass phenomenon (Figure 8.1). In the area surrounding Mega Dereio (Prefecture of Evros), out-migration amounted to reductions of 94.25 percent of the district's original population. Notable is the complete de-Bulgarization of villages in that part of the Sapka region. Bulgarians are gone from the Babalar (Goniko) District. Out-migration of Turkish

Figure 8.1 The State-Military-Church complex is territorialized in the grouping of the Cathedral Church, the municipal police headquarters, City hall, which is adjacent to the local nursing college, and the monument to fallen aviators featuring a F-5 fighter jet set on a concrete plinth in the college's front yard.

Source: Alexandros Ktistakis, with permission.

Table 8.5 Population counts and ethnic designation of settlements in the Mount Sapka region. General Staff of the Army, Office III, Section A, in Athens, 15 November 1938. The report documents high levels of out-migration, including forced migration, and depopulation. Notably, non-Hellenic populations are vastly reduced

Village [Contemporary toponym, when available][2]	*Ethnic identity of prior inhabitants*	*Number of families before the War*	*Ethnic identity of current inhabitants*	*Current number of families*	*Percent change*
SAPKA DISTRICT					
Glyka (Baltiran)	Turkish	42	N/A	0	-100
Monastirion	Bulgarian	350	N/A	0	-100
Pontikia	Bulgarian	300	N/A	0	-100
Tahtajik [Sanidochori]	Bulgarian	25	N/A	0	-100
Pessani	Bulgarian	150	N/A	0	-100
Moukante	Turkish	15	N/A	0	-100
Katrantzides	Turkish	50	N/A	0	-100
Mega Dereio[3]	Bulgarian	350	Turkish	20	-94
Kotronia	Bulg. & Turkish	120	Greek refugees	27	-77.5
Distibaki	N/A	35	N/A	N/A	
[Demir-Ouren] Sidiro	Turkish	100	Turkish	29	-71
BABALAR DISTRICT (Goniko)					
Rousa	Turkish	18	Turkish	4	-78
Armoutli [Apidochorion]	Turkish	45	Turkish	2	-96
Telibiler [possibly, Archontiko]	Turkish	15	Turkish	0	-100
Mylopetra	Turkish	25	Turkish	4	-84
Imamlar [Agrafiotika]	Turkish	20	Turkish	0	-100
İnceköy	Turkish	12	Turkish	0	-100
Dikeli-Tas [Dikella]	Turkish	8	Turkish	0	-100
Soujaklar	Turkish	12	Turkish	0	-100
Ömerler	Turkish	35	Turkish	0	-100
Asia Mahalle [Asia]	Turkish	9	Turkish	2	-78
Tsokekli [or, likely, Çökekli]	Turkish	12	Turkish	2	-83
Tekes [Tavri]	Turkish	0	[N/A	0	
Goniko (Babalar)	Turkish	30	Turkish	7	-77
Sarpidonia	Turkish	20	Turkish	12	-40
Kioseler [possibly, Agnantia]	Turkish	4	Turkish	2	-50
Mesimler [Mesimerion]	Turkish	12	Turkish	5	-58
[Kara Ouren] Ourania (Vamantan Mahalle)	Turkish	20	Turkish	15	-25
Kiousoukli [Mikrakion]	Turkish	8	Turkish	2	-75
Outs-Evler	Turkish	30	Turkish	10	-67
Inagli, Maskarlar, and other very small mahalle	Turkish	50	Turkish	19	-62
TOTAL		1,922		162	-91.5

populations was also significant in Sapka, with the reduction amounting to 77.5 percent of the original population. Ministry surveys report that their Turkish occupants entirely vacated the villages of Telibiler, Imamlar, Injekoy, Dikeli-Tas, Soujaklar, and Ömerler. Concerning the resettlement of exchangee-refugees of the Greek-Turkish War, it is significant that only 27 families were registered as settled in the Village of Kotronia, a village 16.3 kilometers west of the Greek-Turkish border. Information on out-migration of refugees is not provided in table form, although there are references to the difficulties such a phenomenon caused to the financial viability of the state's resettlement, security, and rural development plans.

Clearly, the war had negative impacts on the regional economy, causing population flight, resource wastage, lowered productivity, and increased income inequality.[4] The drawing of an international border on the Evros River in 1923 separated the Sapka Region, the Evros Prefecture, and generally Western Thrace from parts of their hinterland in Eastern Thrace and Bulgaria, causing disruptions to long-standing commercial networks, flows of capital, demographic mobility, and cultural exchanges.

The absence of a hard-surface road network did not help. At the time of the report, the Evros Greek-Turkish borderlands, at the latitude of the Mount Sapka Region, included only 46 kilometers of paved roads, 32 of which – linking Alexandroupolis to Pessani – were constructed by the Army. Most country roads were described as impassable in the rainy season. The report recommended that a 60-80-kilometer extension of the road network would be sufficient to link that vital part of the borderlands to the urban network, and specifically to the City of Souflion (Ibid, 19). Amputated from the broad hinterland to the east, poorly connected to networks, severely depopulated, and featuring poor soils and a subsistence-oriented economic culture, Sapka presented a significant challenge to what was otherwise a reasonable set of planning assumptions. The sober assessment of the Ministry of Agriculture faced these challenges head-on by proposing an ambitious and comprehensive economic strategy: preserve "the best of the past," namely sustainable husbandry practices that integrated the use of forested land; extend forest resources through an aggressive afforestation effort; support existing commercial crops, such as tobacco and vineyards; replant mulberry orchards; and support apiaries and the commercial production of honey and its derivatives (Ibid, 25).[5] Further, there was a strong sense that, in addition to primary sector development, artisanal and cottage industries, and larger-scale commodity production, would be essential to the region's economic uplift. A relaunch of the silk industry that used to flourish in the region in the Ottoman era was desirable. After all, before 1922, the City of Souflion, at a distance of 15-30 kilometers from most of the Sapka settlements, was a renowned center of the silk industry. Cut off from its extensive acreage of mulberry orchards, and from its commercial hinterland, which was dominated by Adrianople and Istanbul, the Souflion-based silk industry largely collapsed. Wishing and planning its return was a vision that was not strongly supported by ground conditions.

However, a dolorous daily life, physical geographical limitations, and inadequate infrastructure were not inviting to the landless and disadvantaged refugees. Any pragmatic resettlement strategy would have had to involve considerable intervention on the part of the national government and, by extension, the Army, which often served as its proxy. In arguing for new settlers in 1937, General Papagos and his Agricultural Bank of Greece advisors suggested that loans with attractive terms could attract settlers, thereby fortifying the borderlands and contributing to regional prosperity. Land grants and loans should have supported the rural settlement strategy. The Ministry assessment, however, was that such instruments, which inspired a similar settlement strategy in Germany (based on land grants of 150 stremmata (15 hectares) and loans of 40,000 Marcs per family), failed to produce small sustainable farmsteads once the government subsidies and other supports were withdrawn (Ibid, 26). Under ideal conditions, the German model in Sapka, would involve the parceling of the 280,000 available stremmata (28,000 hectares) of productive land into land grants supporting 1,866 settler families, or enough new settlers to replace population losses. The best-case prediction was that the region's population would bounce back to pre-war levels. Realistically, the report conceded that Sapka, as a truncated-by-treaty region, could support no more than 500 new settler families (Ibid, 43), a fraction of the envisioned settlements of 3,000 families.

Agricultural bank of Greece's proposed plan for the Mount Sapka region

In the Spring of 1939, the policy was revised and reduced in scope. In his 26 May 1939 report to the Chief of the Army and the Ministry of Agriculture, Agricultural Bank of Greece Section Head, Christos Vasmatzis, proposed a plan for the resettlement and rural development of Sapka. He outlined the plan and laid out the prerequisites for success. His report suggests that new settlement-building in borderlands was an important, perhaps critical, element in the planning for both national defense of borderlands and regional and rural development. The report recommended unified planning and management for all actions under the auspices of a single head and agency. Drawing on ABG's technocratic operational modus, Vasmatzis recommends that the project engage only experienced cadres with local expertise who would extend instead of duplicating the capacities of Athens' based partners. Crucially, the report underlines the importance of financial and political buy-in by all local, regional, and national partners, Agricultural Bank of Greece included.

Vasmatzis laid out an extensive list of preconditions for success (Ibid, 26–37). They concerned planning, project governance and long-term management, transportation and social state infrastructure development, technical support (in the guises of surveying, mapping, and platting of settlements), and financial and material supports to farming households. As a workman-like document free of ideological language, it alluded to the significance of the rural development strategy for border security.

As a first step, and before any settlers arrived, the report recommended constructing a road network to support the construction of new settlements adequately. The plan did not provide budget information. In 1939, the Army was constructing the Alexandroupolis-Pessani road, and it would be reasonable to expect that further road-related expenses in the borderlands would have been folded into the Army's budget. It called for the drawing of detailed topographic surveys, the profiling of soils, the mapping of existing settlements, and the provisional platting of prospective ones. It did not provide budget figures for the work.

Second, much like in a previous iteration of the plan for the Eastern Macedonian borderlands, the report recommended drawing on populations of resource-poor areas in Western Macedonia and Thessaly. They were accustomed to hardship and more likely to adapt well to the rough environment of Mount Sapka. The character and capacities of the settler population had to be suitable to the plan's requirements, and if it failed to integrate fully into the regional economy, the overall strategy would quickly unravel. The concern was legitimate as settlement policy did not distinguish refugees as formerly urban or rural. It was, therefore, entirely possible to imagine urban families settled in the borderlands, which were entirely unprepared and unequipped for a life of farming (Notaras, 1934: 250).

Third, the report called for building at least five new settlements, each accommodating a hundred families. It called cautioned against creating small and widely dispersed villages. The tyranny of soil geography was acknowledged as a limitation.

Fourth, Vasmatzis outlined the plan's financial liabilities for both the ABG and the state. Concerning financial and resource support, each household would receive an amount of 140,000-150,000 Drachmas, a specified number of animals, and animal feed on the initial year of settlement. The budget schema conjured an image of the archetypical farmstead (Table 8.6).

Table 8.6 Projected expenses of model settler household

House, including a barn and storage shed	55,000
80 goats/sheep	32,000
2-3 harness sets	10,000
2 heads of draft animals	10,000
1 carriage	3,000
1 sow	1,000
Agricultural implements	5,000
Planting of a small vineyard, an orchard, a start-up apiary	10,000
Basic necessities, seed stock, a loom and other cottage-industry tools.	14,000
Total	140,000

(Amounts in Drachmas are approximate) (Ibid, 51)

Since the maximal goal was to settle 500 families, the expected financial burden was estimated at 70,000,000 Drachmas (Ibid, 52). Counter to the German model of land grants and loans to settler families, the Ministry recommended the transfer of ownership of public forests and pasturelands in the target region to the communities and the cooperatives they may form, under the proviso that such land resources would be held, managed, and exploited as commons. As for the loans, Vasmatzis is quite frank in stating that, given the unfavorable ground conditions, their repayment in full by the settlers was unlikely; therefore, they should be budgeted as grants (Ibid, 53–54). What is absent from the report is the consideration of informal private credit lending networks, such as moneylending by individuals? Their contribution to the high debt burden of farming households was significant. Operating outside statist frameworks, they remained unregulated and unaccounted for in planning documents (Alivizatos, 1938: 144–146).

Debt service here refers to the system of credit to rural populations developed and managed by and for the Agricultural Bank of Greece (with incidental service contributions in the field by National Bank) with the necessary sanction of the state (ABG, 1929). The proposed plan demonstrated the extent to which state financial resources were commingled with bank reserves. It also strongly supports the view that, first, the banks were not firewalled against political interference, and, second, their solvency, long-term durability, and credit-worthiness in financial markets were of little concern to the state as long as higher policy objectives were being served.

At a more granular scale, Athens reached right inside the rural farmstead to interfere in everyday life. The optics of rural lending was of some concern since the program's inception. Fiscal discipline and rectitude were strongly recommended and regarded as an effect of a conservative value system. The AGB profiled the "ideal farmer"-cum-debt holder as follows: "Farmers must focus on their crops ... They must refrain from playing cards, and must also be indifferent to the speeches of certain menacing individuals ... Farmers must not be wasteful and avoid excessive expenditure ... One must nevertheless know that beer and spirits are not suited to the honest farmer's personality..." (ABG, 1929: 5). This time, surveillance of a loyalist population was consistent with the nature of the regime.

Vasmatzis' fifth recommendation to the state was building an appropriate water management infrastructure to support husbandry activities.[6] The projected cost was 12,000,000 drachmas (Ibid, 52). Additional infrastructure at the community level, such as plant nurseries, clinics, veterinary support, and lumber and forestry resources, were projected to cost 15,000,000 drachmas. The report proposed allocating 1,000,000 drachmas per year, with a budget horizon of ten years, to afforestation and the construction of hillside terraces. For the Region of Sapka alone, then, government expenditure was projected at 107,000,000 Drachmas, a considerable amount of capital at a time that a general European war appeared all but inevitable.

On 1 September 1939, three months after the submission of Vasmatzis' report to the General Staff of the Army, Nazi Germany invaded Poland, thus triggering the Second World War. Greece remained neutral until Italy's declaration of war on Greece on 28 October 1940. Since Italy's occupation of Albania in Spring 1939, Greece was its next territorial objective in the Balkans. Italy launched its land invasion from Albania in late autumn, expecting a short campaign. Instead, Greece repelled the invasion and forced the retreat of Italian forces into Albanian territory. That was the first military setback of Axis forces in the war. British, Australian, and New Zealander reinforcements of the Greek front precipitated Germany's intervention to support its Axis ally. Exposure of its southern flank as it prepared to launch its invasion of the Soviet Union was concerning. Greece lacked the military resources and forces to fight a two-front war. The Metaxas Line fortifications, designed to accommodate a garrison of 200,000 soldiers, were defended at 30 percent strength since the defense of the Greek-Albanian border absorbed a vast majority of Greece's military capacity. Persistent attacks by overwhelming Italian forces on the Greek-Albanian front through the winter and Germany's invasion of Greece through Bulgaria (an Axis ally since 1 March 1941) in April 1941, determined Greece's fate.

Greek strategists' long-standing assumptions that Bulgaria would attempt to reclaim its territorial losses of the Second Balkan War and the First World War were correct. Though ultimately unsuccessful in securing the Greek-Bulgarian border, the Metaxas Line project's strategic logic was sound but politically repugnant as it was based on displacements, loyalty tests, and surveillance. Upon the withdrawal of Greek forces, Bulgaria, as an Axis ally, occupied Western Thrace and the parts of Eastern and Central Macedonia it long-sought to annex. Germany assumed the occupation of the Greek-Turkish borderland zone over Bulgaria's protests. Consistent with classic Balkan statecraft, Bulgaria commenced the re-Bulgarization of occupied territories in earnest to roll back the region's Hellenization. The public and the Greek Resistance fiercely opposed the Bulgarian occupation, which ended on 10 October 1944.

By the onset of the war, Metaxas' strategy had been partially implemented along the Greek-Bulgarian border and was at the planning stage on the Greek-Albanian border. The stratagem entangled the building of defensive fortifications, the ethno-nationalization of the population, resettlement of loyalists, and rural development through an expanded credit economy. Fortifications endured on the landscape and served (and still serve) national sentiment as historic battlefields do. Nonetheless, the significance of population sorting and the expansion of statization through professional services, such as banking, although incomplete, should not be discounted. They concretely contributed to the Hellenization project and the definition of transactional politics between the state and Western Thrace's Minorities. Importantly, Metaxas' stratagem is the starting point of the long period of national garrisoning that defined the oppositional relationship of Hellenism to difference.

Case study III: Didymoteicho on the Greece-Turkey border: the Metropolitan's cathedral, the mayor, and the fighter jet

"Διδυμότειχο μπλουζ" (*Didymoteicho Blues), 1991*

(Excerpt)
...Ethnic rifts bummed cigs and pranks
Same johns like at 'Kino Omonia'
Like cheap cologne;
Anyway, back to mine
to my grubby khakis, the morning bugle,
To my silent watch;
To my transfer, to my crazy shtick
To my raspy wakeup,
This-and-that, anyway you have it;
Off to Didymoteicho; ...
...Didymoteicho blues with a scratchy guitar
Didymoteicho blues my lost dime
Didymoteicho blues my greasy comb
My wet 'rettes, rotten fruit in the pallets
Didymoteicho blues;
Didymoteicho blues its name is 'because',
Didymoteicho blues a hole on the map
Didymoteicho blues empty photo
The service, absurd, the stressful inanity

Yiannis Bach Spyropoulos, *lyrics*
Lavrentis Mahairitsas, *music*

The City of Didymoteicho on Evros' Greek-Turkish border has long been considered a rough place to serve in the military. This popular and much resented by Didymoteichans song, *Didymoteicho Blues,* is in the voice of a young conscript. Cold and wet weather, poor soldier housing, dismal on-base food, boredom, and inanity render the city "a hole on the map." Web lists with titles, "Worst places to do one's military service," almost invariable include Didymoteicho.[7] Per the lore, once in Didymoteicho, one's first thought is how to secure a transfer somewhere else. The lore further suggests that the Metropolitan (the Bishop) can make it happen.

Borderlands cities, like Alexandroupolis, Nea Oresteias, and Didymoteicho, reterritorialized the New Lands as *Hellenic* lands and central places for administration and the economy. Their geopolitical liminality and importance as border settlements also raised the question of national settlement policy-as-strategy we addressed in Chapter 7. The pattern was, ultimately, drawn by the intersection of state intention and capability – a strong desire to Hellenize the New Lands, until 1974, as an illiberal, garrisoned state.

The shaping of the borderlands' and Didymoteicho's geopolitics of difference started in 1919. Deputy Governor-General of Thrace, Kostis Geragas (1925), notes that between 1919 and 1922, the state riding on the successes of the war, set a clear direction for the newly annexed Western Thrace: pacification of the countryside by expelling Bulgarian paramilitaries, adopting confidence-building measures in its dealings with the Muslim minority, and investing in urban renewal and infrastructure.

On 22 September 1922, following the defeat, Greece acceded to the terms of the Treaty of Moudania for the evacuation of Eastern Thrace. Great Britain and France had not intervened to avert the destruction of the Greek forces in Asia Minor. However, they mediated "Moudania" to protect their access to the straits, limit civilian casualties, and maintain a semblance of diplomatic impartiality. The evacuation started on 30 September, once instructions regarding means and methods were communicated to the panicking public, now augmented in size by westbound refugees crossing the straits. Gerakas describes scenes of toil and despair on the roads: "For many days, the road out of Arkadiopolis all the way to Karagac, was covered by a continuous stream of farmers' wagons, each carrying a family and its animals. The drivers were quiet and sickened, their tragic silence broken by the sound of creaking wheels. They moved without stopping toward salvation, west of the Evros" (Geragas, 174–176). Following the Hellenic Army's complete withdrawal on 16 October, the last civil administrators left Kallipolis (Gallipoli) on 11 November. In the meantime, Greek authorities transported across Evros River as much movable property as possible, including cut lumber and other building material, agricultural equipment, including plows and threshing machines, and other supplies that would prove essential to resettlement in the following months.

By Greek government estimates, approximately 185,000 people were evacuated from Eastern Thrace west of Evros River in that short period. Some 68 percent of them were farmers, with the remaining population recorded as urban (Venizelos Archive, 1922: 030–104). Much of this population will shape the demography of the borderlands.

The securitization of the borderlands was of paramount importance. The Hellenic Army was retrenched and reorganized west of the river as the Army of Evros under General Theodoros Pangalos. Preparations were made to resume hostilities if negotiations with Turkey failed (Army Directorate, 1969: 140–143). The Treaty of Lausanne of 24 July 1923, established the status quo still in force.

Since 1923, "Lausanne" obligations set limits nationally to change and development. Lack of capital and the thinness of an institutional and infrastructural scaffold cemented the peripheral status of places like Evros Prefecture. Initially, it was a disaster mitigation strategy rather than long-term, sustainable plan for reconstruction and development. Reduced state capacity necessitated the joint State-League of Nations financing and management of refugee settlement, most of it in a border zone less than ten kilometers deep.

The pattern of post-"Lausanne" borderlands settlement included formally planned fringe belt developments in historic towns, like Alexandroupolis and Didymoteicho, as well as new master-planned towns (notably, Nea Irakleia in Eastern Macedonia and New Oresteias in Western Thrace). Many old Ottoman-era villages were abandoned, others were expanded to accommodate refugees, and many new villages were platted by the Hellenic Army's engineer-corps, most often without planning and development follow-up. There were, therefore, both formal and organic dimensions to urbanization and settlement.

Today, the morphology of these borderland places exemplifies how contests between different kinds of civil Hellenic identities (Helleno-Christian and non-Helleno-lingual and non-Christian identities) are waged. Moreover, it reflects how, and to what effect borderlands networks of the State-Church-Military co-evolving complex drive change at both center and periphery. The output of borderlands networks is continuously modifying perceptions of center and periphery in Athens and the borderlands.[8] As Agnew posits, although state authority may not be fully territorialized as conventionally-defined statization and sovereignty (for example, through policy or administrative agency on the ground), it may become otherwise spatially reconfigured and effective when its power is deployed by network flows (Agnew, 2005: 442). We can add here that, in the case of Didymoteicho, and, more generally the borderlands, the "stop-and-go" network flows of state regulation and public financing, have a decisive impact on how extensive and effective the sovereign reach of Athens can be: routinely, decision-making flows between center and periphery support established national planning sub-routines. However, at other times, flows are intermittent, expressing a "stop-and-go" sovereign reach by the state. When the network flows that affirm state power at "the local" are challenged by local agents, the state's effective sovereignty is challenged. The special-purpose local sovereignty regime crafted by Didymoteicho's City Hall and the Metropolitan Church to build the Cathedral Church did just that. It upended the state's land-use planning schema for Didymoteicho's new downtown public services pole (a relatively low-stakes impact), but, more consequentially, deviated from the state's embracing of a civil state standard of intercommunal politics (κοινωνία πολιτών) – a standard that the state would, unequivocally apply a few years later in the Pomak towns and villages.

In October 2017 and September 2019, we traveled the back roads of the Greek-Turkish and Greek-Bulgarian borderlands to photograph towns and villages at a ten-kilometer depth.[9] Border town landscapes made national statecraft visible to us. They are theaters of asymmetric power relations and geopolitics of difference. The town and village-built forms made us constantly aware of the legacies of regional power geopolitics and the Treaty of Lausanne's minorities regime. Importantly, the City of Didymoteicho's geopolitics of difference bound to its urban landscape, illustrates how Greece's liberal democracy manages the civil society "turn" in the periphery.

A strategic castle town since the Late Roman period, Didymoteicho served as a citadel in the Byzantine period, and, for a short time, as the capital of Emperor Andronikos III and imperial residence of Ioannes VI Cantacuzenos (Ousterhout, 1999: 195). Notable remains of walls and towers of the castellated Old Town – relic urban forms that sustain Paparrigopoulos' doctrine of "continuity of Hellenism" – are prominently showcased on the city's lavishly illustrated official website.[10]

The compact Byzantine castle-town came into Ottoman hands in 1361. Until the Balkan Wars outbreak, Didymoteicho was a majority Christian city with Muslim Turkish, Roma, Jewish, and Armenian minorities. On the eve of the First Balkan War, the Greeks and Turks represented some 95 percent of the region's population of 28,900 at a proportion of 2:1. The Bulgarian community of 1,400 was bolstered during short, subsequent periods of Bulgarian occupation and Bulgarization (1912-1913 and 1915-1919). Jewish and Armenian minorities completed the region's Ottoman-era demography. The town featured seven mosques, three mescids (small places of worship), three Orthodox churches, and a synagogue. Notably, its Jewish community, ranging in size at the turn of the nineteenth century between 600-1,000 per the counts of the Ottoman state and the Alliance Israélite Universelle respectively, had flourished with the opening of the economy following the Tanzimat reforms and, especially, after the arrival of the Edirne to Dedeagac railroad (Rodrigue, 1983: 266-267). However, wars and successive waves of Bulgarization, Hellenization, and population exchanges (the latter under the "Lausanne" treaty regime), dwindled its minority communities (Ibid: 279-281).

Situated on the right bank of Erythropotamos, a short distance from its confluence with the Evros River and two kilometers west of the Greek-Turkish border, the city was cut off from its traditional Eastern Thrace hinterland after the Greek-Turkish War. Didymoteicho has remained a modestly-sized city (population 9,362, 2011 census) with an economy largely dependent on the primary sector.

A martial ambiance permeates Didymoteicho. Admittedly, it does not confirm the worst of "Didymoteiho Blues." In the words of Major General Haralampos Lalousis, "the XVIth Didymoteicho" Army Division is responsible for an exceptionally vital borderland region with uniquely challenging battle theater characteristics in the rivers Evros, Ardas, and Erythropotamos. [Didymoteicho's] inhabitants are, in a sense, conscripted as they have, for many years, identified with the presence of the Army in north Evros … [R]eliant on its soldiers and the exceptional borderlands population (ακρίτες), [the XVIth] will continue to guarantee the defense of national interests and the preservation of peace in the region of north Evros, which is not the ends but the beginning of Greece (*Army Voice News*).[11]

In the Metropolitan Cathedral Church's side yard, in front of Didymoteicho's Nursing College and City Hall complex, a soaring F-5A fighter jet is displayed on a plinth as a further manifestation of the State-Military-Church complex.[12] In 2006, it was made a monument commemorating Evros' fallen aviators. The Didymoteicho Guard Officers' Club, the Army commissary and exchange, the military museum, and the base of the XVIth Mechanized Infantry Division are loosely grouped at a short distance from the city's eastern fringe.

There are three channels for the expression of "the martial" in the Evros borderlands generally and in Didymoteicho, specifically. First, the Major General's speech exemplifies the armed forces' position – and by extension, the national Government's – on the importance of a close relationship between the local population and armed forces units. Historically, the civilian population, urban infrastructure, and the social state, served the illiberal, garrison state. [13] In the Interwar period, the absence of a resident population was considered a strategic liability, as the military depended, at least in part, on civilian infrastructure. Battlefield logistics have radically changed since the 1930s. It is doubtful that local society would be an operational asset in a quickly-unfolding Colonel Blotto-type border conflict. However, the civilian vitality of defended borderlands plays a geopolitical role also in a liberal society: it communicates civil society's consent on the uses of military power.

Continuous, visible interaction between civilian and military leaders in national and civil celebrations and commemorations builds on Didymoteicho's transactional politics. The local state, represented by Didymoteicho's mostly right-of-center political class, is the second channel of Evros' "urban martial." The interlocking leadership networks of City, State/Military, and Church sustain a local economy that significantly depends on the military's enduring presence (Figure 8.1). The Muslim community's leadership, albeit under-resourced and less influential in local affairs, is included in local transactional politics. The Army respects standards established and commitments made by the national Government concerning Western Thrace's Minority Muslims. Official Army language on minority affairs is axiomatically apolitical and non-polemical. Accordingly, the Mufti of Didymoteicho invited to official events open to city leaders, as in the ceremony on 14 March 2019, marking the change of commanders of the nearby "Petalotis" Base (Ibid).

The third channel of "the martial" is, in part, extraterritorial as it expresses current and former Army conscripts' popular, vernacular readings of Didymoteicho as a military service post. It may count for little in the city's transactional politics, although it has become the source of Didymoteicho's imaginary as "the ends of the world," or as Didymoteicho Blues wants, "a hole on the map."

We explore the morphological transformation of two iconic monuments of Didymoteicho – one Islamic, dated to the early Ottoman era, and another

Christian and of recent construction – as exemplary of a borderlands geopolitics of difference.

As in Athens and Thessaloniki, following the region's incorporation in Greece, Didymoteicho's elite and vernacular Ottoman urbanism was gradually supplanted by morphological interventions that followed the Western European canon. Athanasios Gouridis provides an extensive list of lost Ottoman monuments and vernacular buildings that includes numerous mosques and masjid (small mosques), tekes (Sufi meeting places), medrese (religious schools), hammams (baths), fountains, caravanserai (inns), mausoleums and other funerary monuments and cemeteries, and vestiges of secular buildings in the historic market center (Gouridis, 2008: 103–110). At the beginning of the twentieth century, seven of the 12 mosques reported in the city in Evliya Çelebi's *Seyâhatnâme* (Book of Travel, 1630–1672) were still operational. Today, only two remain: the Grand Beyazit Mosque (also known as the Mosque of Mehmet I) and the so-called Red, Small, or Market Mosque (Ibid 105). The Grand Beyazit Mosque, or "Temenos" of Beyazit Yildirim (built during Mehmet I's reign – 1402–1420), is one of Europe's most significant Islamic monuments. The massiveness of Beyazit Mosque's timber roof and the innovative techniques used in its construction, make it "one of the most important wooden monuments in the world" (Bakirtzis, 2009: 72). With the significant exception of the Grand Mosque, therefore, Ottoman heritage morphology is no longer discernable. Didymoteicho's built fabric has been substantially reimagined since "Lausanne."[14]

The Grand Mosque was developed in the early fifteenth century as a fringe-belt development and marked the edge of the urbanized area. It remained the town's central place following the post-"Lausanne" expansion of the city (Karakotsoglu Bros, images ca. 1928). Briefly converted by the Bulgarians into a church in 1912, it was decommissioned after the region's incorporation in Greece and sold to a private concern during in the Interwar period (Εν Διδυμότειχω, 11 March, 2017: 10). It fronts a public square that has long served as the town's main square and is still, nominally, referred to as the "central square" (Bazaar, image, ca. 1970).

Urban morphological modernization could not suppress the visual and topographical dominance of the Grand Mosque. For as long as it stood as the city's seminal monument (the Byzantine castle being marginally integrated into the town plan), it branded the central square a space of Ottoman heritage, and limited the scope of landscape Hellenization. A Hellenic morphological counterweight to it would have to be introduced to rebrand the urban core.

Following the removal of its lead roof tiles and flashing during the Second World War, the monument rapidly deteriorated. The damage allowed water infiltration, which further compromised it (Bakirtzis, 2009: 76). Mostly neglected, concerns about its structural integrity were first raised in 1969. Calls for its stabilization and subsequent restoration remained unheeded by the Colonels' Junta, and following the restoration of democracy in 1974.

It still took several years until the roof was temporarily patched with tin sheeting (1981–1985) (Ibid, 82). In 1995–1999, a protective impermeable layer was applied to the deteriorated roof, and in 2001, more structural supports were introduced in anticipation of a full-scale restoration. By 2005, the support scaffolding and protective roof layer showed signs of wear, making the full restoration of the Mosque ever more urgent and imperative. Finally, in 2014, the state introduced legislation that visioned bundling the restoration project with a broad range of regional development initiatives submitted to the European Union's "Framework Programmes 2014–2020." Such opportunity arose under the *Horizon 2020* Programme: Societal Challenges initiative SC5-21-2016/2017: *Cultural heritage as a driver for sustainable growth: heritage-led urban regeneration* (2016). Nevertheless, the restoration work proceeded at a plodding pace.[15]

All work ceased on 11 March 2017 following an accidental fire started in the roof timberworks.[16] The fire destroyed the monumental roof and significantly damaged the Mosque's interior. On that day, the local, regional, and national states made new strident pronouncements of support for the monument's restoration. Emergency scaffolding currently clads the structure, which otherwise remains a ruin. Two and a half years later, the charred and buckled protective metal canopy erected in the pre-fire phase of restoration is still precariously hanging over the edifice.

The maltreatment of the monument's restoration has had a near-fatal effect on the central square. Urban renewal efforts concurrent with the pre-fire phase of restoration have not reversed the square's gradual decline to the surrounding businesses' considerable distress. The stalled restoration efforts post-fire have littered the square with construction debris and unsightly fencing. The once central square is less of a welcoming positive space and has, since 2017, devolved into a transit space (Figure 8.2).

No ready explanation – other than gross negligence by the national State – can describe the situation. Since meticulous restorations by the state of other Ottoman heritage monuments have been undertaken in Athens, Thessaloniki, Kavala, Rhodes, and Chania, it is unlikely that cultural animus towards all-things-Ottoman-Turkish has been the source of inaction.[17] Further, despite administrative decentralization under the "Kapodistrias" (Law 2539/97) and "Kallikratis" plans (Law 3852/2010), agency and financial control are still substantially in Athens' hands. Therefore, it is Athens through its Ministry of Culture that should have led here. In the absence of national leadership, the regional state – the territorial Department of Eastern Macedonia and Thrace – could have exercised its agency and applied its financial resources. However, despite programmatic commitments to action, no substantive support has been extended to the restoration. The local state, the City of Didymoteicho, could have exercised its own agency and, albeit, limited financial resources to salvage what is without question a world-class monument and, arguably, the centerpiece of its tourism development strategy. Nevertheless, despite, again, forceful public

Figure 8.2 Dydimoteicho's Grand Beyazit Mosque: Debris from the salvage operation clutters the Central Square seven months after the fire that destroyed the Mosque's timbered roof (Source: authors).

pronouncements about collective mobilization, there have not been any meaningful action.

At a more granular scale, we would expect agency on the part of a "machinic" (or non-machinic) urban coalition of local political and business elites, coordinated by City Hall – an application of a classic paradigm of urban politics in, at least, the Western tradition. Drilling still deeper to "right to the city" politics, we would expect to find an on-going contentious politics between the local state and the population on setting local priorities and allocating resources that touch on the Mosque restoration. Such urban regimes of cooperation could bridge levels of democratic governance in urban projects, be they in the interest of elites or the public.

The attendant syllogism would be that inaction stems from Thrace's – and by extension Evros borderlands' – status as a double periphery (marginal with respect to both Greece and the European Union), which would relegate regional development projects unrelated to military security to the bottom of national administrators' priority list. It would be a classic case of center-periphery socio-spatial polarization, whereby the center remains unresponsive to calls to action by peripheral and marginal regional and local states. If the state should be assumed to shape national territory and borderlands according to its own functional and symbolic politics of place, then, in the case of Didymoteicho's development, the state appears disengaged.

Such political scalarities, however, where they are manifested, do not follow the scalar nestedness and conventional forms of transactional urban politics we describe above. Instead, a politics of place is at work in Didymoteicho that reveals the peculiarities and limitations of Hellenic statecraft concerning difference in the borderlands. It describes late-phase Hellenization, which attempts to bridge long-standing borderlands imaginaries of geopolitical peril to Greece's civil society turn in a Europeanist mold.

Following the restoration of democracy in 1974, and especially after the election of Socialist Andreas Papandreou in 1981, Greece's social state grew significantly. Didymoteicho's historic core defined by the central square and the Grand Mosque, could not accommodate the social state's expanding infrastructure. Gradually, administrative and social services moved east of the historic core, forming a new services district (municipal urban plan, 1984). Eleftherios Venizelos Street – Didymoteicho's main commercial artery – connected the historic core and its central square to the new services district. This new hub included an expanded City Hall, the municipal police headquarters, a complex of primary and secondary public schools, the nursing college, and the prefectural government branch. The recently completed Byzantine Museum has extended the district's footprint to the south.

The new district included open public land zoned for future expansion of state services. Much of that open public land became the site for a new iconic and uncompromising Hellenic urban monument, the Metropolitan Cathedral Church – the visual anchor of the expanded downtown. The project illustrates the limitations of the co-evolved State-Church-Military complex in the borderlands in managing the geopolitics of difference.

In 1989–1992, the Metropolitan of Didymoteicho, Oresteias, and Souflion, Nikiforos II, succeeded in building a monumental Cathedral Church on the few remaining lots in the new district, which had been previously earmarked for the expansion of public schools. The Church has since become Didymoteicho's new (alternate) city landmark and the centerpiece of a de-facto new central square. On national holidays, the Cathedral's 33-meter-high bell tower is draped in an immense Greek flag. Only 300 meters distant from the old central square, the Cathedral competes visually with the now fire-damaged Grand Beyazit Mosque – the two structures defining the city's skyline.

The Metropolitan Cathedral of Panagia Eleftherotria (Virgin Mary, the Liberator), inaugurated in 1992, is as much an ecclesiastical as a political project. Its outsize dimensions, for the scale of population and urbanized

Figure 8.3 The Church of Panayia Eleftherotria ("the liberator") is the monumental landmark and epicenter of Didymoteicho's expanded downtown (Source: authors).

area, the exorbitant cost of its construction (estimated at the time to exceed one billion drachmas (EUR 30 million), and the clarion clear political rationale for its construction, describe the outsize agency of the Orthodox Church in the borderlands. It lends Didymoteicho the urban morphology of a "Cathedral City" of old (Figure 8.3).

The placename Virgin Mary, the Liberator bears a double meaning. In Christian Orthodox nomenclature, it celebrates the Virgin Mary's role in liberating humanity of sin by giving birth to the Savior. In the context of the borderlands' politics of place, the Virgin Mary is cast as the liberator of Greece from Turkish rule and the eternal protector of Didymoteicho.

Conceived by the influential Metropolitan Nikiforos II, it serves as the centerpiece of a landscape geopolitics that oversaw the construction of large churches in several borderland villages of the diocese.[18] The ideological motivation behind the building campaign is not in dispute. On the twentieth anniversary of the inauguration of the Cathedral, the Official Site of the Church of Greece published Metropolitan Damaskinos' speech commemorating Metropolitan Nikiforos II (†)[19] work and the events surrounding the Cathedral Church initiative:

> Upon his enthronement in January 1989, the Blessed Metropolitan of Didymoteicho, Orestias, and Souflion, Nikiforos, determined the absence of a church in the city center, which had been dominated for centuries by the Mosque (Temenos) of Beyazit. Didymoteicho's historic churches were situated in the Castle and its surroundings, since that was the city's extent until the beginning of the Twentieth century.[20] Inspired and motivated by patriotic and religious ideals, the Blessed [Metropolitan] could not tolerate the Mosque's domination of the city. Immediately upon his installation, therefore, … he mobilized every one so that a magnificent church would be erected. Despite resistance incited by a group with well-known positions, Metropolitan Nikiforos, distinguished by well-understood stubbornness, and thanks to his persistence, recruited the municipal government, which was led by Mayor Evangelos Papatsarouhas. [The Mayor] adopted and embraced the Metropolitan's vision. His impulse was to place at the disposal of the blessed Shepherd the full resources of the Municipality. And, thus, work begun…[21]

It is unclear here as to who belongs to the "group with well-known positions." The most strident complaints against the Church project were voiced by the Public Schools establishment, which realized that the Cathedral Church's vast footprint would occupy building lots previously earmarked for "educational land-use." The erection of the Church would have precluded any contiguous expansion of existing school facilities [citation from municipal records]. Given the animus exhibited by the Metropolitan toward the Grand Mosque, it is possible, though unlikely, that Didymoteicho's small Muslim community or its leadership (the Mufti of Didymoteicho) would have joined the resistance to the project. Arguably, resistance would have also come from those among the electorate who prioritized quality-of-life over prestige projects. The city has minimal green public space, significant infrastructure deficiencies, long-standing municipal water shortages, gaps in social services, and suffers from a grave and worsening demographic crisis. Reasonably, a church building project, however magnificent, would not address any of these concerns. Further, the Church's report on the Mayor's "impulsive" adoption of the project and his commitment of all town resources to the project, makes resistance likely by some aldermen in the multi-party elected municipal government.

It is notable that the institutional Church of Greece enjoys significant land-regulatory prerogatives over its church buildings. However, it does not enjoy "co-decision" rights with the local civil state on matters involving public land. Nevertheless, the local state prioritized the Cathedral Church's building over all other major projects. It secured the consent of the national state – the Ministry of Education and Religious Affairs had to release the land parcels to the city – which allowed it to provide building permits, staff, material, and financial resources.

Work on the Church began on 17 July 1992, with soldiers of the XVIth Army Division carrying out the foundation earthworks. The building was completed in record time and inaugurated on Sunday, 29 May 1994 – the anniversary of the Fall of Constantinople to the Ottomans in 1453. The Church building is massive for the scale of the city. Its interior – from its iconostasis, which was gilded in 22-carat gold by Romanian artisans, to the liturgical furniture and canonical frescoes – is lavish if not extravagant.[22]

The building project's political intentionality is, again, laid bare in Metropolitan Nikiforos II's celebratory remarks on the occasion of the Cathedral Church's inauguration: with Mayor Evangelos Papatsarouchas at his side, he exclaimed, "We two shall call out that all [who contributed to the project] are entitled to congratulations, and we thank you. But it was you [speaking to Papatsarouchas] and me that toiled… And from this microphone together we shall shout 'We defeated you, Beyazit!'"[23]

A neo-Byzantinist political discourse defines key architectural and artistic flourishes. The Metropolis commissioned artist Pavlos Sarfatis to create a canonical "ktetoric" mosaic in the Byzantine iconographic tradition for the tympanum over the main entrance. It depicts Metropolitan Nikiforos as "ktetor" (founder) presenting to the Virgin Mary and Christ-in-mandorla an unconventionally realistic model of the Church of Panayia Eleftherotria (Figure 8.4). The praying Panayia and Christ extend arms out, palms up, in the "Orans position," a conventional pictorial device in early Christian representation. She is represented as "Platytéra" (Wider than the Heavens), dwarfing the figure of the "ktetor."[24] Beneath them lays castellated Didymoteicho, the full circuit of its walls schematized canonically by representing its full volume on the same plane ("eliminatio angoli"). Ordinary buildings of the city are strewn underneath it. The intention to link Didymoteicho's and Constantinople's imperial past to a neo-Hellenic present and future is clear. The Virgin Mary was the patroness of imperial Constantinople and, now, by the grace of this project, also the protectress of Didymoteicho. In the lower right of the composition, the scaled-down figure of Metropolitan Nikiforos II in episcopal vestments and mitre is represented kneeling, the model of "Eleftherotria" in his hands. In the Byzantine political-historical context, this formulaic pictorial language was a powerful iconographic device that affirmed and illustrated the social dimension of the Empire-Church condominium.[25]

In the modern context, and more specifically, in the borderlands context, it connotes the continuity of Byzantine traditions and the outsize role of the Church as mediator between the Hellenes and the protective divine. In

Figure 8.4 The "ktetor" mosaic decorating the main entrance's tympanum – traditionally, the most important location for monumental religious art in a church's exterior. Church of Panayia Eleftherotria (Source: authors).

that sense, the adoption of canonical iconographic tropes in official art, is fundamental to politically and culturally signifying the Church building, the Metropolitan-as-"ktetor" (and by extension the full Church), the flock of the faithful, the City of Didymoteicho, the Evros region, and Thrace, as co-constructors and repositories of Hellenic cultural continuity.

These are far from run-of-the-mill aesthetic choices made in what is pragmatically a provincial church building. The "ktetoric" mosaic and the Church itself, which is built in the canonical cruciform basilica-with-dome Byzantine style, are more than artistically traditional and exemplary of antiquarian tastes in architecture. After all, the vast majority of church-building projects in Greece are inspired by Byzantine architecture and their interiors decorated in that same canonical iconographic tradition.

The differences here are of agency and place: the Church leader who recruits the eager-to-serve Mayor, both of them securing the services of the Army to dig the foundations, and importantly, receiving the aegis of the national state in the form of licenses and significant financial contributions, all "to defeat Beyazit," which is, ultimately, a shorthand for Didymoteicho's geopolitics of place as expressed by its urban morphology. Theirs is a special purpose local sovereignty regime that leverages the national government's resources for the performance of political tasks of consequence to local political elites but with repercussions that reverberate nationally. Importantly, it also marks the still long way to a civil society "turn" in Greece's periphery.

A monumental bronze statue of Konstantinos Palaiologos XI, the last Emperor of Byzantium, was later installed in front of the Church "as a witness to the unbreakable bonds between Didymoteicho and the 'Reigning City' (Constantinople), and its last dynasty, the Palaiologan [dynasty]." It is a replica of a work created by sculptor Spyros Goggakis in 1978 for Mystra, Laconia, the citadel of the Palaiologos dynasty. In 1990, Center-Right Athens Mayor, Miltiades Evert placed another replica across from the National Cathedral in Athens. On 20 November 2019, a fourth replica was installed in Didymoteicho at the XIXth Army Division's base, "Petalotes."

The signal here is Helleno-Christianity and the signalman the Metropolitan. The Cathedral Church project makes Didymoteicho about the martial and religious "Hellenic Medieval." As a demonstration of Hellenic cultural continuity, it is entirely consistent with the Paparrigopoulos thesis. The absence of either rhetorical or morphological "continuity" references to *pre*-Christian Hellenism or "Romanitas" is conspicuous, although material vestiges from those eras abound.

Why the vastly disparate treatment of the two sites – the Mosque and the Church? We can think of Didymoteicho's politics of place as ideal-typically run by a State-compiled, persistent, formal, visible, regulated political routine that produces two possible outcomes: provided an uninterrupted flow of financing, when the national and the local states' intentions and desired outcomes align, then effective sovereignty is on display. When the national and the local States do not align, effective sovereignty is compromised. In the first scenario territorially adjacent, ephemeral (or special-purpose), informal, less regulated, less visible-beyond-"the local" political subroutines – let us call them "organic" for convenience – are subordinate to national political authority. In the second scenario, national political authority, and by extension effective sovereignty, becomes contingent upon, if not subordinate to, the effectiveness of organic political subroutines territorialized at "the local" and still networked to sources of administrative power in the national core.

Both scenarios are produced in Didymoteicho: the fortunes of the Grand Mosque are significantly at Athens' administrative routine and discretion. Evidence points to Athens' failure to properly managing repair and conservation works despite the locality's support for the project. To wit their priorities and desired outcomes aligned, but in this case, national governmentality was poor. In contrast, while the realization of the Cathedral Church hinges on the Athens Archbishopric's approval and the vast proportion of municipal finances depends upon the national fiscal state, the prestige project is made possible by local agency and political-regulatory authority. Unexpectedly, in this case, the normative State routine falters while the locally situated and networked subroutine produces results.

Figure 8.5 A very Thracian standoff: In the foreground, the fire-damaged volume of the Grand Beyazit Mosque. In the middle ground, the Cathedral Church of Panayia Eleftherotria (Virgin Mary, the Liberator). The Greco-Turkish border lies 2 kilometers to the east (Source: authors).

When the treatment of the two sites is contrasted, the concurrent deterioration of the Grand Beyazit Mosque, officially under restoration, and the building of the Cathedral Church of Panayia Eleftherotria suggest that the restoration misadventures of the former and the rapid realization of the latter cannot be adequately explained as either a temporary incapacity of the national administrative state or a capacity surge in the local administrative state. Both projects enjoyed the approval of some level of the administrative state, receiving permits and licenses as needed to move forward with building works. In the case of the Cathedral, the local state prevailed upon the national state in de-characterizing (effectively, de-nationalizing) land parcels earmarked for public school expansion for the purpose of rezoning them to build the Cathedral Church. Therefore, these stark differences in how political authority is applied point to the divisibility of sovereignty, at least in this borderlands town (Figure 8.5).

The disparate treatment of the two projects cannot be adequately explained in economic terms, by considering the shifting financial capacity of these layers of government. In the last 30 years, both project arcs crossed periods of economic prosperity and deep crisis. To wit, state-institutionalist and economistic explanations for the disparity are inadequate.

An explanation that would cast the Mosque as a cultural amenity in an otherwise "in-decline" urban quarter and the Cathedral project as exemplary of investments made in developing urban quarters, does not satisfy. The Mosque is situated in Didymoteicho's central square. Even if diminished by the fire and the stalled salvage and renovation of the monument, the square still anchors the city's central commercial district. Its repair and renewal are essential to the city's commercial life. The spatial logic of the Cathedral Church project can be found in the functional importance of the services district in which it is situated and the visibility the district affords it. The district land reserved for future social services expansion was of appropriate size for an otherwise very compact city. Therefore, ultimately, both sites define Didymoteicho's downtown and are critical to its development.

The "can-do" story of the Didymoteicho Cathedral Church suggests that the (organic) adjacent, informal, less regulated, less-visible-beyond-"the local" subroutine that entangles productively, the borderlands "local" with the State-Church-Military complex, allows the borderlands a measure of agency to refashion itself – in this case, in the performance of a major prestige project that does not reflect national priorities. That was made possible even in a "doubly peripheral" region (with respect to the national State and the European Union). Accordingly, there is no documentary evidence that the monumental project was formulated by Athens' political establishment or the national administrative state, although, once it was on-going and following its completion, the national political class readily embraced it. It was good national(ist) politics. The initiative was borderlands-based, and political signaling was pulsing from the borderlands to Athens.

Our findings do not suggest national state and Church bias against the Grand Mosque and favoritism toward the Cathedral Church project the Metropolitan's comments regarding Beyazit's defeat notwithstanding. The lesson of Didymoteicho instructs how sovereigntist and "organic" politics of place are both entangled and competitive in the production of different versions of national statecraft.

The case study further suggests that the borderlands' politics of place can reveal how political and ideological distancing between center and periphery can occur. While Athens' administrative entanglement in the Grand Mosque fiasco is overt, it is Didymoteicho's own local elites that author the very political Cathedral Church project that is not necessarily helpful to the national state's civil society "turn."

Concerning the role of the Church of Greece, it is critical to point out the distance between Metropolitan Nikiforos II's polemical views toward the Ottoman past and Islam and the Church of Greece's official position in support of Islam and Greece's Muslim minority. The Church and the Ecumenical Patriarchate of Constantinople are committed to interfaith dialogue.[26] The Church approved the building of, and allocated funds for a needed diocesan church in the region, not a church to "defeat Beyazit." It is notable that Nikiforos II's successor, Metropolitan Damaskinos' interfaith discourse cleaves close to the national Church's.

Beyond Didymoteicho, the morphology of borderland villages reproduces the national landscape politics we encountered in Didymoteicho. Large Orthodox churches have been constructed in several of these settlements – some of them built since 1989 at Nikiforos II's initiative. These church projects can be aptly described as outsize when considering the demographic decline of the region and the borderlands' limited economic resources (Figure 8.6).

Expectedly, the church is the dominant building type in the village. It is commonly co-located with the local war memorial and village services in

Figure 8.6 The Church of Hagia Paraskevi in the Village of Provatonas, Evros. Provatonas is a post-1922, Army-planned settlement with a population of approximately 1,000 inhabitants (Source: authors).

its single public square. Such public spaces invariably host memorial monuments to wars and the fallen in defense of the fatherland. The memorials observe a small range of morphological-artistic conventions: The simplest among them include a base and plain, unadorned pillar, clad in white, veined marble. They, essentially, serve as supports or holders for commemorative wreaths. The majority of memorials are more ornate: Commonly a stele with architectural finishes and a decorative finial (a flame, an acanthus bloom, or an urn, evocative of Classical or Hellenistic Age funerary monuments).

The church, the memorial, and modest public amenities and services (playgrounds, the elementary or high school, the Village Hall, and more often now also the Association for the Care of Elders) would be located in or around that public square. The lack of public investment is evident. The hardscaping is worn out, if extant at all. The village furniture (such as benches, lighting, fences) is in most cases of old vintage and disrepair. In contrast, the churches are in excellent repair, their facades renovated, and the buildings are featuring, in many cases, modern and recently installed amenities, such as air-conditioning units and gleaming furnace chimneys, reception or event buildings, and well-kept gardens (Figure 8.7). Church-centered civic investment has become all the more critical in this less

Figure 8.7 Typical morphology of village cores in Evros' borderlands: The Church of Hagios Dimitrios and war memorial anchor Kavyli, Evros' central square (Source: authors).

developed region. Where the State administration routine falters, the diocesan managerial subroutine supplants it.

The Didymoteicho case study is a striking example of the durability of the nineteenth century's historiographic politics and, specifically, the persistence of the "continuity thesis" as the master narrative of the Church of Greece. The state's historiographic politics is much more nuanced and accommodating, as it is filtered, since 1974, through the experience of democracy, Europeanization, and the gradual development of civil society. By extension, the Hellenic Army's constitutional role in Didymoteicho as apolitical is clear, and it maintains good relations with minority communities in the borderlands.[27] That said, although the Army's construction assistance to the Metropolis may have been informal, voluntary, and temporary, its involvement in a clearly political project demonstrated an aspect of customary transactional politics between Church and state, which we would claim has greater currency in borderlands. The economic footprint of the Army in Didymoteicho and other centers on the borderlands has been diminishing. Although still a citizen's army supported by mandatory universal conscription of males, the armed forces' gradual professionalization has resulted in smaller numbers of boots-on-the-ground. "Didymoteiho Blues" is becoming less current, and Didymoteicho's landmark landscape politics refute its status as "a hole on the map."

In the first two case studies, we found that between 1919 and 1941 historic borderland settlements and new towns in Eastern Macedonia and Western Thrace served as more than an expression of the territorialization of Hellenic statism and nationalist ideology. Newly drawn international borders with Turkey and Bulgaria re-signified existing settlements as *border* settlements: "Hellenic" with respect to the national territorial imaginary and also as places directly implicated in national security. By representing the specter of peril to state and nation, the borderlands' ground conditions in the inter-war period provided context and stimulus for the drawing of internal borders and the crystallization of a garrison-like national political culture – an exclusionary politics of place – well into the last quarter of the twentieth century. In the third case study, the disparate treatment of the historic Grand Mosque of Beyazit and the Cathedral Church of the Virgin Mary, the Liberator in the borderlands city of Didymoteicho tells a story about how state and Church are entangled across scales to produce different politics of place, either consistent with the classic historiographic paradigm of Hellenic cultural continuity, or increasingly consistent with the standards and norms of a European civil state. Importantly, Didymoteicho makes us aware of the trans-scalar interlockedness of networks and routines (and their subroutines): a state-compiled persistent, formal, visible, regulated routine that incorporates one or more "organic" subroutines, that is, territorially adjacent, temporary or special-purpose, informal, less regulated, less-visible-beyond-"the local" subroutines. Both struggle with difference and produce distinct and sometimes contrasting definitions of Hellenic borderlands.

Notes

1. Greece's central bank – Bank of Greece – was founded in 1928 at the direction of the League of Nations and for monetary stabilization in the wake of "Lausanne" (Calvin, 2013; Dritsas 1990). National Bank was a commercial bank founded in 1841. In the absence of a central bank until 1928, National Bank fulfilled that role.
2. We transliterated the toponyms from Greek to English as they appear in the Army report. In some cases, toponyms were not recorded in standard, Latinized Turkish. For clarity, we added, whenever possible, the contemporary Hellenized toponyms. Toponyms were changed en masse between 1908 and 1928.
3. The counts include Mega Dereios' communities of Ferai-Kolymbar [Ferai], Tsoutsour [Thiafi], Semer-Bournou [Samaras], Dobrava [Triada], Tsalil, Kerven-Tsair [Polyvlaston], and Tsiligir).
4. Drawing on Ottoman and Bulgarian sources, the report notes that before the Greek-Turkish War, the region's 1,922 families supported approximately 95,000 heads of goat and sheep (in a two to one proportion), 5,600 heads of cattle, and 1,300 mules and donkeys. The majority of animals were owned by a small number of families in herds of 100-500 heads, while a second cast of families owned 20-80 animals, and many families owned none and were, instead, occupied as laborers in agriculture and husbandry (Ibid, 16). By 1939, husbandry stocks were reduced to 54,000 sheep and 601 horses owned by only 15 families, which employed in total 170 nomadic herders. The most significant herd owner ("tselingas," in the local dialect) was Dimitris Gikas, who employed 25 families of herders, and owned 11,000 sheep and 138 horses in the Imamlar-Mega Dereio-Babalar area (Ibid 17). Notably, local life and economy were described as poor and unequal, defined by daily toil and hardship, and distinguished by thriving black marketeering and animal theft.
5. The pasturelands of Mount Sapka amounted to approximately 285,000 stremmata (28,500 hectares), or a fraction of the region. The survey identifies them as pasturelands interspersed with forested land and scrub primarily appropriate for goatherds, and secondarily for cattle and sheep herding. Both rich in grazing vegetation and supply of surface water were deemed appropriate for the traditional transhumant shepherding practices that distinguished the pastoralist communities in the region. The forested parts of Rhodopi would also have provided additional animal feed for extended winter seasons from branch and leaf harvests (especially from beech, oak, and ash trees, and carpinus bushes). The report noted that an area of mixed forest and an equivalent plot planted with clover would yield equal amounts of animal feed (ibid, 14, footnote) and strongly recommended the judicious exploitation of forest resources, as was practiced at that time, to make husbandry sustainable, despite adverse climatic conditions (Ibid, 24).
6. The terrain was not suitable for commercial-scale cattle husbandry. The report called for the reintroduce of the goat as a permanent feature of husbandry practice in the region, and was quite emphatic that settlement efforts would fail without it. Another challenge was resistance to settlement by indigenous, nomadic herders, who traditionally opposed all settlement efforts in the borderlands. They desired to eliminate new competition for scarce resources and, especially, to maintain their unregulated access to all areas in the region.
7. The ten worst transfer posts to be assigned to in the Army: https://menshouse.gr/antras/47994/10-chiroteres-metathesis-pou-boroun-na-sou-tychoun-ston-strato (accessed on 12 December 2019).

8. The actor-network method has compelling characteristics for a study like ours in which urban materiality, human and community actors, and ideological constructs continuously negotiate the architecture of state and nation networks. This, however, is not a fully developed actor-network-based study. Further fine-granular empirical research on the structure of actor decision making would be required.
9. We did not plan to formally interview officials or members of the public, although, on any visit to a village Kafenio (cafe), we were engaged in conversation. A common theme was population decline as a result of out-migration to Western Europe, notably Germany – a pronounced phenomenon in Xanthi's Pomak villages.
10. The looping images on the top webpage are divided into three themes: "City of castles and legend," "City of artists and creativity," and "City of faith and devotion." The first ensemble features the Byzantine monuments of the city and the Cantacuzene-era castellated mansion in the nearby village of Pytheio. A plan-view of an Ottoman hammam represents Didymoteicho's significant Ottoman built-environment remains. The second ensemble depicts primary school children painting, a cabinetmaker at work, and a scene from a theatrical play. Notably, the third theme juxtaposes in perspective the Metropolitan Cathedral Church and the Grand Beyazit Mosque, with the former dwarfing the latter. https://www.didymoteicho.gr/el/
11. https://www.armyvoice.gr/2019/03/16i-merarchia-didymoticho-teleti-paradosis-paralavis-foto/, accessed on 26 November 2019.
12. The construct begs comparison to the United States' "military-industrial" (and congressional) complex famously referenced by President Dwight Eisenhower in his farewell address of 17 January 1961. In Greece's statist system, the industrial-economic dimension is both limited and less explicitly tied to the military, given Greece's foreign defense procurement. In the absence of Church-State separation, the Ecclesiastical dimension is critical to understanding Greek politics. In the borderlands, the spatial entanglement of all three dimensions is determinative of town and village morphologies: the grouping of town hall - church - police station - war memorial, and in some cases army base).
13. We encountered this strategic understanding in the interwar communiqués between the Metaxas administration and the head of the General Staff, General Papagos. The sparseness of population in the borderlands – a demography that could not adequately support military bases and operations
14. The contemporary town plan incorporates organic street block groups of Byzantino-Ottoman chronology, and planned, regular, and often orthogonal, post-1923 fringe belt developments.
15. The need for external financing of this important project during Greece's Eurozone crisis was unquestionable. The delay of decades, however, cannot be entirely explained in financial terms. Despite the cultural value of the structure and its compromised state, the Didymoteicho mosque restoration was de facto not a State priority. (1) ΕΥΤΟΠ/Α/ΕΣΠΑ/14-20/1013/22-3-2016 Απόφαση της Γενικής Γραμματέως του ΥΠ.ΠΟ.Α έγκρισης του εγχειριδίου διαδικασιών αρχαιολογικών έργων, όπως τροποποιήθηκε με την υπ'αριθ. ΕΔΕΠΟΛ/Α/ΕΣΠΑ14-20/3942/18-11-16 και την υπ' αριθ. ΕΔΕΠΟΛ/Β1/ΕΣΠΑ14-20/909/1-6-2017 όμοιές της. (2) 3362/7-11-18 (ΑΔΑ: ΩΘΑΤ7ΛΒ-ΖΣ0) Απόφαση Ένταξης πράξης "Αποκατάσταση τεμένους Βαγιαζήτ Διδυμοτείχου - φάση Β΄" (MIS 5004072) στο Επιχειρησιακό Πρόγραμμα «Ανατολικής Μακεδονίας και Θράκης 2014-2020.» (3) ΥΠΠΟΑ/ΓΔΑΠΚ/ΔΒΜΑ/ΤΒΜΑΧΜΑΕ/567847/406621/7353/2964/7-12-18 (ΑΔΑ: Ω5Ω44653Π4-ΩΦ2) ΥΑ έγκρισης εκτέλεσης εργασιών του υποέργου 1 «Δομική και -3- μορφολογική

αποκατάσταση μνημειακού κελύφους και περιβάλλοντος» του έργου «Αποκατάσταση τεμένους Βαγιαζήτ Διδυμοτείχου - Φάση Β» (με κωδικό ΟΠΣ 5004072) απολογιστικά και με αυτεπιστασία από την Εφορεία Αρχαιοτήτων Έβρου. (4) ΥΠΠΟΤ/ΔΑΒΜΜ/113732/2812/01-12-2010 υπουργική απόφαση έγκρισης της μελέτης με τίτλο "Μεγάλο Τέμενος Διδυμοτείχου. Μελέτη α) επικαιροποίησης της γεωμετρικής τεκμηρίωσης της στέγης, β) καταγραφής και ανάλυσης του φέροντος οργανισμού ως φορέα, γ) καταγραφής και αποτίμησης της παθολογίας του και δ) διατύπωση μεθόδων επισκευής και ενίσχυσής του." (4) ΥΠΠΟΑ/ΓΔΑΜ ΤΕ/ΔΑΒΜΜ/ΤΜΒΜΜ/370229/ 38237/2196/768/27-7-18 υπουργική απόφαση, όπως ανακοινοποιήθηκε στο ορθό στις 1-8-18 (ΑΔΑ: 6ΔΙΔ4653Π4- Ξ1Ξ), έγκριση μελέτης με τίτλο "Τέμενος Βαγιαζήτ Διδυμοτείχου. Σχέδιο-πλαίσιο αποκατάστασης μνημειακού εμβαδού και περιβάλλοντος. Αρχαιολογική τεκμηρίωση αναγκαίων επεμβάσεων και υποστηρικτικών μελετών."

16. Much like the fire that destroyed the roof of Paris' Notre Dame Cathedral on 15 April 2019, negligence involving electrical installations, welding, and soldering was also likely, responsible for the Grand Beyazit Mosque fire. Police investigations to date have dismissed the possibility of arson.
17. Stefanidou notes that the majority of Greece's Islamic monuments gained landmark status after 1980 (2009: 26).
18. Nikiforos II's strategy scaled-up the post-1923 settlement morphological tradition that cast the church as the epicenter of the village and the anchor of its, most often, sole public space.
19. Deceased, 6 October 2009.
20. In the classic Ottoman urban morphological canon of Castle-City-Suburb, "suburban" (fringe belt) developments would accrue in the periphery of existing town plans. Accordingly, the Grand Beyazit Mosque was the centerpiece of such a "suburban" accretion at the eastern edge of formerly Byzantine Didymoteicho.
21. http://ecclesia.gr/epikairotita/main_epikairotita_next.asp?id=923, accessed on 13 November 2019.
22. Various sources, including the monthly Εν Διδυμοτειχω (15 May 2015).
23. The Metropolitan's words also appear on the City of Didymoteicho's official website: https://www.didymoteicho.gr/el/component/k2/item/227-panagia-eleftherotria.html (accessed on 13 November 2019).
24. Asymmetry is not a commonplace iconographic device in Byzantine art; indeed, it is the exception. It is usually reserved for representations of enemies or sinners. In fact, in the few examples in which we observe asymmetry in "ktetoric" representation, the "ktetor" is depicted larger than other figures, including those representing the divine (Cvetković, 2006: 74–75; Kazhdan and Cutler, 1991:1988–1989). It is one of the intriguing deviations from canonical representation that may be worthy of further examination.
25. An excellent example of Byzantine ktetoric mosaic art can be seen at the Church of Saint Savior in the Chora, Istanbul: Founder Theodoros Metochites offers a model of the church to the enthroned Christ.
26. Speech of the Archbishop of Athens, Christodoulos, to the Synod of the Church of Greece on the need and challenges of interreligious dialogue, 12 May 2007. http://www.ecclesia.gr/greek/archbishop/default.asp?id=638&what_main=1&what_sub=5&lang=gr&archbishop_who=2&archbishop_heading=%CE%95%CF%85%CF%81%CF%8E%CF%80%CE%B7 Accessed on 14 December 2019.
27. One should think that, accordingly, soldiers serving with the XVIth Mechanized Division would be as readily directed to assists in earthworks for an Islamic community center as they did for the Cathedral Church.

Sources

Agnew, John. "The Territorial Trap: The Geographical Assumptions of International Relations Theory." *Review of International Political Economy* Vol. 1 (1994): 53–80.

Agnew, John. "Sovereignty Regimes: Territoriality and State Authority in Contemporary World Politics." *Annals of the Association of American Geographers* Vol. 95, No. 2 (2005): 437–461.

Agricultural Bank of Greece. «Notice to Farmers, Agricultural Cooperatives and their Syndicates.» Athens: September 1929.

Agricultural Bank of Greece. Historical Archives. *Administrative Circulars, 1930-1939.* Athens: 1940.

Alivizatos, Babis. « Le Règlement des Dettes Agricoles. »*Bulletin de la Banque Agricole* 3.3 (1938) : 144–146.

ASKI: KKE Archive, Box 408, F 23/1/177: Resolution of the KKE Macedonia-Thrace Bureau, 7 April 1947.

Brégianni, Catherine. "Echanges de Populations, Organisation de l'Espace et Savoir-Faire des Institutions Internationaux: Le Réseau de la Banque Agricole de Grèce et l'Office Autonome pour l'Etablissement des Refugies a la Grèce du Nord Pendant les Années Vingt et Trente." In Brégianni, Catherine; Cussó, Roser, eds. *Politiques économiques des institutions internationaux.* Projet Politiques et statistiques économiques des institutions internationales durant l'entre-deux-guerres. Paris: Laboratoire de Démographie et d'Histoire Sociale-CRH, EHESS. Paris, 2019.

Brégianni, Catherine. *Les Banques, l'Agriculture et l'Etat. Stratégies de crédit et politique agraire en Grèce de 1861 à 1940.* Lille: Presses Universitaires Septentrion, 2002.

Calvin, Patricia. *Securing the World Economy. The Reinvention of the League of Nations, 1920-1946.* Oxford: Oxford University Press, 2013.

Church of Greece, Official Site (http://ecclesia.gr/epikairotita/main_epikairotita_next.asp?id=923).

Cliadakis, H. "The Political and Diplomatic Background to the Metaxas Dictatorship, 1935-36." *Journal of Contemporary History* 14, no. 1 (1979): 117–138.

Cvetković, Branislav. "Intentional Asymmetry in Byzantine Imagery: The Communion of the Apostles in St Sophia of Ohrid and Later Instances." *Byzantion* Vol. 76 (2006): 74–96.

Dritsas, M. *Industrie et Banques en Grèce dans l'entre-deux-guerres.* Athènes: Fondation Culturelle de la Banque Nationale, 1990: 240–241.

European Union. *Framework Programmes, 2014-2020.* Horizon 2020 (accessed: 12 January 2020) https://ec.europa.eu/assets/eac/culture/library/reports/2014-heritage-mapping_en.pdf

Featherstone, K.; Papadimitriou, D.; Mamarelis, A.; Niarchos G. "On the Path to War." In: *The Last Ottomans.* New Perspectives on South-East Europe Series. London: Palgrave-Macmillan, 2011.

Kazhdan, A.P; Cutler, Anthony, et al. "Symmetry." *The Oxford Dictionary of Byzantium*, 3 Vols. Oxford. New York: Oxford University Press, 1991: 1988–1989.

Notaras, M. *L'établissement Rural des Réfugiés.* Athènes, 1934.

Ousterhout, Robert G. "A Byzantine Chapel at Didymoteicho and its Frescoes." (http://repository.upenn.edu/histart_papers/3, 1999).

Rodrigue, Aron. "Jewish Society and Schooling in a Thracian Town. The Alliance Israelite Universelle in Demotica, 1897-1924." *Jewish Social Studies* Vol. XLV, Nos. 3-4 (1983): 263–283.

Αγροτική Τράπεζα της Ελλάδος (Agricultural Bank of Greece). *Το έργον μιας δεκαετίας, 1929-1939* (Works of the decade 1929-1939). Αθήνα, 1940.

Γουριδης (Gouridis), Αθανασιος Ι. *Διδυμοτειχο. Μια Αγνωστη Πρωτευουσα* (Didymoteicho. An Unknown Capital). Δημος Διδυμοτειχου, 2008.

Καύκουλα, (Kafkoula) Κική. «Εξευρωπαϊσμός και Μεταρρύθμιση σε Δύσκολους Καιρούς. Το Πρόγραμμα Ανοικοδόμησης της Ανατ. Μακεδονίας το 1919» (Europeanization and reform in difficult times. The program for the reconstruction of Eastern Macedonia). *Εφημερίδα του Τεχνικού Επιμελητηρίου Ελλάδας/Τμήμα Κεντρικής Μακεδονίας* [Vol.] 442 (1 [June] 2012): 10–11.

Καύκουλα, (Kafkoula) Κική. *Η Περιπέτεια των Κηπουπόλεων. Κοινωνική και Περιβαλλοντική Μεταρρύθμιση στην Ευρώπη και την Ελλάδα του 20ού Αιώνα* (The saga of garden cities. Social and environmental reform in 20th Century Europe and Greece). Θεσσαλονίκη: University Studio Press, 2007.

Μπακιρτζής (Bakirtzis), Αργύρης. «Το Τέμενος Μεχμέτ Τσελεμπή Διδυμότειχου. Περιγραφή, Παθολογία, Επεμβάσεις και Ενέργειες για την Αποκατάσταση της Ξυλοστέγης του» (The Mosque of Mehmet Çelebi, Didymoteicho. Description, pathology, interventions and work for the conservation of its timber roof). In Αιμιλία Στεφανίδου (Stefanidou) [ed], *Η συντήρηση και η αποκατάσταση των οθωμανικών μνημείων στην Ελλάδα.* Θεσσαλονίκη: University Studio Press, 2009, 71-86.

Σπυρόπουλος (Spyropoulos), Γιάννης Μπαχ [lyrics]; Μαχαιρίτσας (Mahairitsas), Λαυρέντης [music]. *Διδυμότειχο μπλουζ* (*Didymoteicho blues* song), 1991.

Στεφανίδου (Stefanidou), Αιμιλία [ed.]. *Η Συντήρηση και η Αποκατάσταση των Οθωμανικών Μνημείων στην Ελλάδα* (The conservation and restoration of Ottoman monuments in Greece). Θεσσαλονίκη: University Studio Press, 2009.

Σωτηρόπουλος (Sotiropoulos), Δ. Π.; Παναγιωτόπουλος (Panagiotopoulos), Δ. «"Ειδικοί" Διανοούμενοι και Θύλακες Χειραφέτησης στο Μεσοπόλεμο. Μεταρρυθμιστές Γεωπόνοι και Μηχανικοί στην Υπαιθρο και το Αστυ» ('Specialist' intellectuals and pockets of emancipation in the Interwar Period. Reform agronomists and engineers in the countryside and the city). *Μνήμων* 29 (2011): 121–150 (file:///C:/Users/User/AppData/Local/Temp/7716-13692-1-PB.pdf Accessed on 18 August 2020).

9 Contesting the building of the Athens mosque

Worship and placemaking in the margins

Historically, a source country of international labor migration, and very recently a destination of, and transit country for economic and forced migrants, the Greek State confronts in the twenty-first century an unprecedented challenge: how to manage increased flows of migrants from the developing world, how to administer to their needs, and how to socially integrate those among them who stay in-country. These new populations join significant immigrant communities of Albanian, Middle Eastern, and North African origin that were established in the 1980s and 1990s.

The vast majority of immigrants residing in Athens come from Muslim countries. Muslim migration and settlement in Athens are relatively recent and geopolitically unrelated to Greece's "Lausanne" Muslim minorities regime. They represent "New Islam," as distinguished from Western Thrace's legacy Muslim communities. The contestation of the building of a new mosque in Athens provides us with an opportunity to explore how Hellenic statecraft manages difference in an age of global flows of migration and reconciles conservative public views on immigration to Hellenism's brand of cosmopolitanism. The politics of Muslim worship in Athens reveals the shifting balance between the public's desire for preserving national religious privilege and the state's aim of advancing a civil society in the Europeanist mold.

How have subaltern communities in Greece achieved rights in the past, and how do they do so today? In the telling of these stories, we are alerted to the primacy of a statist – "politics from above" – mode of inter-communal politics. Through a liberal lens, Prime Minister Eleftherios Venizelos' minorities policies of the 1910s and 1920s reveals a functioning statist, deliberative democracy attempting to engage productively with religious difference. In the aftermath of the Balkan Wars, Venizelos's Liberal Party legislated in defense of the lives, rights, and property of Thessaloniki's Jewish community (1912–1913), albeit by geopolitical exigency. During the post-First World War French Mandate period and following Greece's annexation of Western Thrace (1918–1919), his Liberal Government negotiated a model regime for the protection of the region's Muslim communities. Later, Venizelos' decision to expand the country's architectural preservation regime to include

Ottoman antiquities (1920), ended earlier practices (by plan or through neglect) of ethnic landscape scrubbing. "Lausanne" (1923) formalized in a durable manner regionalized religious and social rights for Muslim communities, albeit imperfectly concerning their integration into Hellenism. More recently, the decision of Minister of Defense, Gerasimos Arsenis, and the third Andreas Papandreou Government, to abolish all internal borders in Western Thrace's Pomak region and engage in a systematic politics of inclusion, is evidence of a still statist but increasingly contestatory liberal democracy. The making of the new Athens mosque is the latest example of Greece's civil society turn in a Europeanist mold.

On the flip side, historically, there has been significant social resistance to the idea of expansive and inclusive definitions of Hellenism and Hellenicity, and the inclusion of heterodox and heterolingual communities of Greek nationals. Exceptionally, as long as the community spoke some version of Greek, the default strategy for heteroglossic, and, in some instances, bilingual communities, was assimilation to Hellenic culture. In the past, the state guardedly accepted different cultural norms, as in the case of legislation beneficial to Thessaloniki's Jewish community (Law 147/1914, Law 180/1914, Law 231/1914), only to dilute it, three years later, with new law and policy that favored the Christian charter group. Under Ioannis Metaxas' hermetic police regime, exclusionary policies forcibly sequestered "suspect populations" in the northern borderlands – policies that, to varying degrees, endured well-past Greece's transition to democracy in 1974. By legislating, establishing processes, and promulgating policies, the state, by its sovereign prerogative, established what was legal, albeit not necessarily legitimate with respect to international norms for the treatment of minorities.

Collectively, such peaks and troughs of liberty demonstrate how the state and opposing stakeholders competed, contested, contended, and transacted asymmetrically in their efforts to, respectively, establish or challenge authority. Dissidence against the state by violent means on the part of minority groups did not occur – a phenomenon that begs further research.

In the nearly 100 years since "Lausanne," the trend line of liberal statecraft bends toward the Europeanist goal of a civil, liberal constitutional democracy that gives space and extends protections to communities of difference. Openness to difference and the development of a well-functioning civil society are, arguably, both the source and the outcome of a contestatory democracy.

In his treatise on republicanism and the sources of freedom, Philip Pettit posits that the benchmark of liberty is an ideological commitment to non-domination by the state (Pettit, 1997). To approximate that normative state of relations, three general preconditions need to be met: a basis for contestation (defining the desideratum), a channel or voice by which decisions may be contested, and a forum for contestation – what, according to Pettit, makes a "the responsive republic." If for Pettit liberty is a political ideal, as opposed to an end-goal, critics point out that his genre of liberty is made possible by an absence, summarized by Charles Larmore as follows: "We are free to the extent that we are not under the domination of others, subject to their will, and exposed to the vicissitudes of their desires" (Larmore, 2001: 229-230). Freedom, therefore, is predicated on non-interference by the state as opposed to an action program undertaken by the State in the service of liberty.

In this case, the main critique is that freedom cannot be served when the state eschews concerted actions in support of freedom as long as it fosters conditions and an institutional framework for the contestation of its policies. Accordingly, it is not sufficient to rely on organic deliberations "from below" over what may be a proper basis for contestation, how channels for contestatory speech and action can be opened, and forums established where this transactional politics would occur. In sum, liberty, critics claim, requires as much a rights-minded activist base as an activist, liberty-minding State. Contestatory politics is where they come together to arbitrate the contours and limits of liberty.

In the Greek experience, until 1974, the record of the state in defense of liberty was inconsistent, with Venizelos' liberalism at the apex of the arc and Metaxas' and the Colonels' Junta's oppression at the bottom of the trough. In all instances, the pursuit of liberty – in this case, minority rights – entangled opposing and differently empowered stakeholders in a process of asymmetric contestation. In such an environment, Pettit's expectation of vigorous contestatory agency by those whose liberty is at risk (in this case, religious minorities) is unlikely to be met. Such would be a political environment where immigrant/non-citizen status translates into less social and political capital – a position from which it would be difficult to challenge the state – unless plaintiffs can secure the assistance of civil society or international organizations. The non-interfering (meaning, non-activist) state in Pettit's republican mold would thus be, contributing to the political disability and social marginalization of immigrant populations. The danger in this case is that such a normative paradigm of liberty justifies inaction (putatively, neglect) as an expression of "non-domination," thereby perpetuating the conditions of exclusion of subaltern communities.

Antje Wiener articulates a theory of contestation according to which state and non-state stakeholders engage in the activist challenge of norms, processes, or procedures. In the Greek case, the national state sets the norms for establishing and maintaining all places of worship in the country based on the Hellenic Constitution's Article 13. Regional (Prefectural) and local states apply laws on the principle of subsidiarity and, since the implementation of the decentralization strategies "Kapodistrias" (Law 2539/1997) and "Kallikrates" (Law 3852/2010), enjoy a broad range of devolved competences. Subsidiarity and devolution, therefore, create opportunities for contestatory politics at different scales of administration.

Wiener describes contestation as a formal or informal political or social process of meta-organizing expressed through language (legal and political speech). Contestation takes different forms and can proceed in stages: *arbitration* (legal mode of contestation, whereby stakeholders consider pros-cons); *deliberation* (political mode, which addresses the formal application of rules and regulations; *contention* (critical interrogation of societal rules, regulations, or procedures); and *justification* (moral mode questioning principles of justice) (Wiener 2014) – forms and processes that one would expect to describe interactions among stakeholders in the Athens mosque case.

Until 2019, Athens remained the only European capital without an official public place of worship for its Muslim inhabitants. Muslim immigrants mainly from Afghanistan, Pakistan, Bangladesh, Iraq, Syria, Egypt, Iran, and Tunisia, congregate and worship by necessity in places such as private apartments, stores, storage facilities, either abandoned or idled industrial plants, building basements, and parking garages, which serve as makeshift, unlicensed places of worship. Between 60 and 100 such unofficial places of worship are spread across the Prefecture of Attica and, primarily, the Athens and Piraeus combined metropolitan region, mapped on clusters of co-national co-religionists.

Under that layer of informal Muslim places of worship lies a small constellation of Ottoman-era mosques, medreses, and tekes (convents or lodges). The majority of Ottoman mosques were demolished in the nineteenth century in the course of urban development and renewal. The desire to morphologically re-Hellenize the Capital city produced a regulatory environment for antiquities, which was (and continues to be) substantially oriented toward preservation and axiomatically opposed to the functional re-integration of historical sites into the living fabric of cities. That preservation ethic took Ottoman legacy mosques and medreses out of consideration for restoration for community use. Their status as museums reflects a similar preservation regime concerning particular Byzantine-era churches in Turkey – a cultural politics that strives to balance popular and state sentiment on heritage religious architecture on both sides of the border.[1]

The two geographies (the preserved but inactive legacy mosques layer and the vernacular layer of informal Muslim places of worship), are now complemented by the first state-licensed mosque to be built outside Western Thrace, since 1830. As a place of Islamic worship, the new mosque is non-denominational, welcoming Athens' different Muslim communities to use it for services and community events. The road to its completion is a lesson on the nature of contestatory politics in Athens.

Our study of the geopolitics of the building of Athens' new mosque is based on three stories of contestation of the prerogative of the state to establish norms, processes, and procedures for the regulation of places of worship – a critical dimension of what constitutes freedom of religion in like constitutional liberal democracies. We ask what caused the much-delayed licensing of this significant place of Islamic worship; how this building project relates to the preservation politics that regulate Ottoman legacy mosques and contemporary informal places of worship; and how the new mosque becomes entangled in the Muslim communities' geography of settlement, mobilities, and networks. We problematize, by extension, the geopolitics of the new mosque and make connections to other cases in our study where the intersection of religion, national and minority politics, and the built environment yielded insights on the nature of Hellenic statecraft and nationcraft.

Islamic places of worship in Athens: regulation, contestation, and placemaking

Upon Athens' elevation to national capital in 1834, Ottoman-built heritage was at best neglected and at worse targeted for demolition. The segregation of religious communities in different quarters distinguished ottoman urbanism. Accordingly, the castle and the mansion of the local governor, vakifs (charitable organizations and their services), ordinary homes, mosques, medreses, hammams, and some portion of market activity would have collectively constituted the "Turkish quarter." *Millet* Greeks, Jews, and Armenians would, commonly, be segregated to their quarters, enjoying a certainly degree of self-government. As we observed in the urban morphology of Ottoman Thessaloniki, "Turkish"-built fabric was veined throughout the city – a reflection of privilege.

Athens, however, did not exhibit classic Ottoman sectarian urban structure. Accordingly, Athens did not exhibit a high degree of population segregation based on religion. Instead, as Dimitris Karidis notes, Ottoman political and economic power, the former held by Muslim elites and the latter shared with elites of resident non-Muslim populations, was mapped on the central marketplace – the bazaar – and its immediate surroundings. That core became the locus of built forms of consequence: residences of the wealthy, both Muslim and Christian places of worship, and the center of civil administration (the voyvodalik, in this case). Karidis writes that "[p]roximity to the bazaar, while not necessarily desirable for living, was a source of much prestige and could be afforded by the very wealthy... Athens closely followed this 'principle.' ... In other words, the boundaries between urban space and social structure were not drawn on any dichotomy between the residential districts of Muslim and non-Muslim citizens" (Karidis, 2014, 72–73).

At the time of its incorporation in the Hellenic State in 1833, Athens was a modestly-sized city (population estimates range between 5,000 and 9,000 inhabitants). It did not have a significant Muslim population, as demonstrated by the modest number of Muslim places of worship. Drawing on the archaeological record and the 1668–71 travel accounts of Turkish geographer Evliya Çelebi, Ioannis Travlos reports that in addition to the mosque built inside the roofless cella of the Parthenon there were only another four mosques in Athens: the Fetiye Mosque (Mosque of the Conqueror), dated to 1458, expanded in 1670, and preserved to this day; the Küçük (small)

Mosque (possibly the mosque Çelebi describes as the Ula Bey Mosque) in the immediate periphery of the bazaar; the so-called Mosque-of-the-Column (possibly, Hadji Ali Mosque in Çelebi's account), situated presently in the intersection of the Hadrianou and Flessa Streets; and the Yeni Mosque, also known as the Beina Mosque, located on the land parcel delimited by Voulis, Nikodemou, Thucydidou, and Apollonos Streets (Figure 9.1). Evliya Çelebi references a small mosque on the path leading to the Acropolis' Propylaea, which were used in the Ottoman period as a barracks. Çelebi also records three hammams, seven medreses, two teke, one study "of sages", three mektebs, or boys' schools, two inns, and a great number of public fountains (Travlos, 1960/2005: 184–86). Two more mosques would be constructed in the eighteenth century: the Mosque of the Upper Fountain (also known as the Softa Mosque) and the Mosque of the Lower Fountain (built in 1759 by Athens' Voyvod Tzisdarakis), the latter still extant, and in a prominent location on Monastiraki Square, at the beginning of Pandrosou Street. Outdoors worship on the grounds on the Temple of Olympian Zeus, presumably by Ethiopians or Arabs, was memorialized in the 1819 Louis Dupré lithograph Le Temple de Jupiter Olympien et l'Acropolis d'Athens.[2]

The future of Muslim places of worship in independent Greece was politically fraught. Following the departure of the Muslim population in 1833, Athens' Muslim places of worship closed. Under Ottoman rule, some Christian churches had been transformed into mosques. Under Greek rule, they were quickly re-sanctified as churches. In a central district in the throes of structural change and landscape Hellenization, mosques were placed in a particularly precarious position. In contrast, built fabric with clear connections to Hellenism (pre-Classical, Classical, Hellenistic, and Roman buildings) were earmarked for preservation and integration into the mosaic Athens-as-national-capital. Non-culturally compliant buildings (for a time, including Byzantine-era built fabric) were targeted for demolition. The new canon precipitated the demolition of the Ottoman-era (described as "post-Byzantine" (μετά-Βυζαντινή)) Christian Orthodox metropolitan church and its replacement by a new, vastly larger National Cathedral of the Annunciation.

Although Ottoman-built heritage has enjoyed formal protections for the last 100 years, no heritage mosque has been open for worship since 1833. Preserved and restored mosques and other Ottoman-era buildings are under the Ministry of Culture jurisdiction, as are all archaeological sites. By law and regulation, therefore, legacy mosques could not (and cannot) serve as places of worship for Athens' growing Muslim population. On that principle, therefore, reopening Athens' two largest legacy mosques – the Fetiye and Tzisdaraki Mosques – as Muslim places of worship would endanger them as archeological sites. Moreover, their modest size would not have addressed the needs of the city's Muslim communities. A different site in another part of the metropolis would have needed to be developed to that purpose. Following decades of deliberation, in 2006, the national state

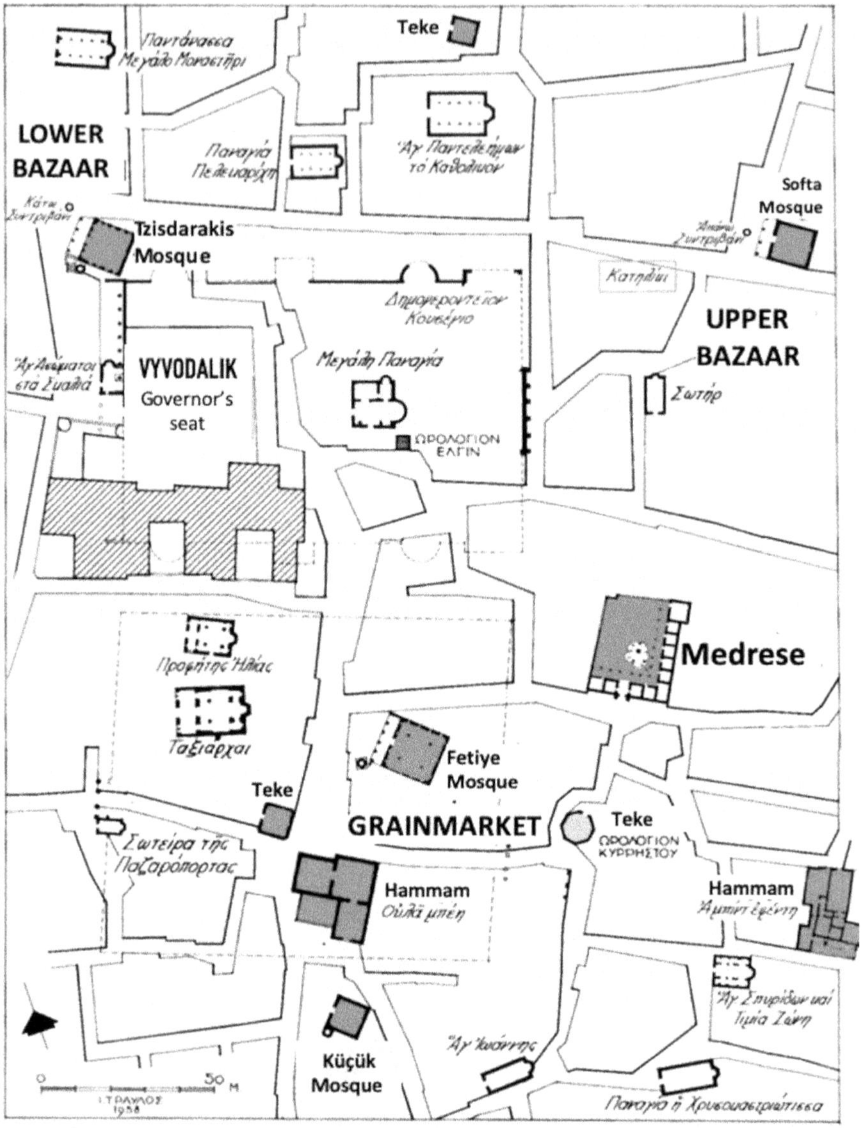

Source: Adapted from Travlos, 1958, 1993: 211.

Ottoman buildings in the urban core

Pre-Ottoman buildings adapted by Ottomans

Figure 9.1 Athens' core at the beginning of the Nineteenth Century. The co-location of mosques and churches in the same district at such close quarters is striking (approximate area: 35 hectares or 85 acres) (Adapted from Travlos, 1952, 1993. In Karidis, 2014).

decided to build a new official Muslim place of worship in Athens. It was inaugurated in 2019 after long delays. In the interim, Muslim communities had to devise their own solutions to the problem.

Athens' informal mosques – placemaking in the margins

Muslim communities are often conceived in Greece's public discourse as a unitary whole, their broad diversity concerning country of origin, sect or denomination, legal and occupational status, and the material, and geographical circumstances of their daily life obscured. The geographic provenance of Muslims in Greece includes Pakistan, India, Bangladesh, Afghanistan, West Africa (mainly, Senegal and Nigeria), North Africa (mainly Egypt and the Maghreb), Palestine, Iraq, Syria (the vast majority of the latter group arriving in Greece after the onset of the Syrian civil war in 2011), and Sahelian-zone countries (mainly Chad and Sudan).

Concerning immigrant community access to services, Greece finds itself where Western European countries with Muslim immigrant populations were 20 or 30 years ago. Despite the ebb and flow of migration over decades, Greece has only recently come to terms with the reality of significant, long-term co-ethnic and third-country resident immigrant communities. The consequence has been a relatively slow pace of integration of these populations in the Greek social state and Greek society (Gropas and Triandafyllidou, 2009). The emergence of a broadly-missioned and vocal civil society has been addressing some of these deficits. More recently, and with the encouragement of, and financial support of the European Union, networks of civil society organizations dedicated to human rights and the protection of refugees and other immigrant groups have proliferated (Ibid, 11; Papadopoulos et al., 2013: 348-349). Immigrant labor and religious associations have also proliferated, and immigrants have been made welcome in Greek trade unions (Fouskas, 2012). Such associational politics, although necessary for community agency, enjoy limited recognition by the state and are, thus, at a disadvantage in contestatory proceedings involving the state. Ultimately, they cannot adequately address alone the needs of immigrant communities for protection, social development, and cultural expression. In coalition with civil society partners, such associations have been more successful in their pursuit of resources.

Outside Western Thrace, Muslim communities in Greece experience limits to economic and educational access and political representation. They also experience structural impediments to the practice of religion: in the licensing of Islamic cemeteries and, where that is not possible, reserving sections of Christian cemeteries for their use, empowering muftis to give legal rulings on matters of religion (as in Western Thrace) and, finally, securing the licensing and permitting of mosques (Ioannou, 2013: 180).

The study of informal, vernacular, places of Muslim worship in Athens and the communities that brought them into being presents significant challenges. Estimates on the number of such places range between 60 and 100. Informal mosques are invisible to the public. On the one hand, this visual illegibility is the result of communities' desire for security and privacy, made in part necessary by the simmering menace of "Golden Dawn," the anti-immigrant extreme-Right party, now suspended by the Courts. On the other hand, the absence of, or limited zoning provisions for the establishment of non-Christian religious places limits their visibility in the built environment. Signage and Islamic epigraphs on facades are rare and, where present, they are discrete. Other religious communities – including Hindu, Buddhist, Sikh, and a range of evangelical and charismatic Christian sects that remain unaccounted for in Greece's legal and planning frames – share this state of invisibility in the landscape.[2]

"All known religions shall be free…"

The ability of religious communities to build official places of worship depends on their legal status. Official status cannot automatically be granted to a religion as the expression of a fundamental right, primarily because of the privileged position of the Eastern Orthodox Church as a quasi-state religion. Therefore, legal ambiguity about the status of a religion and the influence of the Church in Greek politics, produce paralysis in liberally extending recognition and rights to all religions.

On the one hand, the Constitution states that "[t]he prevailing religion in Greece is that of the Eastern Orthodox Church of Christ. The Orthodox Church of Greece, acknowledging our Lord Jesus Christ as its head…" (Section II, Article 3). This can be reasonably read as the recognition of a State religion.

On the other hand, Article 13 of the Constitution protects freedom of religion (ΦΕΚ 120 Α΄/27-6-2008):

> "Paragraph 1. Freedom of religious conscience is inviolable. The enjoyment of civil rights and liberties does not depend on the individual's religious beliefs.
>
> Paragraph 2. All *known religions* [our emphasis] shall be free and their rites of worship shall be performed unhindered and under the protection of the law. The practice of rites of worship is not allowed to offend public order or the good usages. Proselytism is prohibited…"

Accordingly, then, one should conclude that all religions would be equal before the law and would have the right to establish formal places of worship. The key here is the granting of protection to "known religions" only. Per Greek law, a religion is "known" (meaning, recognized as having legal status) if it has been granted a permit for the erection and operation of a place of worship (ναός) or a house of prayer (ευκτήριος οίκος)

(Law 4301/2014, Article 17). Simply put, to gain official status as a religion, a license to build and operate a temple needs to be secured first. Ergo, the refusal to grant such building and operator permits, often on technical grounds, creates structural impediment for communities observing an "unknown" religion. And while under the terms of "Lausanne," the Muslim communities of Western Thrace can readily secure such licenses and building permits for mosques, that privilege is not extended to Muslims outside Western Thrace.

Overturning this regulatory rule would be challenging for subaltern communities, despite the rule's *prima facie* incongruity with Greece's Lisbon Treaty obligation to protect the freedom of thought, conscience and religion. The EU Charter of Fundamental Rights (EUCFR) created by Regulation EC168/2007 guarantees freedom of thought, conscience and religion (Article 10). It came into force at the signing of the Lisbon Treaty in 2009 and it is, therefore, binding on all Member States. It is also aligned with the European Convention on Human Rights. Any departure from that blanket protection must respect Article 9 (2) of the Convention, which reads as follows:

> Freedom to manifest one's religion or beliefs shall be subject only to such limitations as are prescribed by law and are necessary in a democratic society in the interests of public safety, for the protection of public order, health or morals, or for the protection of the rights and freedoms of others. (EUCFR, Title II, explanation).

The Greek and the European legal articulation of religious rights do not overlap fully, the Greek Constitution's being succinct and not inclusive of all conditions stated in the European Charter. Ambiguity, therefore, allows for divisible sovereignty exercises, possibly contestable at the European level. Such multi-phase contestation involving courts at national and European levels would require significant community resources and support by civil society organizations. It is not impossible but unlikely.

Yet, Muslim communities do not appear at a dead end on their quest to secure their own places of worship. Their founding of as many as 100 informal mosques, some of which are now licensed as official places of worship, sheds light on the workings of Hellenic statecraft at the local scale.

Home-informal mosque spatial match

In 2020, the demography and geographies of immigrant communities in Athens remain unclear. The latest authoritative information is found in the *Population and Housing Census* of 2011; moreover, the state does not release population projections at a level granular enough for meaningful analysis.

Thomas Maloutas reports that between 1991 and 2011 the migrant population of the Athens metropolitan region increased from 38,351 (1.06 percent of the total population) in 1991 to 367,903 (8.27 percent) in 2001, and reached 398,599 (9.46 percent) in 2011. Notably, a substantial percentage of that population resides in the Municipality of Athens (the metropolitan region's historic core): 41.1 percent (1991), 39.9 percent (2001), and 38.47 percent (2011) (Maloutas, 2018: 141–142). There is a strong expectation that the 2021 Census will reveal yet larger immigrant population counts. New flows stem from the Libyan, Syrian, and Yemeni civil wars, the rise and fall of Islamic State in Syria and northern Iraq, the on-going war in Afghanistan, and economic precarity in parts of West and South Asia and Africa. These events have substantially increased the flows of migrants and refugees forced on the road by war, political oppression, and poverty. Situated at the crossroad of Southeastern Europe and the Eastern Mediterranean, Greece has emerged as both a reception country and a major European gateway destination for migrants wishing to reach Europe's most prosperous countries.

The growth and spatial distribution of Athens' immigrant populations can be understood in terms of the changing morphology of city and region, the mobilities of Athenian Greek nationals, and Muslim communities' income and occupation profiles. At the end of the 1970s, upper- and middle-class Athenians started out-migrating to new, fringe belt municipalities that did not suffer from the urban core's structural ills: poor town planning outcomes, very dense-built fabric, limited green and open spaces, under-resourced and poorly managed municipal services. Between 1991 and 2011, the Municipality's Greek nationals' population was reduced by approximately 290,000 (36 percent) (Ibid: 143). This demographic, fiscal, financial, and occupational hollowing out of the urban core undermined the local state's efforts to engage in extensive urban renewal there. Instead, the attention of the regional and national states was reterritorialized in the newer ring municipalities.

The spatial distribution of these populations varies greatly: Albanian immigrants (Christian and non-Christian) are substantially integrated in communities across the Prefecture of Attica. The community's high Greek language acquisition rate, not being subject to the level of casualized and structural racism black and brown immigrants encounter, the existence of a Greek-Albanian legacy population in Albania, and Albania's political support, has allowed Albanian emigrants to settle across the region. In contrast, non-Christian majority ethnic groups cluster mostly in the western, poorer quarters of the Athens Municipality and adjacent community areas to the west of it, including the old industrial core. Also different is the suburban and peri-urban settlement pattern of Pakistani and Indian immigrants, commonly employed in commercial agriculture (Figure 9.2).

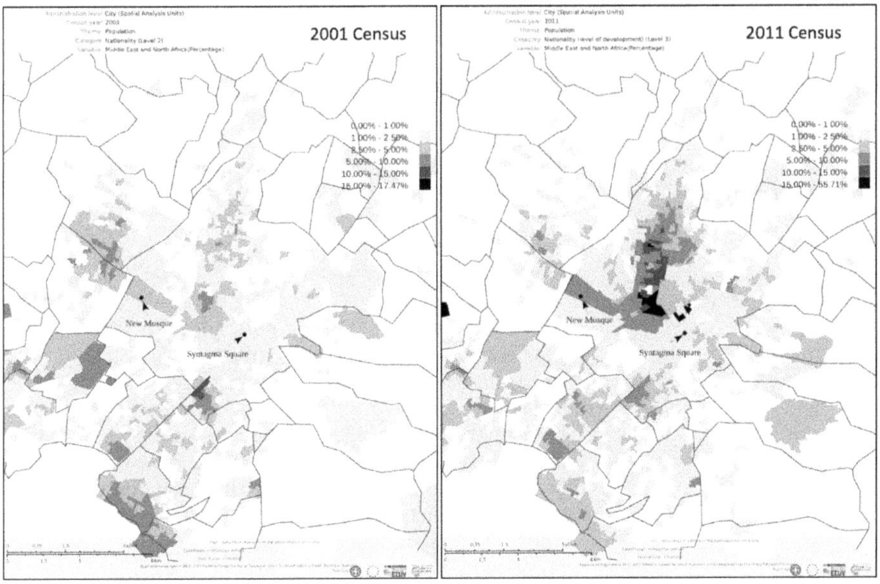

Figure 9.2 Cumulative visualization of the spatial distribution of non-Christian populations from North Africa and the Middle East, 2001-2011.

Source: ELSTAT, ESPA/EU; cartography: Vagelis Papadias).

By international standards, the rate of Greek-Athenian-immigrant residential segregation is relatively low (Kandylis et al., 2012: 268–269). Low-income immigrants from developing countries occupied residential space in a range of low-to-middle land rent neighborhoods. In these city quarters, they often occupied (and still occupy) the least desirable housing units in Athens' vertically class-stratified residential apartment buildings: ground floor, basement, and sub-basement units (the 2011 census records the floor on which the respondent resides).

During the GREXIT crisis (2009–2017), income polarization between the city and its ring municipalities increased further. Many households were consolidated to reduce housing costs, thereby, further hollowing out Athens' housing market and creating low-rent niches and pockets. Proportionately younger, male, with low educational attainment, living collectively in non-family households, occupied in low-paying sectors of the economy, and with limited access to goods and services, such as access to the internet and automobile ownership, immigrant populations, generally, and Muslim populations, specifically, experience degrees of segregation and the diseconomies and inconvenience of high-density occupancy in a municipality under fiscal and financial stress (Table 9.1).

Athens' informal mosques are spatially embedded in the Muslim communities they serve and are frequently co-located with the seats of immigrant

Table 9.1 Muslim minority groups' social characteristics by country of origin

Country of origin	*Egypt, India, Syria, Bangladesh, Pakistan*	*Afghanistan, Iraq, PR China, Nigeria*
Age & household	Younger; majority males; non-family households	Younger; majority males; non-family households
Occupation	Manufacturing; construction; agriculture; trades; unemployed	Commerce, unemployed
Educational attainment	Low; no access to education for 15-19-year cohorts	Low; no access to education for 15-19-year cohorts
Housing	Per capita residential space substantially below national average; majority in street level, basement, or sub-basement apartment units; low percentage of homeownership	Per capita residential space substantially below national average; majority in street level, basement, or sub-basement apartment units; low percentage of homeownership
Community places	Urban core; neighborhoods in recessive phase; peri-urban agricultural communities	Urban core; neighborhoods in recessive phase
Access to goods and services	Limited access to the internet; no personal vehicle	Limited access to the internet; no personal vehicle
Segregation & community pop. density	High; high	High; high
Residency status	Documented (as refugee or other); undocumented	Documented (as refugee or other); undocumented
Sequestration	N/A	Incarceration

(Source: Adapted from Maloutas, 2018)

associations. Since the 1980s, such mosques have developed organically in neighborhoods with clusters of co-national co-religionists. They are mainly found in the urban communities of Kypseli, Renti, Peristeri, Tavros, Kerameikos, Ampelokipi, Nea Ionia, and Neos Kosmos, and in peri-urban agricultural communities, such as Marathon, in the southeastern fringe of the metropolitan region, and Lavrio on Attica's east coast (Papadopoulos, 2017: interview). They serve as places of worship and prayer, religious schools, mess halls, community centers, and event spaces. Some are hosted in residential flats that temporarily transform into communal spaces for one of these uses. The majority of informal places of worship are constructed in building basements, ordinary stores, closed factory buildings, and garages (Figure 9.3). They are also temporarily put together in public places, such

Figure 9.3 A morning in Gerani Street. The environs of Omonia Square – the heart of the Nineteenth Century core of the city – are home for many minority businesses, and cultural amenities, including a number of informal mosques.

Source: Christos Paridis, with permission.

as neighborhood squares and soccer fields. Some of them are accommodated in shanties (Argyropoulou and Tzirtzilakis, 2015: XI). Altogether, they define a geography of care. By necessity, informal mosques blur the boundaries between public and communal, public and private, and sacred and lay spaces.

Significant donations can produce upgraded, special-purpose facilities like the Hellenic-Arab Educational and Cultural Center, funded by a Saudi businessman in 2007 (Figure 9.4). The Egyptian-born president, Mohammed Ibrahim, remarks, "[The converted factory building that also serves as a school] is much more comfortable than where we used to pray, but it cannot serve hundreds of thousands of Muslims in Athens" … "We help Arab-speaking citizens to teach their children Arabic but also to understand the Greek mentality" (Kitsantonis, 2007: NYT Online).

Photographer Tassos Vrettos' Τ(ρ)οποι λατρειας (Wor(th)ship)

Muslim communities are very welcoming. However, personal access to informal places of worship is difficult because of security concerns and the desire of religious communities for privacy. These are not clandestine sites. Importantly, the mosques-cum-cultural centers are known to the police.

Figure 9.4 In the foreground, the yard of *Hellenic Metal Scrap*. In the middle ground the *Hellenic-Arab Educational and Cultural Center*. A Saudi businessman funded the nonprofit community organization. In 2007 it converted an old factory building into a special-purpose facility that includes a school and space for worship (Corner of Kyprou and Dimitsanas Streets, Ano Moschato)

Source: Christos Paridis, with permission.

Since they could not be licensed as mosques until recently, they have long been licensed as cultural centers. The state's awareness of these informal sites of worship runs counter the often-made security argument that they could potentially shield and propagate Islamic extremism. To wit, one argument made in support of the new state-licensed mosque is that it provides greater transparency on Muslim community activities. Ground truth does not support this argument. The new mosque's modest capacity and its impractical siting at an old deindustrialized zone do not satisfy demand for community spaces by Athens' populous Muslim communities. Ergo, neighborhood mosques will remain a feature of Muslim community life in Athens.

Since 2012, photographer Tasos Vrettos has enjoyed privileged access to such places. We had the opportunity to interview him in December of 2017, to draw on his first-hand observations of Athens' "religious-vernacular." His extensive body of work provides fine-grained information on the topographies, landscapes, and rites of worship of Athens' less-visible religious communities.

Vrettos describes his work as an impetus for self-analysis and self-discovery. His approach to photographing the human subject is not explicitly ethnographic. It has a strong narrative character (or in Henri Cartier-Bresson's terminology, it is a picture story). In Vrettos' own words, his art-photographic reportage is instinctive, unplanned, uncontrived, unstaged and unposed, spontaneous, and seeking to capture the unexpected (Paridis, 2015: *LIFO*).

Tassos Vrettos' diverse work explores the daily vernacular (studiovrettos.gr). It is at times documentary; it has visually chronicled and recorded kitsch landscapes and material culture (a long-standing interest), and it studies the demimonde and the transgressive. Much of this work is collected and curated in anthologies and exhibitions. Vrettos' commercial work includes fashion photography, theater, cinema, advertising, and magazine covers, including exclusive work for composer Manos Hatzidakis' influential art and music magazine *Tetarto*. As a crossover "art-commercial" photographer, Vrettos' work expresses a high degree of awareness of class and capitalism's contradictions, culture, and politics in Greece.

Henri Cartier-Bresson influenced Vrettos' body of work Τ*(ρ)*οποι λατρειας (2015) ("Topoi" and ways of worship): "To take a photograph is to align the head, the eye, and the heart. It's a way of life." The work reflects on Bresson's importance of the "decisive moment," when the photographer manages to capture a geometric pattern that provides clarity about what is essential in life (Bresson, 1952). For Roland Barth, the photographer may capture that one, unique, emotively expressive frame, which belongs to "the studium" (loosely translated as the study or the field of action, also the focused interest of the photographer) in an otherwise ordinary world of "things-as-they-are."

Vrettos' work embraces Barthes' idea of "punctum": that which makes the image memorable, if not also critical and important: the element in the photograph that slays the viewer like an arrow through the heart: "...it is this element which rises from the scene, shoots out of it like an arrow, and pierces me" (Barthes, 1980: 26; Fried, 2015). Whether visually sparse, as in the case of large, warehouses modified for worship, or baroque in aspect, as in the case of intimate, crowded, private apartments, with riotous décor used for services, Vrettos' photographs of informal Christian and Muslim places of worship reveal the vast variety of approaches and solutions to the absence of official mosques and churches for religions not "known" to the state.

Vrettos suggests that nothing could substitute time-space presence in that genre of photography. He describes his experience of a Paschal liturgy in an informal Ethiopian church in the neighborhood of Polygono, as "a complete shock to the senses;" "it was as if I had been transported to an alternate space-time;" "as if I had pulled up the cover to a manhole and slipped into an underground city" (Papadopoulos interview, 2017) (Figure 9.5).

Figure 9.5 Ethiopian Orthodox Church Kidane Miheret, Polygono, Athens, Greece (photograph by Tassos Vrettos, courtesy of the photographer – original photograph in color).

Whether temporary or transitory, or long-standing community places, such sites range in appearance and temperament. Vrettos describes his visits to mosques in the suburban town of Marathonas, home to extensive commercial agricultural enterprises: “In Marathonas, I visited three mosques; I am not sure how many more there might be: a large one, which serves as the main mosque and is housed in an underground garage; one in the inland reaches of Marathonas, where I encountered three or four individuals, all very friendly; and a smaller, private [mosque], which was built by an individual for his personal use as prayer space (Figure 9.6). They were all Pakistanis” (Bratsiakou, 2015: Interview).

Vrettos narrates his visits to the Senegalese Brotherhood of the Murids on Agiou Konstantinou Avenue (Figure 9.7):

> They do not call their place of worship either a mosque or a church. They hold their meetings every Wednesday evening… Women participate, albeit, in small numbers. The place is stark; not iconic [of Islam], and includes only two or three pictures of the Brotherhood founder. The space is strewn with carpets of all sorts, which creates a peculiar kind of a mosaic. The day I photographed them, most were dressed in traditional garb; each [worshiper] held a sacred text in photocopy and read or chanted in a low voice – They created a very devotional atmosphere,

Figure 9.6 Fatima-Zahra Mosque, Marathon, Greece (photograph by Tassos Vrettos, courtesy of the photographer – original photograph in color).

Figure 9.7 Dahiraa Mouride (Senegalese), Agiou Konstantinou street, Athens (photograph by Tassos Vrettos, courtesy of the photographer – original photograph in color).

> where the sound dominated the space. During the service, they passed around coffee, and at the end of it, we ate a meal of spiced meat with carrots. They expected around that time the arrival of an important cleric from Senegal, and on the day of his arrival, and because they could not all fit in their place of worship, they congregated at a night club near Karaiskaki Square (Ibid).

Vrettos' photographs bridge the gap between the sublime-experiential and his desire to make visible to the majority Greeks the worlds of Athens' religious-vernacular. This alter-Athens (a subaltern city of many faiths) is not visible to the majority of Athenians. Vrettos noted that over the three years of the original endeavor (2012–2015), building of mutual trust was what made these photographs possible. He was grateful to be allowed to "make them visible to us" (Athens' majority society) and to introduce them to national and international audiences via his exhibition of the photographs at Athens' Benaki Museum (Vrettos et al. 2015; Papadopoulos, 2017: interview). The project is ongoing.

Athens' informal places of worship have organically proliferated since the late 1980s to meet the demand of small, clustered communities of Christian and Muslim co-ethnic co-religionists. Unlike Athens' new official Athens Mosque in Votanikos, they are not the product of overtly public or collective contestation of state power by the impacted communities, although, continuous lobbying and trust-building measures were essential to their preservation. Their existence, proliferation and durability in Athens' cultural geography despite their long-standing unregulated status, suggests that there is space between the state's *de jure* deference to a national religion, as expressed in the Hellenic Constitution, and more liberal (or at least not illiberal) *de facto* State practices concerning religious minorities. While inadequate in meeting Athens' Muslim communities' needs for full recognition, resources, and space, the state has judiciously avoided closing informal mosques as part of enforcing zoning regulations or applying the constitution's standard, which endows with rights only religions "known" to the State. Importantly, the state is also ahead of mainstream Greek society, which retains conservative views of immigrants (DiaNEOsis survey, 2016), by declaring its commitment to the building of a civil society (κοινωνία πολιτών) and gradually converging with European standards of openness concerning difference. Ergo, there is an exclusionary national way of doing things, to which subscribes the majority of Greek citizens, a constitutional frame for society that increasingly bends towards Europeanization, in part out of political necessity, and then there is the state's pliable, variable, and not always effective strategy of bridging these divergences.

The operative components of this strategy are subsidiarity and deliberate non-enforcement of select laws. First, with subsidiarity the national state cedes initiative to the local state on matters of compliance to urban regulations. As the most multicultural and cosmopolitan of Greek cities, Athens has long had to accommodate difference in its politics and manage the outcomes of socio-spatial exclusion. Critically, it is more eager than the national state to engage in dialogue with resident communities. The second dimension of the state's practice is the decision to selectively defer the application of laws of consequence to inter-communal relations (as in informally allowing the operation of places of worship of religions not "known" to it). It is, in some important sense, exemplary of Pettit's non-interventionist state catering to liberty.

This pliable and conditional regime of Muslim community accommodation on matters of religious worship could never sufficiently address the substantial demand for sacred and community space. The obvious solution has long been the licensing and building of new mosques of the type, size, and in locations most suitable for underserved immigrant Muslim communities. That has been an incredibly fraught scenario.

Building the Athens mosque in the industrial fringe

The Athens Mosque's symbolic and political significance as the first new mosque built in Athens in nearly 200 years cannot be overstated. After several false starts between 1880 and 2006, Greece passed the necessary legislation to build a mosque to "address the religious needs of the Muslims living in [the Prefecture of] Attica (για την κάλυψη των θρησκευτικών αναγκών των μουσουλμάνων που διαβιούν στην Αττικη) (Law 3512/2006). The modest-sized mosque cannot possibly address the needs of Athens' populous Muslim communities, the population of which may be as high as 250,000 out of a 2015 population estimate of 3.052 million (ca. 8.2 percent). However, the passing of the law was without question, a point of inflection in Greece's political engagement with difference.

The road to law 3512/2006

Although a decision was finally reached in 2006, to build an official Muslim place of worship in Athens, inertia and drift on the part of the state, and resistance to the project by extremists and religious conservatives delayed its completion thirteen years. The latter groups consider it a threat to the Hellenic character of the State and its capital.

The need for such a facility became all the more acute in the last two decades as refugees from the Middle East, North Africa, and West Asia joined the growing numbers of economic refugees from these regions and also sub-Saharan Africa. The 2001 and 2011 census data report that demographic growth. The 2021 census is likely to record further growth. Therefore, the pressure to legislate the building of a significant new mosque for the Greek capital increased over time.

The question of new mosque foundations dates back to the creation of the Hellenic State. Urban development and renewal were often a proxy for landscape Hellenization, and landscape Hellenization meant the promotion of the built environments of the Hellenic past, often at the cost of others' heritage landscapes. As a result, many mosques were either decommissioned and, ultimately, turned into museums or demolished. Then, in 1880, a decision was made to erect a mosque in the Athens-Piraeus municipal area, on land allocated close to the Port of Piraeus. The plan was never implemented. With the conclusion of the Balkan Wars, Greece acquired lands in Macedonia that included a Muslim minority population. According to Article 4 of the Third Protocol of Athens of 1913, Greece assumed the responsibility to erect five mosques in its national territory, one in the capital and the remaining four in other Greek cities. That also proved to be a dead-end, primarily because the First World War and the Greek-Turkish War that followed it created new ground conditions: population exchanges and the resettlement of minorities.

More than 20 years later, Law 6244/1934 created the legal basis for the construction of a mosque and Islamic Cultural Center for the use of members of the Egyptian Government undertaking scholarship-supported studies at the University of Athens. That law also fell victim to indecision and inaction.

Stasis persisted in the years of the dictatorship of nationalist Ioannis Metaxas, the Second World War, the Greek Civil War, and years of great civil and geopolitical instability culminating with the fall of the Greek military Junta in July 1974. Neither did the issue of Muslim worship outside Western Thrace appear to be a priority, nor was there a possibility to engage in contestatory politics to compel the Government to action.

Greece's diplomatic and economic isolation ended in 1974 with the return of democracy and a resumption of Greece's efforts for the internationalization of its economy. International pressure soon mounted for redress of the situation. In 1976, Arab and other Muslim states renewed their call for the erection of a mosque in Athens. The project gained some traction following the election of Andreas Papandreou, a Socialist, to Greece's premiership. His Government reintroduce in 1983 a plan for a mosque in Athens. The proposal was for its erection in the Community of Marousi (Amarousion), a newer middle-class suburb with a population of 48,283 (Census, 1981, Volume V), approximately ten kilometers north of the city's historic core. However, since the 1981 census recorded only internal migration and returns to Greece of nationals of the diaspora, it could not reasonably support location-specific community-based proposals (Census 1981, Volume IV).

Consequently, the Government's locational logic in selecting Marousi as the site for the mosque is unclear. Moreover, the 2001 and 2011 censuses that counted immigrants would show that upmarket Marousi had not been a site of notable Muslim settlement in that period. The Muslim community's settlement geography, therefore, did not appear to support the Marousi plan. Ultimately, the Marousi proposal was met with great resistance by the local community, and the plan stalled in Parliament at the committee level for several years.

Inaction on this issue was met with embarrassing diplomatic demarches against Greece. Notably, in 2002, the European Commissioner for Human Rights, Alvaro Gil-Robles, put the Government on notice and demanded immediate action. Legal exigencies associated with Athens' bid for the 2004 Olympic Games compelled the Government to reanimate the debate on the new mosque. Obligated to provide a place of worship for Muslim athletes, the Government proposed the construction of a mosque in the working-class community of Paiania, at the extreme eastern fringe of the metropolitan region and the site of a sizable Muslim population, primarily employed in commercial agriculture. The proposal was made all the more attractive by Saudi Arabia's commitment to fund the project. This plan was also derailed. The influential Archbishop of All-Greece, Christodoulos, opposed it. Incredibly, he successfully argued that the building of a mosque in Paiania and in close proximity to the under-construction Eleutherios Venizelos International Airport, would "confuse and alarm" visitors to a Christian country. The claim echoed Metropolitan Nikiphoros II's indignation statement in 1988 about the prominence of the Grand Beyazit Mosque in the City of Didymoteicho that precipitated the building of the large Cathedral Church of Panayia Elefterotra. Ultimately, and for the 2004 Olympic Games' duration, the Government constructed a temporary modular building for the worship needs of Muslim athletes.

It is in 2006 when, finally, an effective law – Law 3512/2006 – was passed by the Center-Right Government of Prime Minister Konstantinos A Karamanlis. All political parties voted in favor of the new mosque, except for the extreme rightist party LA.O.S (Λαϊκός Ορθόδοξος Συναγερμός - Popular Orthodox Alarm). The neo-Nazi "Golden Dawn" Party did not have a presence in Parliament at the time, but supplanted LA.O.S following the parliamentary elections of 2012, as the leading voice of opposition to immigration and the mosque project.

With legal obstacles removed, planning for the Athens mosque was set to start. The Ministry of Education and Religious Affairs, which was vested by the Law to oversee the project and manage the mosque affairs following its opening, requested a Hellenic Navy property of 5.2 hectares in Athens' industrial Votanikos community area were ceded for the siting of the new mosque. Nevertheless, drift and inaction persisted. The European Commission's 2009 *Report on Racism and Discrimination in the Union* condemned the delays. Of great concern for the Government was sustained pressure from Turkey, under the polemical premiership of Recep Tayyip Erdoğan, who threatened to make the continuing operation of Orthodox churches in Turkey contingent upon the opening of a mosque in Athens – a political gaffe that served Turkish strategy until the opening of the mosque ten years later.

In 2010, a year after the European Commission's scathing report, Socialist Prime Minister George A Papandreou and by Presidential Edict, directed the Ministry of Defense to cede 1.7 hectares of the original 5.7-hectare plot in Votanikos initially earmarked for the project.

It will take nearly another ten years of uncertainty and delays to complete the rezoning of the parcel and its surroundings, invite public procurement bids, announce and hold a competition for the design, select a design, then hold two rounds of bids for the contracting of the project (the first did not attract any bids) until, in 2017, construction began.

Law 3512/2006 – ΦΕΚ 264/A'/5.12.2006

The Law created a not-for-profit legal entity in the public interest with the name Administrative Committee of the Mosque of Athens," chartered in the Municipality of Athens and supervised by the Minister of Education and Religious Affairs (ME&RA). It is described as "not belonging to the broadly-defined public sector" (Article 1).

The scope of the Committee is to administer, manage, and maintain the mosque, which would be built to "address the religious needs of the Muslims who live in [the Prefecture of] Attica" (Article 2). The Committee's broad responsibilities would make its composition – outlined in Article 5 – of paramount importance. Its disproportionate staffing with government officials would, effectively, make it a parastatal entity. Further, the second part of the article alludes to the mosque's capacity to "address the religious needs of Muslims across Attica" – an overstatement that leaves the door open for future demands.

Article 3 describes the division of labor between the Ministry of Education and Religious Affairs and the Ministry of the Environment, Planning, and Public Works (MEP&PW). ME&RA would build the mosque on public land on parameters set by Presidential Decree, by mobilizing its educational infrastructure branch (effectively, the Ministerial department responsible for building and maintaining public schools). MEP&PW will be responsible for financial planning through the Direct Investment Program. The two ministries will advise the Presidency on the matter.

Article 4 outlines the property and financing regime: first, the mosque building and the associated land are the property of the not-for-profit legal entity created by the law. Second, the mosque's primary source of financing is the ME&RA's annual budget. Bequests, gifts, and other kinds of donations from physical and legal persons are acceptable as incidental sources. The last articulation opens the door to endowments by Islamic foundations (*wakf/vakif*) that, again, suggests future non-state sources of support and, possibly, the state stepping back from wholesale management of like religious projects.

Article 5 lays out the composition of the seven-member Administrative Committee (appointments were announced on 21 August 2017 (Regulation No. 135952/Θ1): the Committee includes seven voting members, including the President and the Vice-President, who are appointed by the state. A non-voting Secretary is also state-appointed. All three, including alternates, hold high administrative office in the Ministry of Education, Research, and Religious Affairs (formerly (2006), ME&RA). Three additional voting members and their alternates are drawn, one each, from Athens' municipal administration, Athens' City Council, and the

Areopagus, the Supreme Court (a separate power from the executive branch under the Constitution). The remaining two seats are reserved for members of the Muslim community and are drawn from the largest ethnoreligious communities in Athens (Pakistan, Afghanistan, Egypt, Bangladesh, India, and Syria, Census, 2011).

Officers of the Administrative Committee of the Mosque of Athens

President: Deputy Head of the General Secretariat of Religious Affairs of the Ministry of Education, Research, and Religious Affairs.
Alternate: Head of Directorate of the Ministry of Education, Research and Religious Affairs.

Vice president: Deputy Head of the General Secretariat for Fiscal Policy of the Ministry of Finance
Alternate: Head of Department of the Ministry of Finance.

Secretary (non-voting): Administrative Officer of the General Secretariat of Religious Affairs of the Ministry of Education, Research and Religious Affairs.
Alternate: Education Officer of the General Secretariat of Religions of the Ministry of Education, Research and Religious Affairs.

Members

Head of Directorate, Municipality of Athens
Alternate: Head of Directorate, Municipality of Athens

City councilperson, Municipal Council, Municipality of Athens
Alternate: City councilperson, Municipal Council, Municipality of Athens

State's Attorney at the Supreme Court
Alternate: State's Attorney at the Supreme Court

Representative of Muslims residing in Attica
Alternate: Representative (as above)

Representative of Muslims residing in Attica
Alternate: Representative (as above)

The minority participation of Muslim community representatives is striking. The parastatal character of the mosque's administration is hard to refute.

Article 6 states that the Athens mosque's Imam, the head religious officiant, is appointed by the state, is a public civil servant, and his salary is paid by the Education Ministry at a rate set by the Finance Ministry. Importantly, the duties and responsibilities of the Imam are strictly circumscribed to religious and community activities in the facility.

The law is paradigmatic of Greece's statism and *dirigisme* in its intent to fully encapsulate the processes of building and administering the mosque within mechanisms of the state. It would be hard not to characterize such insistence on control as a reflection of Greek governments' geopolitical anxieties, their desire to placate conservative constituencies and, no less, the Church of Greece, and more generally, avoid introducing unnecessary stressors in Greece's precarious clientelistic political system, which, incidentally, the economic crisis of 2009–2017 all but extinguished. It can also be reasonably explained by the political establishment's and, more broadly, Greek charter society's ambivalence about broadening the definition of Hellenism and reclaiming the essence of much-talked-about but often unpracticed Mediterranean cosmopolitanism.

The big story here is not only the long and difficult road of contestatory politics in defense of the mosque project. As Pettit posits about the prerequisites to pursuing liberty, there was a long-established *basis* for contestation and the articulation of a concrete objective, there were *voices* and *channels* through which decisions were contested (the Church of Greece, Parliamentary political parties, the State itself in its democratic incarnations, other states (Turkey, Saudi Arabia, Egypt) that claimed stakeholder status, international organizations, such as the European Union, international and domestic civil society, and a broad range of Muslim community group associations. Finally, there were public *forums* where contestation took place: the Hellenic Parliament, Athens' city government (via its public participation planning outlets and lobbying by community organizations, the courts, and the press.

The critical point here is that the power positionalities of the stakeholders in the mosque's contestatory politics reveal the struggle to have been asymmetric. State and intergovernmental stakeholders have disproportionate control and influence over norms, processes, and procedures (Table 9.2). Immigrant groups, such as the Muslim Association of Athens, lobbied politicians for years. Its Director, Naim el-Ghandour, an Egyptian businessman living in Greece since 1974 who administers the Association with his wife Anna Stamou, remarks, "I try to stay romantic about Greece and not feel that [politicians] manipulated me… My children go to school, to university, live equally with Greek children. But when they go to pray, one goes to church and the other goes to an underground garage. This proves to the young Muslim that he is not an equal citizen" (Speed, 2019: FT online).

The Athens Mosque

The mosque is modest in scope. The choice of site does not readily reflect its functional and symbolic significance, nor does its architectural design. The mosque accommodates a maximum of 500 worshipers – far fewer places

Table 9.2 Asymmetric contestation across scales and stakeholders. An Athens model

Stakeholder interaction	*International Organization(s) (EU, ICC, CoE, ILO)*	*Other state(s) EU Member States, or with special interest)*	*The State in contestation (national, regional, and/or local)*	*Civil Society B (Domestic: national, local, mission-specific)*	*Minority group B (Sub-national, Transnational, Diasporic, difference, special interest)*
International Organization(s) *(EU, ICC, CoE, ILO)*	(formal) -Policy making -Collaboration -Advocacy -Opposition -No jurisdiction		(formal) -Policy-making (i.e., EU) -Defend policy -Appeal decision		Charter of rights and protections Observatory Support (financial, legal)
Other state(s) *EU Member States, or with special interest)*		(formal) (covert) -Supporter -Adversary/ Demarches -Arbitrator -Bandwagoning -Sanctions	(formal) (back channel) -Diplomacy -Gauge leverage -Soft power politics		Patronage Support (financial) Diplomatic representation Support in IOs
The State in contestation *(national, regional, and/or local)*	-Contesting/Defending norms, processes, procedures -Arbitration -Demarches -Legal action (international court) -Reporting non-compliance to treaty obligations	-Demarches -Good offices -Bilateral arbitration -Tit-for-tat politics -Bargaining/ Arbitration Minority advocate or patron -Non-interference	(formal)(informal) -Norms, processes, procedures setting/ (following – i.e. EU *acquis*) -Contesting from below (i.e. EU *acquis*) -Contesting from above (national scale stakeholders)	-Contracting -Intermediation -Consortial arrangement -Support (financial, in services, in material) -Lawsuit in national court	(formal) (informal) -Setting and/or contesting/ Defending -Norms -Processes -Procedures … -Negotiation/arbitration -Contesting to preserve right -Expand right - Win new right

(Continued)

Table 9.2 Asymmetric contestation across scales and stakeholders. An Athens model (Continued)

Stakeholder interaction	*International Organization(s) (EU, ICC, CoE, ILO)*	*Other state(s) EU Member States, or with special interest)*	*The State in contestation (national, regional, and/ or local)*	*Civil Society B (Domestic: national, local, mission-specific)*	*Minority group B (Sub-national, Transnational, Diasporic, difference, special interest)*
Civil society A *Transnational, International, Diasporic, mission-specific i.e. difference, civil rights, climate)*			(formal) -Collaboration -Subcontracting -Funding -Compliance -Assessment/evaluation	(formal)(informal) -Contesting **from below** -Contesting **from abroad**	-(formal)(informal) -Community-based -Individual -Funding -Services -Advocacy
Minority group A *(Sub-national, Diasporic)*			- Same as w/ Minority B -New regime		(formal) (informal) -Associational politics (intra-, inter-group) -Collaborating to preserve right - Expand right - Win new right

Figure 9.8 The newly-inaugurated Athens Mosque in Votanikos.

Source: Photograph by Demetrios Ioannou courtesy of Middle East Eye

than the large Muslim minority requires – and includes a smaller space for the exclusive use of 50 female worshipers. The budget amounted to approximately one million Euro (Figure 9.8).

The competition for the mosque's design yielded high-quality architectural projects (Tsioutsiou, 2007). However, the chosen form was the outcome of lowest-common-denominator politics – a design by subtraction.

Following long deliberations in which the state appeared to capitulate to conservative voices, a decision was made not to include a minaret in the plan. The unexceptional style of the exterior eschewing all traditional Islamic architectural forms, gives the mosque the air of a car dealership – certainly no avant-gardist homage to the industrial quarter in which the mosque has been embedded. Arguably, the state missed an opportunity to display its cosmopolitan cred and prove its international critics in the Islamic world wrong. In its part, Athens failed to live up to its reputation as Capital of the Arts.

Practical considerations determined the mosque's location. Building it on public land (the parcel includes a depot of the Hellenic Navy), the State avoided compounding "NIMBY" contestatory politics upon the fraught politics of approving and building the mosque. Also, systematic resistance by politically influential entities – especially the Orthodox Church and loud ultra-rightist conservative media, like the newspaper *Eleftheros Kosmos* – significantly limited potential locations (Figure 9.9).

Figure 9.9 "I don't want a mosque in my country," thinks the towheaded boy, countenancing the raging masses of Muslims – a montage of classic Orientalist and Islamophobic tropes, equating Islam with violence and fanaticism and contrasting it with the Aryan child-hero.

Source: Eleftheros Kosmos, 24 June 2015, online edition.

The mosque is located in Votanikos, a thinly-populated quarter in the city's old industrial core that experienced significant capital divestment in the last twenty years. Concerning the population it was built to serve, geographic marginality made industrial Votanikos an initially poor choice of a site for a public amenity. The mosque's locational logic is weakly related to Athens' Muslim communities' spatial distribution. The strongest argument would be that it is close to the historic core, which is a zone of dense settlement for several Muslim communities. However, it is not geographically embedded into any one Muslim community, nor is it located within convenient walking distance from any community cluster. Currently, Athens Metro Line 3 (Eleonas Station) and buses serve it – a practicality that may place burdens on Muslim communities' gendered patterns of urban mobility.

Economic expansion associated with Greece's accession to the European Communities in the 1980s raised the prospect of further industrial development. Elias Kourliouros (1997) notes that despite attempts for urban-industrial planning in the 1980s, plans faltered. "[T]he lack of a comprehensive and deliberate industrial strategy for the development and effective location of manufacturing" compounded by "the role of speculative land and housing interests" and a general anti-industrial posture based on a fictitious 'environmental protection-versus-industrial development' discouraged corporate investment in manufacturing in Athens (Figure 9.10).

Investors' turn to the service sector further marginalized Athens' historic industrial zone close to the downtown. Votanikos' life as an industrial district may be ending, but significant investment in corporate headquarters a few blocks away on Athenon Avenue is rapidly transforming the urban morphological profile of the Mosque's surrounds. What appeared in 2007 a

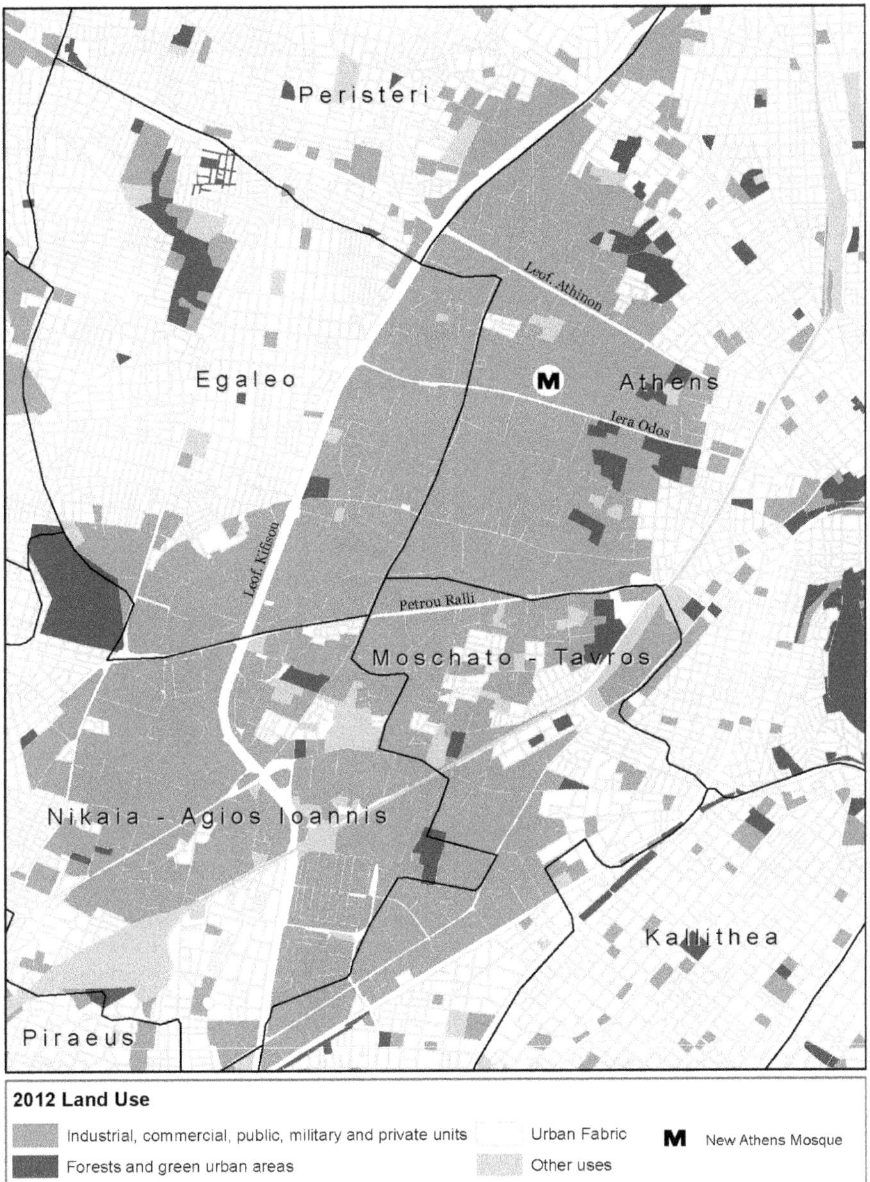

Figure 9.10 Land-use profile of Athens' historic industrial core.

Source: Census, 2011; cartography: Vagelis Papadias.

Figure 9.11 (Above) The closed "Autocar" auto parts plant. Tour bus companies use the alley as an informal parking area. (Below) Two blocks north of the new mosque's location, Athenon Avenue has seen extraordinary investment levels in services headquarters (WIND Telecom, Aegean Oil, OPAP Lottery, ALPHA Bank, and Global Aviation SA (Source: Authors).

remote, frankly unlikely location for a mosque, is now gradually recast as a growing and dynamic urban quarter (Figure 9.11).

With legacy manufacturing in full retreat, these improvements appear to have a net positive effect on the community area's physical plant. Significant transportation and utility infrastructure were required to accommodate the incoming services industry. In 2017, Eleonas Metro Station's transportation prospects improved dramatically. The European Commission in consultation with the national, regional, and local states, proposed the building of a Central Interurban Bus Station in the Votanikos community area: the station will serve all Greece's Interurban Bus Lines. At the same time, there will also be provisions made to serve international bus lines. The plan envisions other facilities and improvements, such as retail stores, dining and coffee areas, spaces for cultural activities, tourism and hospitality units, and foresees an underground parking area (EC Investment Project EIPP-20160108). The new transportation hub is another unexpected boon for the Athens mosque, as it makes it highly accessible to travelers across the region and beyond.

Commercial outlets are proliferating along Athenon Avenue and also, gradually, in the tributary streets. There is some evidence of gentrification in the form of dispersed new construction of midrise residential buildings. It is too early to discern the effect of significant corporate investment and gentrification upon the Athens Mosque's campus and the facility's use.

Conclusions

The making of the new Athens Mosque tells a story that links Athens' rebirth at Hellenism's national awakening in the 1830s to its urban futurity as a multicultural capital city. Deliberations for, and the contestation of the Athens Mosque project took place in a multiple-speed Europe that is guardedly embracing racial, cultural, and sexual difference. Despite the presumed discipline the European Union *acquis* imposes upon the Member States, the path to liberal Europeanization is neither linear nor relentless. In that sense, the state's dilemmas that underwrote the asymmetric contestatory politics of the Mosque project between 1983 and 2019 signpost where Greece, historically an affirmatively national state, is on the road to becoming a civil society and polity in the Europeanist mold.

There is much that will be made apparent in the next few years: the measure of the Mosque's contribution to Muslim communities' legibility; any improvement to the State's legitimacy among minority communities; the use-value of the institution and its facilities for different Muslim communities spatially dispersed across Attica, and the discursive urban effects of the Mosque and the nearby globalizing Athenon Avenue. Relatedly, some measure of gentrification is expected at the scale of the Votanikos community area triggered by on-going State and EU-financed infrastructure improvements. The sort of gentrification this investment would produce in the longer term is difficult to foresee, especially since the COVID-19 pandemic now convulses the world economy. European priorities may change. Nor should we preclude that the old industrial zone has seen the last of manufacturing investment. Green investment that balances environmental sustainability, economic development, and social equity or "green-enough" investment, driven by private sector priorities (Hamilton and Curran, 2017), can reshape the community area's employment profile and make it attractive for residential and mixed-use investment. Lastly, the Mosque is now embedded in this changing part of historic Athens, possibly influencing future Muslim community mobility.

The successful opening of the Athens Mosque should have measurable geopolitical consequences as long as the state follow-up affirms a broader commitment to freedom of conscience and religion. Foremost, such openness undermines Turkey's argument that Greek society and state are inimical to religious difference. It communicates to the European Union Greece's civil society turn and earns Greece political capital in Brussels and Strasbourg. It satisfices long-standing Arab demands for a mosque in Athens, although it is doubtful that it is sufficient, as a gesture, to earn Greece substantial political capital in the Islamic World. Importantly, the building of the Athens Mosque serves as a model of a new Hellenic geopolitics of difference produced by contestatory democratic politics. It establishes new norms and procedures that can support further change in the country in a Europeanist mold.

Notes

1. President of Turkey, Recep Tayyip Erdoğan's push to convert the UNESCO-listed Church of Hagia Sophia to a mosque was approved by the court on 10 July 2020. By annulling its museum status, established in 1934 in a spirit of reconciliation between Greece and Turkey, Erdoğan serves Neo-Ottomanist objectives while he draws a sharp line between Turkey and the EU. The inability of Greece to build a mosque and Turkey's reconversion of historic sites into mosques are exemplary of the politicization of built heritage in both countries: Associated Press. "INESCO statement on Hagia Sophia" https://en.unesco.org/news/unesco-statement-hagia-sophia-istanbul (accessed on 17 July 2020).
2. Dupré, Louis. "*Le Temple de Jupiter Olympien et l'Acropolis d'Athens.*" [The Temple of Jupiter Olympus and the Acropolis of Athens] Paris: Imprimerie de Dondey-Dupré, Rue St Louis, No 46, Au Marais, ca. 1825-1837.

Sources

Barthes, Roland. *Camera Lucida*. New York: Hill & Wang, 1980.

EUCFR Title II: Freedom of Thought, Conscience and Religion can be found at https://fra.europa.eu/en/eu-charter/article/10-freedom-thought-conscience-and-religion (Accessed on 10 June 2020).

Fouskas, Theodoros. "Low-Status Work and Decollectivization: The Case of Bangladeshis in Athens." *Journal of Immigrant & Refugee Studies* Vol. 10, No. 1 (2012): 54–73.

Fried, Michael. "Barthes's Punctum." *Critical Inquiry* Vol. 31, No. 3 (Spring 2005): 539–574.

Gropas, Ruby; Triandafyllidou, Anna. "Immigrant and Political Life in Greece: Between Political Patronage and the Search for Inclusion." Report prepared for the Project: *EMILIE - A European approach to multicultural citizenship: Legal, political and educational challenges,* ELIAMEP, 2009: 1–29.

Hamilton, Trina; Curran, Winifred. *Just Green Enough: Urban Development and Environmental Gentrification.* London: Routledge, 2017.

Ioannou, Nasia. "The Role of Migrant Integration Councils at the Coexistence of Different Religious Groups at Local Level: The Case of Islam in Greece." *Social Cohesion and Development* a publication of the Ministry of Interior, General Secretariat of Population and Social Cohesion Vol. 8, No. 2 (2013): 177–185.

Kandylis, George; Maloutas, Thomas; Sayas, John. "Immigration, Inequality and Diversity: Socio-Ethnic Hierarchy and Spatial Organization in Athens, Greece." *European Urban and Regional Studies* Volume 19, Issue 3 (2012): 267–286.

Karidis, Dimitris N. *Athens from 1456 to 1920: The Town under Ottoman Rule and the 19th-Century Capital City.* Athens: Archaeopress Archaeology, 2014.

Kitsantonis, Niki. "Athens Muslims Turn Factory into a Mosque." *New York Times* Online, July 4, 2007.

Kourliouros, Elias. "Planning Industrial Location in Greater Athens: The Interaction Between Deindustrialization and Anti-industrialism During the 1980s." *European Planning Studies*, Vol. 5, No. 4 (1997): 435–460.

Larmore, Charles. "A Critique of Philip Pettit's Republicanism." *Philosophical Issues: Social, Political, and Legal Philosophy* Vol. 11 (2001): 229–243.

Maloutas, Thomas; Spyrellis, Stavros Nikiforos. Ethnic Groups: "Inequality and Segregation in Athens: Maps and Data" [December 2019]. In Thomas Maloutas and Stavros Spyrellis, eds. *Athens Social Atlas* https://www.athenssocialatlas.gr/en/ (accessed 12 June 2020).

Maloutas, Thomas. *The Social Geography of Athens: Social Groups and the Built Environment of a South European Metropolis.* Athens: Alexandria, 2018.

Papadopoulos, Alex. Tasos Vrettos interview. 13 December 2017.

Papadopoulos, Apostolos G; Chalkias, Christos; Fratsea, Laukia-Maria. «Challenges to immigrant associations and NGOs in contemporary Greece." *Migration Letters* Vol. 10, No. 3 (2013): 342–358.

Paridis, Christos. Tasos Vrettos interview. *LIFO*, The Athenians. 2015 (https://www.lifo.gr/articles/the_athenians/80517).

Pettit, Philip. *Republicanism: A Theory of Freedom and Government.* London: Clarendon Press, 1997.

Speed, Madeleine. "The Battle to Build a Mosque in Athens." *Financial Times* (01 February 2019) (https://www.ft.com/content/ae4fa654-2416-11e9-8ce6-5db4543da632 (Accessed on 13 June 2020)).

Vrettos, Tasos; Argyropoulou, Nadja; Tzirtzilakis, Yorgos. *T(r)opoi latreias (Wor(th)ship)* exhibition catalogue. Mouseio Benakē, 2015.

Wiener, Antje. *A Theory of Contestation.* New York: Springer, 2014.

Βρεττός (Vrettos), Τάσος; Αργυροπούλου, (Argyropoulou) Νάντια; Τζιρτζιλάκης (Tzirtzilakis) Γιώργος (eds.). *Τ(ρ)όποι λατρείας* (Wor(th)ship), Κατάλογος έκθεσης (bilingual exhibition catalogue). Αθήνα: Μουσείο Μπενάκη, 2015.

διαΝΕΟσις (DiaNEOsis), *Έλληνες, Ευρώπη, Κόσμος και Εθνική Ταυτότητα: Τι Πιστεύουν οι Έλληνες Πανελλαδική Ερευνα – Έκθεση Αποτελεσμάτων* (Greeks, Europe, the world and national identity. What the Greeks believe nationwide survey –

report of results), [December] 2016 (https://www.dianeosis.org/wp-content/uploads/2017/03/ti_pistevoun_oi_ellines_final_version.pdf Accessed on 10 June 2020).

Κούρου (Kourou), Καλλιρόη, Σακκά (Sakka), Νικολέτα. «Γεωπολιτικές του Ιερού Χώρου στη Σύγχρονη Αθήνα: το Παράδειγμα των Ατυπων Τζαμιών» (The geopolitics of sacred space in Athens. The case of the informal mosques), Διάλεξη (public lecture). Εθνικό Μετσόβιο Πολυτεχνείο, Σχολή Αρχιτεκτόνων Μηχανικών, Τομέας Πολεοδομίας και Χωροταξίας, 9 [October] 2017.

Μπρατσιάκου (Bratsiakou), Ιωάννα. «Που Προσεύχονται οι Μετανάστες της Αθήνας; Εικόνες ενός Αθέατου Κόσμου» (Where do immigrants worship? Images of an invisible world), *News 247*, 21.11.2015 (https://www.news247.gr/weekend-edition/poy-proseychontai-oi-metanastes-tis-athinas-eikones-enos-atheatoy-kosmoy.6388035.html Accessed on 18 August 2020).

Παρίδης (Paridis), Χρήστος. «Τάσος Βρεττός. Συνέντευξη» (Tasos Vrettos. Interview). *Lifo*, 11 November 2015 (https://www.lifo.gr/articles/the_athenians/80517 Accessed on 18 August 2020).

Τραυλός (Travlos), Ιωάννης. *Πολεοδομική Εξέλιξις των Αθηνών* (Athens' urban planning evolution), Αθήνα: Καπόν, 2005.

Τσελεμπή, Εβλιγιά (Evliya Çelebi), *Οδοιπορικό στην Ελλάδα (1668-1671). Πελοπόννησος - Νησιά Ιονίου - Κρήτη - Κυκλάδες - Νησιά Ανατολικού Αιγαίου* (Travels in Greece (1668-1671). The Peloponnese – the Ionian Islands – Crete – the Cyclades – the Islands of the Eastern Aegean). Αθήνα: Εκάτη, 2005.

Τσιούτσιου (Tsioutsiou), Βασιλική. *Ετεροτοπίες: Προσεγγίσεις για το Νέο Τζαμί στην Αθήνα* (Heterotropias: Approaches to Athens' new mosque), Πτυχιακή εργασία, Πανεπιστήμιο Θεσσαλίας, Πολυτεχνική Σχολή, Βόλος, 2007 (file:///C:/Users/User/AppData/Local/Temp/P0004399.pdf Accessed on 18 August 2020)

Index

For Product Safety Concerns and Information please contact our EU
representative GPSR@taylorandfrancis.com
Taylor & Francis Verlag GmbH, Kaufingerstraße 24, 80331 München, Germany

www.ingramcontent.com/pod-product-compliance
Ingram Content Group UK Ltd.
Pitfield, Milton Keynes, MK11 3LW, UK
UKHW020153200926
475702UK00016B/2685
9781032006741